AF830909

# Applied Spatial Modelling and Planning

This book highlights the extraordinary range of areas to which geographical analysis and spatial modelling can bring lessons and insights. It shows how these techniques have been used to address 'real world' issues that are of concern to international organisations, public agencies and businesses, as illustrated by actual funded projects that geographers have developed collaboratively with end-users.

*Applied Spatial Modelling and Planning* shows how much geographical research is policy relevant to a wide variety of agencies through the use of GIS and spatial modelling in applied geography. The book's chapters contain a cross-section of innovative applications and approaches to problem solving within five major domains of the dynamics of economic space, housing and settlements, population movements and population ageing, health care, and the environment. Using a number of case studies on the use of GIS and spatial modelling, this book demonstrates the fact that much of what is done by quantitative geographers is not only relevant within academia, but also has use in policy work.

This book will appeal to an international audience interested in cutting-edge spatial modelling to better understand the processes involved in solving real problems.

**John R. Lombard** is an Associate Professor in the School of Public Service at Old Dominion University, USA.

**Eliahu Stern** is Professor of Geography and Planning at Ben-Gurion University of the Negev, Beer Sheva, Israel.

**Graham Clarke** is Professor of Business Geography at the University of Leeds, UK.

## Routledge Advances in Regional Economics, Science and Policy

For a full list of titles in this series, please visit www.routledge.com/series/RAIRESP

12. **Knowledge Economy in the Megalopolis**
    Interactions of innovations in transport, information, production and organizations
    *T.R. Lakshmanan, William P. Anderson and Yena Song*

13. **Regional Disparities in the Enlarged European Union**
    Geography, innovation and structural change
    *Valentina Meliciani*

14. **Tourist Clusters, Destinations and Competitiveness**
    Theoretical issues and empirical evidences
    *Edited by Francesco Capone*

15. **Regional Development in Australia**
    Being regional
    *Robyn Eversole*

16. **Knowledge Intensive Business Services and Regional Competitiveness**
    *Edited by João J. M. Ferreira, Mário L. Raposo, Cristina I. Fernandes and Marcus Dejardin*

17. **Urban Recycling Cooperatives**
    Building resilient communities
    *Jutta Gutberlet*

18. **Situated Practices of Strategic Planning**
    An international perspective
    *Edited by Louis Albrechts, Alessandro Balducci and Jean Hillier*

19. **Applied Spatial Modelling and Planning**
    *Edited by John R. Lombard, Eliahu Stern and Graham Clarke*

# Applied Spatial Modelling and Planning

Edited by John R. Lombard, Eliahu Stern and Graham Clarke

Routledge
Taylor & Francis Group
LONDON AND NEW YORK

First published 2017
by Routledge
2 Park Square, Milton Park, Abingdon, Oxon OX14 4RN

and by Routledge
711 Third Avenue, New York, NY 10017

*Routledge is an imprint of the Taylor & Francis Group, an informa business*

© 2017 selection and editorial matter, John R. Lombard, Eliahu Stern and Graham Clarke; individual chapters, the contributors

The right of John R. Lombard, Eliahu Stern and Graham Clarke to be identified as the authors of the editorial material, and of the authors for their individual chapters, has been asserted in accordance with sections 77 and 78 of the Copyright, Designs and Patents Act 1988.

All rights reserved. No part of this book may be reprinted or reproduced or utilised in any form or by any electronic, mechanical, or other means, now known or hereafter invented, including photocopying and recording, or in any information storage or retrieval system, without permission in writing from the publishers.

*Trademark notice*: Product or corporate names may be trademarks or registered trademarks, and are used only for identification and explanation without intent to infringe.

British Library Cataloguing in Publication Data
A catalogue record for this book is available from the British Library

Library of Congress Cataloging in Publication Data
Names: Lombard, John R., editor. | Stern, Eliahu, editor. | Clarke, Graham, 1960- editor.
Title: Applied spatial modelling and planning / edited by John R. Lombard, Eliahu Stern and Graham Clarke.
Description: Abingdon, Oxon ; New York, NY : Routledge, 2017.
Identifiers: LCCN 2016012523| ISBN 9781138925700 (hardback) | ISBN 9781315683621 (ebook)
Subjects: LCSH: Spatial analysis (Statistics) | Geospatial data–Mathematical models. | Geographical information systems.
Classification: LCC HA30.6 .A67 2017 | DDC 001.4/22–dc23
LC record available at https://lccn.loc.gov/2016012523

ISBN: 978-1-138-92570-0 (hbk)
ISBN: 978-1-315-68362-1 (ebk)

Typeset in Times New Roman
by Cenveo Publisher Services

# Contents

## PART II
## Housing and settlements



## PART III
## Population dynamics and population ageing



## PART IV
## Health care planning and analysis

# Plates

# Figures

# Tables

# Contributors

**Nick Addis** is a PhD student in the School of Geography at the University of Leeds. His current research seeks to understand the heterogeneity of burglary within Leeds, using both primary and secondary data to develop and inform an agent-based model. During the course of his research he has worked with a range of criminal justice agencies, including the police, Safer Cities Partnerships and HM Prison Service. As a former prison psychologist, his research interests lay within the broad field of crime analysis and how quantitative and modelling approaches may be used to support crime prevention efforts.

**Robert G.V. Baker** was an Associate Professor, now Adjunct, in Geography and Planning at the University of New England. He was a pioneer in the development of social physics in the late 1970s and 1980s and has been published in over 50 refereed journal articles in diverse fields of applied modelling and public policy, such as, retailing, gambling studies, physics, traffic engineering and climate change. He was Chair of the International Geographical Union's Commission on Modelling Geographical Systems between 2004–2008 and won the UNE University Research Prize in 2001 for work in the above fields

**Joana Barros** is a lecturer in GIS at Birkbeck University in London which she joined after completing her PhD at the Centre for Advanced Spatial Analysis (CASA). Joana's background is in Architecture and Urban Planning. She holds an MSc in Urban and Regional Planning (UFRGS, Brazil) and a PhD in Planning Studies from University College London. Her areas of expertise are urban planning and modelling, more specifically agent-based and cellular automata models applied to urban systems and urbanisation in developing countries (with a focus on Latin America).

**Paul Bidanset** is a PhD student at Ulster University (Newtownabbey, UK), where he is completing a dissertation on modelling techniques for hedonic real estate valuation models used for property tax purposes. He is concurrently working as the Real Estate CAMA Modeller for the city of Norfolk, Virginia (United States), where he is responsible for the annual valuation of over 74,000 residential properties.

**Mark Birkin** is Chair of Spatial Analysis and Policy in the School of Geography at the University of Leeds, UK. He is also chair of the University of Leeds' Consumer Data Research Centre (CDRC). His research interests include modelling societies and their behaviour, the use of information and communication technologies in the public, industrial and commercial sectors, and the development of Geographical Information Systems for human and environmental analysis.

**Michael Bithell** is Assistant Director of Research in computing at the Department of Geography, University of Cambridge. After a degree in Natural Sciences and a PhD in astronomy at the University of Cambridge, he worked on the dynamics of ozone at the Rutherford Appleton Laboratory before moving to the School of Geography at the University of Oxford to work on climate change and its human impacts. At present, his current research is based on computational techniques for the representation and analysis of systems consisting of interacting discrete components. The aim is to work towards understanding the interaction of environmental processes with human systems, and the adaptation of social dynamics to environmental change.

**Janet Cade** is Professor of Nutritional Epidemiology and Public Health at the University of Leeds, UK. She is leader of the Nutritional Epidemiology Group, which she established at the University of Leeds in 1994. It is a research-intensive group to create, advance and disseminate nutrition research to make an impact on population health. She is also a founder member of the new professional organisation, the Association for Nutrition, which has responsibility for the registration of nutritionists in the UK. Her main areas of research are: nutritional epidemiology, the role of diet in chronic disease aetiology, diet–gene interactions, assessment of nutrition interventions, and the development of dietary assessment methods.

**Graham Clarke** is Professor of Business Geography at the University of Leeds. He has worked extensively in various areas of GIS and applied spatial modelling, focusing on many applications within urban/social geography. A major research interest has been spatial modelling, especially spatial interaction modelling and spatial microsimulation (especially for estimating small-area patterns of income and wealth, and in applications relating to retail, crime and health). Currently he is Chair of the IGU Applied Geography Commission.

**Jonathan Corcoran** is a Professor in Human Geography within the School of Geography, Planning and Environmental Management at The University of Queensland, Australia. His research interests lie in the fields of Human Geography, Demography, Spatial Science and Regional Science. His publications cover a broad suite of topics, including human mobility and migration, human capital, and urban fires, each of which has a focus on quantitative methods. He has worked closely with a range of government agencies both to inform and evaluate operational and strategic planning through the development of geographic-based tools. He is co-editor of *Papers in Regional Science*.

**Kim Edwards** is a Course Director in Sports and Exercise Medicine in the Faculty of Medicine & Health Sciences at the University of Nottingham, UK. She has expertise in research methods, including ethics, study design and medical statistics, with significant experience of managing large datasets and undertaking advanced statistical analyses, including multi-level modelling, geographically weighted regression, geographic information systems, spatial analyses, cluster analyses, structural equation modelling and spatial microsimulation modelling in the fields of obesity, sport science and diet.

**Alessandra Faggian** is Professor at the Ohio State University AED Economics Department. Her research interests lie in the field of Regional and Urban Economics, Demography, Labour Economics and Economics of Education. Her publications cover a wide range of topics, including migration, human capital, labour markets, creativity and local innovation and growth. She is co-editor of *Papers in Regional Science.* In 2013, she was appointed by the European Commission as one five jurors to select the European Capital of Innovation 2014.

**Daniel Felsenstein** is a Professor in the Department of Geography and Director of the Center for Computational Geography, Hebrew University of Jerusalem. His research interests include economic geography, spatial econometrics and urban simulation. He serves as a consultant to the OECD in the area of local employment and economic development.

**Robert Fligg** works as Senior Surveyor (Arpenteur principal) in the Ontario Regional Office (Bureau régional de l'Ontario), Natural Resources Canada (Ressources naturelles Canada). Natural Resources Canada seeks to enhance the responsible development and use of Canada's natural resources and the competitiveness of Canada's natural resources products. It is a federal institution that leads the 'Natural Resources portfolio' and works with its partners.

**Victor M. Garcia-Barrios** is product manager at a large e-commerce solutions provider in Graz (Austria) as well as lecturer and consultant at Graz University of Technology, University of Applied Sciences Joanneum, Carinthia University of Applied Sciences (all in Austria) and Galileo University (in Guatemala). Victor studied Telematics at Graz University of Technology where he received his Doctorate. After that, he was research assistant at Graz University before he was appointed a professorship for Informatics at Carinthia University of Applied Sciences. Victor's research interests focus mainly on e-learning, Web and mobile development, usability engineering and geoinformation science.

**A. Yair Grinberger** is a PhD student in the Department of Geography, the Hebrew University of Jerusalem. His research interests include spatial behaviour, urban dynamics, GIS and agent-based simulation.

**Kingsley E. Haynes** is a professor at George Mason University in Washington, DC. He built the School of Public Policy out of The Institute of Public Policy, which he founded in 1990 while still Dean of the Graduate School.

He served as Dean of the School of Public Policy until July 2010. He also holds appointments in the departments of Decision Sciences, Geography and Public Affairs. His research interests focus on infrastructure investment, transportation and regional policy analysis. His is Fellow of the Academy of Public Administration, Regional Science Association International, and the International Geographical Union, the author of 20 books and 200 academic articles, and an international consultant on regional economic development.

**Claire Hulme** is Professor of Health Economics and Head of the Academic Unit of Health Economics at the University of Leeds. Her research is on the economic evaluation of community programmes spanning the health and social care sectors and economic evaluation alongside clinical trials. Her current portfolio includes trials in oncology, dentistry, musculoskeletal, stroke and complex interventions; community-based research includes dementia, physical activity and rehabilitation. In addition to evaluation work, her research interests lie in the areas of health and employment and informal care.

**Peter Kinderman** is professor of Clinical Psychology at the University of Liverpool and an honorary Consultant Clinical Psychologist with Mersey Care NHS Trust. His research interests are in the psychopathology of psychosis, psychological formulations and the interface between psychological theory and public policy. His work is collaborative with other disciplines (principally psychiatry) and uses a variety of approaches. He has been instrumental in developing theoretical models of paranoid thought and mania. These models have clarified the roles of maladaptive self-regulatory processes, involving the self-concept and causal attributions in the aetiology and maintenance of psychopathology. These models are widely cited in international research and form the basis of recognised therapeutic interventions.

**Yan Liu** is a Senior Lecturer in Geographical Information Science at the School of Geography, Planning and Environment Management of The University of Queensland. Her research interests lie in the fields of: (1) complex systems modelling and geo-simulation focusing on cellular automata (CA) and agent-based modelling (ABM) to describe, understand, model and predict environment and human change processes and interactions; and (2) social geographic studies using Big Data and GIS technologies, including human travel behaviours, spatial accessibility modelling and social inequalities associated with access to resources and service.

**John R. Lombard** is Associate Professor and Interim Chair of the School of Public Service at Old Dominion University. He served as Executive Director of E. V. Williams Center for Real Estate and Economic Development. He has presented and published papers on a variety of real estate and economic development topics. His current research examines the interplay of real estate and economic development policy and applied spatial analysis.

**Leslie Mayhew** is part-time Professor of Statistics at Cass Business School, London, UK. He is Honorary Fellow of the Institute for Actuaries and a member of the Royal Economic Society. He is a former senior civil servant and graduate of the government's Top Management Programme. He is a member of the Office for National Statistics expert panel that advises on population projections. His research interests include health, pensions, and health and long-term care, for which he has received various research grants. He is twice a winner of the Cass research prize. He is also managing director of Mayhew Harper Associates Ltd (MHA), a research consultancy specialising in the use of large administrative datasets and Geographical Information Systems.

**Michelle Morris** is a postdoctoral research fellow in the Consumer Data Research Centre, University of Leeds. Her primary research interests are in spatial variations in diet, lifestyle and health. Michelle is an interdisciplinary researcher with a background spanning spatial analysis and policy, nutritional epidemiology and health economics. Michelle plans to use quantitative methods to investigate geographies of consumption and how consumer behaviours are linked with health outcomes.

**Karyn Morrissey** is a Senior Lecturer in the European Centre for Environment and Human Health at the University of Exeter Medical School. Karyn has two main areas of research interest: population health and natural resource valuation. Karyn has a specific interest in using quantitative methods to examine both these research areas.

**Paul Norman** is an applied demographer/population and health geographer in the School of Geography, University of Leeds, with expertise in time-series and longitudinal analyses of both area- and individual-level data derived from census, survey and administrative records. Paul is highly experienced at teaching and researching on demographic, statistical, GIS and area type methods.

**Jitendra Parajuli** holds a BE degree (Electrical and Electronics) from Kathmandu University, Nepal, an MSc (Communications Technology) from the University of Ulm, Germany, and a PhD (Public Policy) from George Mason University, US. His research interests include infrastructure, entrepreneurship and economic development.

**Eleanor (Ellie) Pontin** is a Research Associate in the Institute of Psychology Health and Society at the University of Liverpool, UK. She is a qualitative researcher on the 'Enhanced Relapse Prevention MRC trial' and a researcher on the 'Relatives Adaptation to Psychosis study'.

**Philip Rees** is Professor Emeritus of Population Geography at the School of Geography, University of Leeds, UK. Recently he has been part of a European network (DEMIFER) and implemented policy scenario projections for EU regions. He has led projects (one current) to forecast the local authority populations of England by ethnicity, measuring the speed and diffusion of the third demographic (ethnic) transition, driven by a continuing stream of immigrants. He has investigated the consequences of population ageing, in particular for health.

**Francisco Rowe** is an Economic Geographer and Lecturer in Quantitative Geography within the Department of Geography and Planning at the University of Liverpool in the UK. Previously he was appointed as a Postdoctoral Research Fellow at the University of Queensland Australia. His research focuses on three core areas of human and economic geography: human capital mobility, spatial labour markets and statistical economic modelling. He has co-authored 26 academic publications. His articles appeared in journals such as *Applied Geography, Environment and Planning C* and *The Australian Geographer*. Francisco is currently a guess editor of a Special Issue in Youth and Graduate Mobility in the Journal The Annals of Regional Science and works closely with the United Nations' Latin American Demographic Centre (CELADE).

**Matthias Schwannauer** is Professor of Clinical Psychology and Head of Clinical and Health Psychology at the University of Edinburgh. His current research interests include the application of attachment theory, reflective function and psychological processes of affect regulation to further our understanding of the development, adaptation to and recovery from major mental health problems in adolescence, in particular psychosis and recurrent mood disorders. He is particularly interested in the psychology of onset and recovery of severe mental health problems in young adults with regard to current developmental models of psychiatric disorders and the advancement of specific psychological interventions.

**Holly Shulman** received her PhD at the University of Leeds from the Department of Geography in the study of patient hospitalisation rates in Yorkshire, UK. Dr Shulman has more recently worked for the Canadian Institute of Health Information in Ottawa, Canada, where she worked on a variety of health-related (GIS) projects and reports.

**David Smith** is a Senior Lecturer in Actuarial Science at CASS Business School, City University London. He is an expert in demography with a special interest in health care issues (especially health and non-life insurance). Over the last three years he has mainly published papers on developing new methodologies of predicting changes in population sizes and in addition looking at developing products to tap into the housing wealth of the UK population to help fund long term health and care needs.

**Eliahu Stern** is a professor of geography and planning in the Department of Geography and Environmental Development at Ben-Gurion University of the Negev, Israel. He chaired the Israel Planners Association, the Israel Geographical Association, Israel Association of Transportation Research & Planning and the Israel World Heritage Committee for UNESCO. Currently he is the Secretary of the IGU Applied Geography Commission. His research interests and expertise include transportation geography, urban and regional planning, spatial behavior, conservation and preservation of cultural heritage.

**Robert (Bob) Stimson** is Director of the Australian Urban Research Infrastructure Network (AURIN), Faculty of Architecture, Building and Planning, and Adjunct Professor, Department of Resource Management and Geography, The University of Melbourne and Emeritus Professor in Geographical Sciences and Planning, The University of Queensland. His substantial research expertise and experience spans numerous fields, in which he has published, including regional economic development, urban and regional planning, human spatial behaviour, locational analysis, housing studies and housing policy including seniors housing, urban quality of life, spatial decision support systems and survey research methods.

**Tal Svoray** is a Full Professor in Geography and Head of the Geography and Environmental Development Department in Ben-Gurion University of the Negev. Tal's research interests are in understanding how environmental and climatic factors and processes affect land degradation and ecosystem functioning in terms of primary production and species richness and diversity. The methods he uses include field studies and remote sensing data analysis with an emphasis upon spatially and temporally explicit modeling at several scales and multi-criteria decision support systems development.

**Sara Tai** is a Senior Lecturer in Clinical Psychology at the University of Manchester and Consultant Clinical Psychologist at Cheadle Royal Hospital and Greater Manchester West Mental Health NHS Foundation Trust. She currently works within a multi-disciplinary team in an acute adult in-patient unit. Sara is an experienced practitioner, trainer and supervisor of transdiagnostic approaches for working with people with serious mental health problems and their families, providing one-to-one interventions, group interventions and family work.

**Melanie Tomintz** is a Postdoctoral Fellow within the GeoHealth Laboratory at the University of Canterbury in New Zealand. She was formerly Senior Researcher and Lecturer at the Carinthia University of Applied Sciences (CUAS), Department of Geoinformation and Environmental Technologies, in Austria. Her main research areas include GIS, spatial modelling and simulations, especially in the area of health and environment, usability evaluation and health information systems. Her PhD is from the School of Geography, University of Leeds, UK.

**David Wadley** is a senior lecturer in the School of Geography at the University of Queensland, Australia. His research interests are in Philosophy of Planning and Development, Industrial and Retail Planning, Economic Geography and Futurology. David Wadley received his PhD from the Australian National University in 1975. His current research projects are in the fields of Visual Impact Assessment in Planning, Effectiveness, Efficiency and Equity in Retail Planning, Trust in Business, and Social and Work Experience in the Future.

**Sir Alan Wilson FBA, FAcSS, FRS** is Professor of Urban and Regional Systems in the Centre for Advanced Spatial Analysis at University College London. He is also Chair of the Home Office Science Advisory Council and of the Lead Expert Group for the Government Office for Science Foresight Project on The Future of Cities. He has been responsible for the introduction of a number of model-building techniques which are now in common use internationally, such as the use of 'entropy' in building spatial interaction models. He has rigorously deployed accounts' concepts in demography and economic modelling and is now working with dynamical systems theory to model the evolution of urban structure.

**Pia Wohland** is a lecturer in Health Inequalities, Hull York Medical School. She is a quantitative Health Geographer with strong expertise and keen interest in health inequalities across geographical areas, time, age and population sub-groups, population projections and future composition of the UK population and population ageing. She has worked on a wide range of projects, including research in ethnic mortality in the UK and health differences for local areas in the UK and across European countries. She is also a keen data scientist and R programmer and developed software for ethnic population projections in the UK.

# 1 Introduction

*John Lombard, Eliahu Stern and Graham Clarke*

This is the fourth volume of applications of geographical modelling and planning produced by members and friends of the International Geographical Union's Commission on 'Applied Geography'. Each year the Commission holds two conferences and invites contributions from those geographers (and related disciplines) who work actively with public or private sector planners on real-world issues. We believe these contributions help to show how much geographical research is policy relevant to a wide variety of agencies. In this particular volume we explore the use of GIS and spatial modelling in applied geography. Although still heavily criticised by many 'critical' human geographers, the field of GIS and spatial modelling has been transformed since the quantitative revolution of the 1960s. The field is now much more policy relevant and the 19 chapters in this book show research undertaken with many different types of policy makers.

Before the individual applications are presented, Sir Alan Wilson gives an overview of applied spatial modelling in a short paper relating to his keynote address to the Commission in 2013. He has been a leading figure not only of spatial modelling in geography and planning but very much of applied modelling work, setting up the consultancy company 'GMAP' with colleagues in the early 1990s. He reflects on the progress we have collectively made in applied modelling through a variation on the traditional SWOT analysis. He concludes that urban modellers need to tap into the current policy environment around big data, arguing that we need 'big science' to tackle effectively many of the biggest problems in our cities and regions. The paper provides a great deal of food for thought.

The first main section that follows provides a set of three studies dealing with the **dynamics of economic space** in order to find, or to formulate, the best policy for their given situations. Starting with the analysis of spatial disparity of the rental housing market, Jitendra Parajuli and Kingsley E. Haynes examine the distribution of new firm formation in New England from 1999 to 2009 using entropy and entropy decomposition. Their study provides an understanding of the distribution of single-unit firm births in a region, over a period in which the US experienced major economic disruptions resulting from recession, terrorism and global financial crisis. In addition, their study also provides the distributional patterns of new firms in various industry sectors as well as comparing the distribution of new firm births with other economic indicators. The findings are useful to regional planning agencies that are formulating policies to attract new businesses, to create jobs and to promote economic growth. At the same time, the study also demonstrates the application of the entropy-based method to examine spatial variation of single-unit firm births, which could be replaced by other spatial variables such that their distributions could also be studied.

The second study in this first section investigates the human capital implications of a policy change resulting from the introduction of the 485 graduate visa in Australia in 2007. The authors, Jonathan Corcoran, Francisco Rowe, Alessandra Faggian and Robert Stimson, provide an exploratory analysis of a specific aspect of the skilled immigration phenomenon in Australia resulting from this recent visa policy change. This change precipitated a sharp increase in the number of international students who then stayed-on post-graduation and entered the workforce. The study enabled an analysis and comparison of the 'before' and the 'after' patterns of regional distribution of those immigrants as they transit from higher education to employment. It thus sheds light on the fields of employment, working conditions and inter-regional migration of patterns of those stay-on graduates and their potential impacts on Australian regions. The main interested party has been the Australian Government's Immigration team.

The last example for the 'dynamics of economic space' section is provided by Robert Baker, who deals with the laws of spatial interaction modelling. He argues that Tobler's law of distance decay can be deconstructed into two further 'laws' relevant to the digital economy, namely, that 'all things can be, but not necessarily will be, connected globally' and, secondly, 'distance always matters, but time matters more'. The application of these new 'laws' to origin-destination trips are shown in a number of examples, such as the time-sensitivity in stock market transaction trades, satellite reception lags and filtering retail call-centre operators. In a more applied context Baker considers the implications of a new set of spatial interaction models (especially relating to e-commerce) on the traditional location models of Australian retailers.

The second section provides examples of applied spatial modelling and planning in the **housing and settlements** arena. Yan Liu, David Wadley and Jonathan Corcoran bring a new perspective to the analysis of the spatial disparity in the rental housing market. Its purpose is to advance understanding of affordability which often emerges as a major issue in developed countries. They bring a new perspective to the issue by linking graduated household income and rent outgoings data at the neighbourhood scale. A set of spatial metrics is developed to map and visualise the geographical disparity in the supply of, and demand for, private rental housing stocks. Principal Component Analysis, followed by a two-step cluster analysis method, is employed to define the spatial typology of the private market at a localised scale known as the State Suburb. The methodology is innovative since it brings a new perspective to the analysis of current, not previously linked, income and rent data at the neighbourhood scale to identify the spatial and structural grouping or clustering of suburbs. The results reveal four distinctive spatial clusters, representing a range of socio-economic and demographic outcomes at the local neighbourhood level. The divergence of the private rental market among the different suburbs indicates the pertinence of community and people-based policy responses and solutions to housing problems. It is also found that the extent of affordability over many suburbs can be reduced when households of more than sufficient income move down market in their choice of rental housing.

A deeper look into the housing market problems is provided by Paul Bidanset and John Lombard, who improve the performance and the predictability power of a mass real-estate appraisal model. They are using a geographically weighted regression across Gaussian and bisquare kernels with both fixed and adaptive bandwidths. They found that continuous kernels achieve superior city-wide dispersion levels, particularly exponential kernels with fixed bandwidths. Additionally, neighbourhood disaggregation of dispersion and price-related differential levels reveal that exponential kernels with fixed bandwidths significantly reduce the amount of sub-markets suffering from regressivity and relatively lower uniformity. The application of spatial modelling therefore aims to promote equity, uniformity and overall government accountability within the realm of *ad valorem* property tax.

A different tool of applied spatial modelling is presented by Eliahu Stern, who provides an example of the never-ending practice of spatial search. It is a GIS-based search aimed at finding locations for housing the growing ultra-orthodox population in Israel. Bounded by several sectorial restrictions like proximity, and/or high accessibility to Jerusalem, and low-rise buildings, the Ministry assigned the author to find feasible locations to accommodate a minimum of 50,000 dwelling units. Feasibility is defined by several different objective and subjective indicators. A paired-comparison analysis of location criteria is also used to examine the rank-orders of feasible locations by the leaders of the ultra-orthodox population and the community of urban planners. An agreeable solution was achieved. However, despite the comprehensive spatial search procedure, at the end, one should expect, to some extent, public objection to a near-by ultra-orthodox settlement, a common NIMBY phenomenon.

The final example in the housing and settlements arena is the joint work of Yair Grinberger and Daniel Felsenstein, who simulate the long-run impacts of an earthquake on the urban system. They present the first of a number of dynamic agent-based models in the book. This model simulates the disaster outcomes in two different urban contexts in Israel: Jerusalem and Tel Aviv. Attention is paid to the effectiveness of urban policies aimed at restoring the urban equilibrium. The policies relate to land use regulation, public provision of shelter and the restoration of damaged urban services. Results show that a similar shock in two different locations results in very different outcomes. Policy simulations imply that interventions directed at rebounding to pre-shock state do not do well and may even inhibit urban stability.

In the next section we have three very diverse papers but all relating to the broad topics of **population movements and population ageing**. In the first chapter, Philip Rees, Pia Wohland and Paul Norman review the context in which population projections are needed in applied and planning analyses. Countries experiencing below replacement fertility and high international immigration are experiencing an ethnic transition from a state of low diversity to high diversity. This process is illustrated by the authors, drawing on a set of ethnic population projections for the UK which show dramatic growth in most Minority Ethnic populations, contrasting with stagnation or decline in the White British population. However, because the inputs to such projections are difficult to estimate and because there is more uncertainty because of the multiplicity of groups, they argue that it is vital to evaluate their accuracy. They compare the ten-year projections from a 2001 base with 2011 Census results. The projections seem to overestimate the White British and White Irish populations because of over-optimistic mortality assumptions. As the groups with the highest share of the elderly, these two groups gain most from better survival into old age. The projections also under-project the growth of Black and Asian Minority populations because the official projections of all group immigration were much lower than the actual inflows from 2008 to 2011. So the UK's multi-ethnic future was arriving faster than the analysis suggested. The authors thus worked to revise the inputs and assumptions of new projections of the UK's local ethnic populations, based on the 2011 Census. This is invaluable information for a whole swathe of Government offices and service providers.

The second chapter in this section, by Les Mayhew and David Smith, also looks at the future population in the UK, in this case exploring different scenarios around population ageing and life expectancies. Many applied policy issues surround population ageing. A major Government concern is how to pay for a larger retired population. The authors speculate that people will be required to take additional steps in planning for their own financial needs in old age and to become less dependent on the state. However, they argue that geographers and policy makers should be wary of taking official forecasts of the older population at face value. Given the fact that people will retire with very different financial resources, rapidly rising life expectancy will mean new coping mechanisms are required to help individuals and families understand the true costs of ageing for them. It will mean that available resources, for example pension savings, need to last longer. There will also be implications for service providers. Mayhew and Smith argue that the increasing uncertainty over how long a population or an individual will live will impact services affecting the whole care economy, types of employment and also health and social care services. This chapter argues that geography is well equipped to address the details but only if it is able to forecast ageing populations with reasonable accuracy and, in turn, the social and spatial ramifications are clearly understood.

In the next chapter by Nick Addis, population movements are captured through the application of agent-based models. As noted in the Grinberger and Felsenstein chapter, this is an important and growing spatial modelling technique (which will also be applied again in the chapters by Bithell and Fligg and Barros – see below). In this instance, the focus of enquiry is the movement of burglars through the urban landscape. The evolution of this new type of computational modelling approach has coincided with a culture whereby criminal justice agencies find themselves subjected to increasing levels of scrutiny over their performance. This has contributed to an increased awareness among these agencies of how such modelling approaches may be used to support crime prevention efforts. Addis's chapter illustrates through case studies how agent-based models have been effectively applied to crime phenomena to help understand the underlying dynamics of existing crime systems, and how this technique may be supported through collaborations with criminal justice agencies.

The next section includes a series of chapters on **health care planning and analysis**. This is an area of geography which has always been very applied. Spatial simulation models and analysis can contribute hugely to the area of health care and health care planning.

The first chapter, by Holly Shulman, Graham Clarke and Mark Birkin, looks at the impact of opening new community hospitals in the West Yorkshire area of the UK. Since radical reforms were made to the UK National Health Service (NHS) in the 1990s, policy makers have tended to favour the promotion of economies of scale in health care provision and funding and have sought to concentrate surgical facilities in large, mainly urban hospitals. However, it is widely recognised that as service provision becomes more centralised, accessibility for patients becomes a more crucial consideration. Many geographical studies have shown that as distance from a service location increases, utilisation generally decreases. To counter increasing centralisation, the UK NHS is currently considering the policy of opening more local hospitals – smaller community hospitals. These are deemed to be especially important for serving older patients with greater mobility problems. Shulman, Clarke and Birkin explore access to hospital treatment for one major type of condition which could easily be treated in local (day care) community hospitals or ambulatory centres – cataracts (especially because surgery today has routinely fallen from a five-day inpatient stay to a day case). In order to evaluate the impact of new community hospitals they build two types of model. The first is a morbidity model for cataracts. This allows them to compare current geographies of hospital usage against potential demand (to explore whether there are areas of the city-region where they suspect patients are not getting the treatment they might need). Second, they then build a suite of spatial interaction models to reproduce the known flows of patients between an origin (patient residence) and destination (hospitals) to model the impacts on access of adding new community-style hospitals.

In the second chapter on health care, Melanie Tomintz and Victor Garcia-Barrios introduce simSALUD, a freely available and user-friendly web application that allows the simulation of health-related issues for small geographical zones. They demonstrate the power of this application through the estimation of small-area smoking rates. The application platform includes spatial microsimulation algorithms to allow users to produce synthetic data by combining different input data sources (normally a survey combined with small-area census-type data). Requiring no programming skills, the simulation model is run through a wizard-based web user interface, which guides the user through each step of model building. This allows non-experts in spatial microsimulation modelling to run powerful models for estimating a variety of health-related variables: that is, potentially smoking patterns, diabetes, obesity, and depression, variables which are not routinely available for small-area geographies. SimSALUD also provides a visualisation tool that allows the model results to be mapped, showing hot-spots of the simulated variable of interest. The visualisation tool is also designed in a way that health care planners, regional planners or non-experts from other sectors can access and use it easily, thus contributing to their future decisions on where to spend resources more effectively.

The third chapter in this section provides another example of the power of agent-based modelling. In this chapter Mike Bithell models infectious diseases at the scale of the individual, allowing him to include a number of factors, such as spatial structure and individual heterogeneity, which have often been neglected in more traditional aggregated approaches. In this chapter, an agent-based model of a primary school (children aged 4–11) is derived directly from observation of a real school. Laser-scanner measurements of the school building are used to determine its size and shape, the location of social spaces, and of obstacles to movement. Observations of classroom activity and the school timetable determine daily activity, and a fine-scale social-force model routes teachers and pupils through the school, while retaining the collective aspects that group together pupils in shared classroom spaces, or see them dispersed in the playground. All activities are set by a timetable that uses real-time units and real-space co-ordinates, so that the spread of a notional proximity-based infectious disease can be matched to observations. Preliminary results show how it is possible to separate disease-specific factors from social activity that mediates disease transfer.

In the fourth chapter in this health care section, Michelle Morris, Graham Clarke, Kimberley Edwards, Claire Hulme and Janet Cade (another multi-disciplinary team, this time at the University of Leeds) examine the growing problem of obesity. They provide a cross-sectional analysis using weight status data from a large cohort of UK women to explore the geographies of overweight and obesity at four different spatial scales, including a geodemographic classification, in order to identify areas with a higher likelihood of persons being overweight and obese. Statistical analysis is carried out using Stata statistical software and results are visualised using ArcGIS. Higher prevalence of obesity is observed in the North compared with the South of England and also in urban compared to rural areas. Significant differences are also shown to exist between the nine Government Office Regions in the UK. Once demographic characteristics are accounted for through the use of a geodemographic classification, it can be seen that those living in 'Constrained by Circumstance' and 'Blue Collar Communities' (the lowest income groups) have the highest odds of being overweight or obese. They conclude that such analysis (when undertaken for a combination of spatial scales) could be the best way forward when investigating the geographies of obesity and for producing meaningful results of use to policy makers.

The final chapter in this health care section is provided by a team of experts from a variety of different disciplines (Karyn Morrissey, Peter Kinderman,

Eleanor Pontin, Sara Tai, Matthias Schwannauer) and addresses a growing concern in UK health planning – an increasing number of persons with mental health problems, particularly depression. There is considerable evidence of a social gradient in the prevalence of depression that may lead to lower levels of wellbeing among the population. Including the Index for Multiple Deprivation for England (2010) with the Stress Test, a survey launched on *All in the Mind*, a BBC Radio 4 programme, and developed by psychologists in Liverpool and Manchester Universities, Morrissey et al. explore the association between depression and area-level deprivation, controlling for demographic and socio-economic factors. This research shows that individuals in lower income categories have higher rates of depression, and within income categories, individuals with depression reside in relatively more deprived areas relative to those without depression.

In the final section we turn to **environmental modelling** and applied spatial modelling. In the first chapter Robert Fligg and Joana Barros present another agent-based model, this time for the Limpopo River Basin Area in Mozambique. More specifically, the study focuses on the evacuation procedure of a fast-flooding area in the vicinity of Xai-Xai, in the province of Gaza. First, Fligg and Barros simulate the phenomena of a fast-flooding area and evacuation behaviour; second, they develop a neural network designed to simulate an agent's cognitive ability to sense, learn and adapt when travelling over a landscape during a flooding episode. The evaluation process undertaken at the end of the modelling exercise demonstrates there is potential for using this design of hybrid model of neural networks for further research about the evacuation behaviour of people in the fast-flooding area of Xai-Xai, which can be possibly extended to other geographic areas where models integrate human decision-making and land use/land cover.

Finally, Tal Svoray explores the issue of land degradation, an important topic in environmental management given that it is such a destructive phenomenon that causes damage to agricultural fields and neighbouring man-made infrastructures. Svoray provides a methodological framework to decrease land degradation based on the following three procedures: (1) the physical understanding of the processes by using GIS and spatial modeling; (2) identifying areas under risk; and (3) using the risk layer as an input to a spatial decision support system for prioritising actions. The framework is partially demonstrated on wheat fields in a semi-arid zone of Israel, an area suffering from intensive water erosion processes.

We hope this book provides a number of interesting case studies of applied spatial modelling and planning, relevant to a wide range of policy makers and advisors.

# 2 Applied spatial modelling

## 'Big science' and 'best practice' challenges[1]

*Sir Alan Wilson*

### 2.1 Introduction

'Applied spatial modelling' is a long-established field with great strengths, but it is under-exploited and hence ripe for more extended use. But it also has scientific challenges to meet to reinforce its value. These issues are explored in this chapter through a SWOT analysis, with a final emphasis on the considerable opportunities for future development. To facilitate this, the SWOT is re-ordered as SWTO. The argument will be developed using a framework derived from Brian Arthur's (2010, 2014) book *The nature of technology*. His argument relates mainly to hard technologies but it can be applied much more broadly. Cities, the planning system and modelling can each be viewed as 'technologies' so that, to look ahead, we can apply Arthur's idea of 'combinatorial evolution' as the basis of our analysis.

### 2.2 Combinatorial evolution

A technology can be seen as a system with a set of subsystems, each of which in turn may have further subsystems – indeed a complex system will have a hierarchy of subsystems with many levels. Technologies can then be seen as evolving through innovations at various levels in the hierarchy – and indeed through new combinations of these systems – hence Arthur's concept of 'combinatorial evolution'. We can view a city as a technology – using Arthur's definition of a technology being 'phenomena put to use'. We can also see the planning system – and indeed other domains of application – as a 'soft' technology. The argument is also put that the evolution of science can be seen as combinatorial in the same way, so if spatial modelling is seen as the science of planning, then in turn we can examine the hierarchy of its subsystems to understand how they can be combined in different ways to generate that science – and, again in turn, to see how this science can be most useful in applications.

### 2.3 Cities, planning systems and models as technologies

The combinatorial evolution concept resonates very strongly with ideas about cities and regions (seen as 'technologies'), the planning systems (as a soft

*Table 2.1* Cities as technologies – urban 'phenomena put to use'

- geology and soils – agriculture, forestry, mineral extraction, biodiversity
- water – irrigation, flooding, potable water, treatment
- energy production, stocks and flows (all sources)
- food
- landscape
- heritage
- buildings
- networks (all modes)
- social infrastructure (numerous sub-systems)

technology) and urban modelling as the underpinning 'science'. We can tease out these resonances.

The elements of Table 2.1 are all 'put to use' in cities. We can identify the significance of technological invention in each of these areas: in traditional planning, we focus on a subset of these lower level systems, especially land use, while in urban modelling, we tend to focus on the human use of technologies. Perhaps it would be appropriate to build a more complete picture to help us build better models of urban evolution? In particular, for example, we could model urban utilities and their interdependencies.

In Table 2.2, we list the elements of the planning system when viewed as a technology. We can look for innovations in any of these areas or through combinations; it is certainly an argument against silos. This simple presentation alone is a strong argument against silos and for what might be called interprofessionalism.

The focus in the rest of this chapter will largely be on the development of urban modelling through combinatorial evolution; and its applications. In the light of the preliminary argument to date, we need to see the field in a hierarchical way; we need to identify the main subsystems – building on Table 2.3; we need to be able to look for new combinations, always bearing in mind the uses of these models in planning and applications.

## 2.4 The SWTO analysis

We consider in turn: the strengths – in the case of urban modelling, as a 'technology', we need to think subsystems and hence the idea of a toolkit; the

*Table 2.2* The planning system as a technology: the main subsystems (and players)

- the planning system itself
- planning law
- regulatory – officers and elected members
- developers, businesses
- designers/engineers
- the community
- analysts and urban modellers

*Table 2.3* Urban modelling as a technology – the main modelling elements

- demographic account-based models
- input-output models
- spatial interaction models of the main flows
- an ability to integrate these into transport models and to load them on to networks
- retail and service 'revenue' and 'use' models
- residential location models

And then, with more difficulty

- retail centre models
- economic activity location models

weaknesses – not enough resources? Too many silos? A disconnect between the 'science' and the 'planning'?; the threats – can modellers get their act together?; and the opportunities – an enormous canvass.

### *2.4.1 Strengths: the toolkit*

We have reached a position where we have an enormously strong toolkit for urban models with elements that are good enough to apply. This can be characterised in terms of the main subsystems and how they link, the theories that we represent in our models, and the methods we use to implement these models. See Figure 2.1 for a sketch.

We have reasonable theoretical underpinnings for model development at different scales, as shown in Table 2.4.

We have appropriate elements of theory: of individuals and households; firms and organisations: utility maximising, profit maximising, innovation – inventing

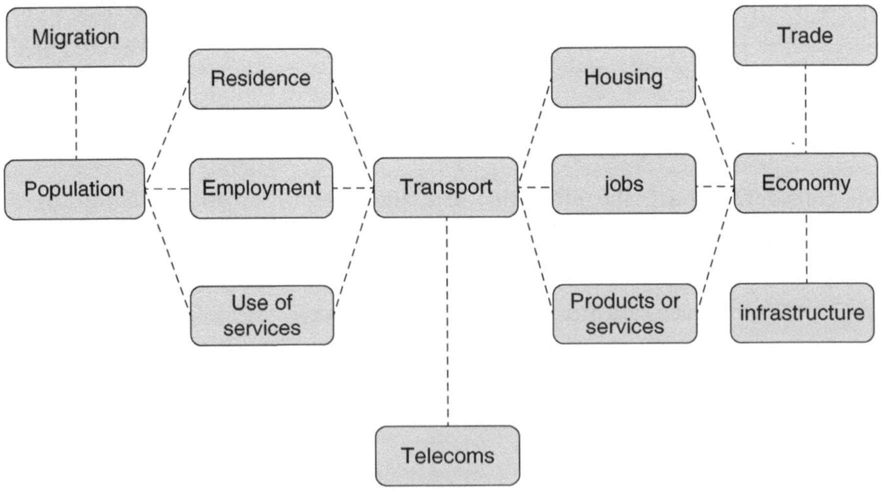


*Figure 2.1* The main elements of an urban model

*Table 2.4* Key elements of models at two scales

- the aggregate scale:
    - demographics: birth, death and migration
    - economics: input-output-interdependencies
- the 'zonal' scale:
    - residential location
    - service location
    - location of economic activities
    - spatial interaction: journeys to work etc.; accessibilities

new production functions and business models. This makes it sound easier than it is! It is the interdependencies, both within and between scales that make modelling difficult. So we then seek appropriate methods to build our models to represent our theories, and again, the toolkit is pretty good. We aggregate upwards through scales (with some losses of information). We decide what we can 'predict' in our models – the endogenous variables – and what has to be exogenous. As we become better modellers, more of the exogenous can become endogenous. What holds our models together are accounts. In the aggregate demographic and economic cases, the accounts are fundamental and provide the basis for model building through 'rates'. Interaction models also function as accounts since they must add up to the same totals at each 'end'; and the 'ends' must be mutually consistent: housing and resident population in an area must (more or less) match. In many cases, the 'ends', of course, can be the subject of independent sub-models. These interactions provide the links that represent the functioning of the city.

In practice, some parts are easier than others. We are pretty good at statistical averaging when large numbers of individuals are involved. It is much harder, indeed in many cases impossible, when small numbers are involved – so, for example, the location decisions of major companies have to be exogenous. (We can identify conditions that might be favourable but we can't determine the outcomes.) And we shouldn't underestimate the difficulties in these areas where we do have models: there are competing approaches in many cases with no consensus on what is 'best'; and in many cases, the models are dependent on 'rates' which are difficult to estimate – migration in the case of demographic models; coefficients, which certainly can't be static, in input-output models; average trip rates in spatial interaction models. And this, of course, is only half of the story. So far, we are modelling the functioning of a city in equilibrium or in a steady state. So the next step, starting from these states as initial conditions, is to model the dynamics, and this is very much harder – but much progress has been made and I would count the present state of knowledge as a strength, albeit in terms of what is now visible potential.

Notwithstanding these comments which imply future model-building challenges, we do know a huge amount and we have models that work well in a variety of situations – certainly in population forecasting, retail analysis and transport; possibly public services. The present state of knowledge rests on the base of an extensive (mathematical) 'methods' toolkit, for example: statistical averaging, whether entropy maximising or random utility theory – one way or another 'blurring' implausible optimal or rational behaviour; mathematical programming; nonlinear dynamics; complemented by excellent computer visualisation.

### *2.4.2 Weaknesses*

In spite of the strengths, it is not difficult to identify weaknesses. There has been a failure to apply modelling systematically in town planning or urban development planning. The planning systems' technology is under-developed. We have not designed models in such a way that they can represent the most serious problems – for example of social disparities. Some major modelling challenges have not been met – for example the high-dimensionality problem; not being comprehensive enough to model interdependencies; and working in methodological, problem and disciplinary silos – particularly a problem with economics. In general, perhaps we have not been ambitious enough – at least in part, this is because it is a small field within science; it should be 'big science'.

### *2.4.3 Threats*

Probably the most significant threat is the failure to deliver what has been promised in the past (notwithstanding the successes in retailing and transport) – especially in real-world situations. It could go the way of 'general systems theory' in the 1950s. We can add failure to deliver the interdisciplinarity resource needed for future development and application – in practice, a reversion to disciplines – in many cases without much modelling, as in Geography. Necessary skills have been lost while at the same time there has been too rapid a shift of resources to fashionable fields like agent-based modelling in situations that are not yet ripe for application. In terms of applications more broadly, if appropriate resources have not been available, potentially good analysis will be replaced by second or third best.

### *2.4.4 Opportunities*

The analysis of weaknesses and threats demonstrate the existence of many new opportunities – challenges to be met! We need a research agenda that seeks models of real-world complexity and real areas of application together with education opportunities at both undergraduate and especially postgraduate levels that offer breadth as well as depth. We need a broader understanding across relevant disciplines and professions of the available tool kit – avoiding the myopia of silos. We need to rise above inappropriate competitiveness between 'paradigms'. In particular, we need to tackle the comprehensiveness problem to represent interdependencies. This almost certainly means tackling the high-dimensionality problem; and this in turn almost certainly means integrating microsimulation with a specification of conditional probabilities (and hence 'chains of causality'). Finally, we need to be in a position to tackle the most serious (wicked) planning problems.

*Table 2.5* Topics for a city leader's agenda

| |
|---|
| • economic development – GPD per capita, growth/decline trajectory, sector mix, skills mix; possibilities of inward investment |
| • demography, including migration – in and out |
| • the population in terms of income, skills, those in particular need (the elderly, the sick, the unemployed, NEETS) – availability of employment – locally or commuting – in relation to the skills base |
| • housing – private, social, rented |
| • 'public' services – education, health, security, social care, environment |
| • private services – retail, leisure, … |
| • infrastructure – utilities, transport |
| • land use |

Can we put a broad agenda together systematically? In Brian Arthur's terms, widen planning's domain. We can look at it from a city leader's point of view; what would he or she need to understand for their city – see Table 2.5.

The 'design' question then is: what are my action possibilities in each of these areas and do I understand the interdependencies? Can the analysts and the professions combine, bringing their expertise together? The Brian Arthur argument is that this would be an engine for innovation.

## 2.5 Summary: a modelling Utopia

We are not doing enough work on the most important problems and we are not deploying the full urban science toolkit, although, in our defence, relative to numbers employed on conventional technologies, globally, we are a very small research community which may mean that we need to engage more effectively with areas where more resources are found – conventional planning and government agencies.

So what can we do? What would a modelling Utopia look like? We need to produce an 'offer' that enables us to be recognised as 'big science' – that is, put the proper resources in place to ensure that we can make effective contributions. Imagine a properly funded research institute (ideally several, scattered around the world). What would it do? In terms of structure, it would be truly interdisciplinary, building a science of cities and regions and applying this science to the most challenging problems: no silos, task-oriented. The elements would include information systems (big data, visualisation for science and planning); a modelling skills base; experimental models team; planning project teams – national, regional and local – all organised as project teams on an 'adhocracy' basis, within each division. This effort would be complemented by some research teams, many applied project teams contracted out as consultants – staffed by computer scientists, mathematicians, statisticians, geographers, economists, sociologists, planners and designers, engineers and ecologist. Specialist skills would be available, but all would need to understand the basics of modelling and planning – having both depth and breadth – embracing the principal systems of interest (at various scales) – national (e.g. UK infrastructure, housing land), regional and local together – with the main subsystems –

transport, retail, public services, and so on. Comprehensive models would provide the basis for recognising interdependencies and contributing to master planning.

In terms of modelling research priorities, we can take what might be called the new basics as step 1: tackling the high-dimensionality problem by integrating most (all?) possible approaches with microsimulation; full integration with economics: discrete space; realism rather than Cobb-Douglas and 'variety of goods' concepts; resurrection of cost-benefit analysis; systematic application of input-output modelling (with new sectors?), combined with demographics, at all scales from the global to the neighbourhood.

For step 2, we can be more ambitious and seek to meet new challenges: dynamics (path dependence and high-dimensional 'cones of possible development'; possibilities of 'genetic planning'); explore a wider range of structure-generating mathematics – Turing and beyond; ABM – progressing to realism and usefulness; network 'science' – ditto; economics again. There are also opportunities to apply models through extending the range, for example in new disciplines (for modelling), such as history and archaeology.

We can then focus on modelling priorities for planning. We need to establish the basics on a broader footing: in effect, 'best practice' models for current practice. These include demographic and input-output accounts and models, a full range of application of appropriate subsystem spatial interaction and activities' model together with a range of comprehensive models based on decision about levels of disaggregation, choice of submodels, etc. Doing the basics properly would include developing effective performance indicators such as catchment populations, indeed a whole battery of pi's, linked also to real-time data, with some kind of intelligent scanning. A wide range of demonstrators should be available, including new model development for the 'wicked problem' challenges.

Put very broadly, we need to be ambitious in model development and to argue the case for urban and regional science as 'big science', to ensure that 'best practice' is routinely deployed in applications and to make sure that we contribute to the 'wicked problems' agenda.

## Note

1 This paper was presented as a lecture and is therefore more informal than the traditional academic paper. On the same basis, there are few references as the subject matter is well known to the audience. However, Boyce and Williams (2015) is a model history of one branch of applied spatial modelling, Wilson (2012) provides a brief overview and Wilson (ed.) (2013) a much more detailed one.

## References

Arthur, B. (2010) *The nature of technology*, Allen Lane, The Penguin Press, London; Penguin edition, 2014.

Boyce, D. and Williams, H. (2015) *Forecasting urban travel: past, present and future*, Edward Elgar, Cheltenham.

Wilson, A.G. (2012) *The science of cities and regions*, Springer, Dordrecht.

Wilson, A.G. (ed.) (2013) *Urban modelling* (5 vols), Routledge, Abingdon.

# Part I

# Dynamics of economic space

# 3 An exploratory analysis of new firm formation in new England

*Jitendra Parajuli and Kingsley E. Haynes*

## 3.1 Introduction

Entrepreneurs play a crucial role in economic growth (Schumpeter, 1934, 1942), and through new firms they create new jobs and distribute wealth (Kirchhoff, 1994). The US has the right mix of factors for entrepreneurial activities (Acs and Armington, 2006; World Economic Forum, 2012; Fairlie, 2013). However, studies in economic geography suggest that economic activities are unevenly distributed across regions and time and that this should also be true with the formation of new firms. For example, according to Glaeser and Kerr (2009), variables, such as costs and the number of suppliers, often determine the heterogeneity of manufacturing entrepreneurship at the local level. Similarly, Renski (2009) found that there is a significant spatial variation in firm formation, survival, and growth in the US.

The purpose of this study is to examine the distribution of new firm formation in New England from 1999 to 2009 using entropy and entropy decomposition. On the one hand, this study should provide an understanding of the distribution of single-unit firm births in a region over a period in which the US experienced major economic disruptions resulting from recession, terrorism, and the global financial crisis. In addition, it should also provide the distributional patterns of new firms in various industry sectors as well as compare the distribution of new firm births with other economic indicators. The findings will be particularly useful to the regions that have been formulating policies to attract new businesses to create jobs and to promote economic growth. On the other hand, the study should also demonstrate the application of the entropy-based method to examine the spatial variation of single-unit firm births, which could be replaced by other spatial variables such that their distributions could also be studied.

This chapter is arranged in the following way. The next section presents theoretical background and research questions followed by the methodological framework and data. Thereafter, the results are presented. Finally, the concluding section summarizes the findings and provides policy implications and the future direction of research.

## 3.2 Theoretical background and research questions

Recession is the contraction of the business cycle and is associated with the fall of real gross domestic product and employment. Since there are various types of business cycle – Juglar, Kitchin, Kuznets, and Kondratiev – and fundamental differences exist among macroeconomic performance and business cycles, it is a challenge to identify the short-run and long-run economic behavior and predict the factors that cause economic fluctuations (see Abarmovitz, 1961; Tvede, 2006).

Economic crises with dire consequences were not uncommon in history (see Kindelberger and Aliber, 2011), but a unifying theory that is helpful in understanding the causes and remedial measures is missing. Keynes ([1936] 1957) posited the rise and fall of aggregate demand as the source of economic fluctuations. Although even if the supply and demand did balance out at equilibrium, it would not still deliver full employment. Thus, government needs to deliberately finance public works to create jobs. On the contrary, Ludwig von Mises suggested that the intervening banking policy – underbidding the interest rates and expanding credit – is the source of business cycle. Policies that establish interest rates, wage rates, and commodity prices as opposed to the market mechanism not only invite destruction and chaos, but also hamper economic progress (see von Mises, 2006). According to Schumpeter (1934, 1939), an economy is a self-perpetuating system that experiences booms and busts. This cyclical phenomenon is mainly caused by innovations and is non-monetary. Other theoretical frameworks proposed by Friedman (1970), Lucas (1975), and Kydland and Prescott (1982) are also adept at explaining aspects of economic fluctuations and their implications.

Minsky (1976) proposed the theory of financial crises which suggests that hedge finance, speculative finance, and Ponzi schemes are vulnerable to the market condition and become liabilities to firms. Because of their vulnerabilities, any shock in the economy induces financial disturbances and thus affects it through feedback mechanisms. In addition, there are also other factors, such as coordination failures, moral hazards, regulatory mechanism, and liquidity crunches, that contribute to financial crises (see Brunnermeier, 2009; Crotty, 2009; Goldstein and Razin, 2013).

After the 2007 financial crisis, the US government created a bipartisan committee, the *Financial Crisis Inquiry Committee*, to study its causes. According to the Committee, the crisis was a result of human action and inaction that was tied to issues, such as financial regulation and markets; failures of corporate governance and risk management at financial institutions; a combination of excessive borrowing, risky investments, and lack of transparency; and a systemic breakdown in accountability and ethics (Financial Crisis Inquiry Commission, 2011). Similarly, the *Levin-Coburn Report* of the US Senate attributed high-risk lending, regulatory failure, inflated credit ratings, and investment bank abuses as the major drivers of the financial crisis (US Senate, 2011). Others argued that both the financial sector and the government were the enablers of the turmoil (see Krugman, 2009; Reinhart and Rogoff, 2009; Kindelberger and Aliber, 2011).

Entrepreneurs play a crucial role in the economy and in economic growth dynamics (Schumpeter, 1934, 1942), and through new firms they create new jobs and distribute wealth (Kirchhoff, 1994). Empirical studies in the field of entrepreneurship research have suggested that population growth (Audretsch and Fritsch,

1994; Guesnier, 1994; Reynolds, Miller, and Maki, 1995), income growth (Armington and Acs, 2002; Lee, Florida, and Acs, 2004), human capital (Armington and Acs, 2002), financial capital (Sutaria and Hicks, 2004), social capital, and cultural diversity (Saxenian, 2002; Lee et al., 2004; Audretsch, Dohse, and Niebuhr, 2010; Hart and Acs, 2011), in general, are positively associated with new firm births. However, the relationships between establishment size and new firms (Armington and Acs, 2002; Fritsch and Falck, 2002; Sutaria and Hicks, 2004), and employment size and new firms are not consistent (Storey, 1991; Reynolds et al., 1995; Fritsch and Falck, 2002; Sutaria and Hicks, 2004).

A well-functioning financial system is important for economic growth and development (see Levine, 1997; Aghion, Howitt, and Mayer-Foulkes, 2005). According to Schumpeter (1934), since innovations are important for the growth of a capitalist society, the banking system is crucial for identifying and funding promising innovations. In today's economic environment, capital sources extend beyond the conventional banking system and entrepreneurs obtain financial capital from a range of sources, such as venture capitalists, personal savings, and social networks. However, Rajan and Zingales (1998) suggested that while financial development can enhance innovation and growth indirectly, imperfections in the financial markets can have detrimental impacts on investment and growth.

A number of regional development theories on the spatial distribution of economic activities have been proposed. The theory of location by Johann-Heinrich von Thunen, in which agricultural activities are located in the proximity of transportation facilities, has been generalized to locational factors related to the theory of rent and land use. Isard (1949), on the other hand, pointed out that traditional general equilibrium analysis failed to account for the spatial dimension and neglected transportation and spatial costs on the distribution of economic activities in space and should be operationalized to consider the spatial array of economic activities with the geographic distributions of inputs and outputs and prices and costs. More recently, Krugman (1991, 1998), Markusen (1996), Ellison and Glaeser (1997), and Fujita, Krugman, and Venables (1999) also explain the importance of configurations and regional variations of economic activities.

According to Arthur (1990), new firms that spin-off from existing firms are found in close geographic proximity to their genetic parent. Porter's theory of competitive advantage, following the Marshall-Arrow-Romer knowledge spillover view, emphasizes the importance of geographic clusters and suggests that new businesses locate themselves in clusters that supply specialized inputs and require specialized infrastructure (Porter, 2000). In addition, Arbia (2001) noted that there is a higher chance of a new firm locating in the neighborhood of existing firms. Thus, whether in West Germany (Bade and Nerlinger, 2000) or in the US (Renski, 2009), new firms locate themselves in regions that are beneficial for growth and survival.

The purpose of this chapter is to understand the distribution patterns of new firm formation in New England – Connecticut, Maine, Massachusetts, New Hampshire, Rhode Island, and Vermont – from 1999 through 2009. Based on the theoretical background and empirical evidence, the study proposes to explore the following questions:

1. How are new firm (single-unit) births distributed in New England from 1999 through 2009?
2. Are the distributions of new firms comparable across industry sectors?
3. Are the distributions of new firms comparable to the distributions of other economic indicators?

Answers to these questions are important for policy makers and practitioners alike. From the policy perspective, regions that base their economic growth policy on entrepreneurship can understand the spatial distribution patterns of firm births at both the aggregate and sectoral levels in a period of financial turmoil. These distribution patterns will help them to determine whether the firm births would be concentrated or dispersed within and/or between states. Moreover, the comparison of the distribution single-unit firm births with the distributions of other economic variables should show how the distribution of new firm formation behaves with respect to the distributions of other economic trends. Thus, policy makers will not merely benefit from understanding the distribution of firm births during financial crisis, but will also have the opportunity to compare such a distribution to the distribution pattern of other economic indicators of the region. A practitioner, on the other hand, can apply a method based on entropy to examine the distribution of single-unit firm births as well as compare its distribution trend with the trends of other economic indicators. In addition, single-unit firm births can be replaced by other spatial variables and their distributions can also be studied.

## 3.3 Methods and data

The Gini coefficient, the Herfindahl index, the ogive index, and the entropy measure are some of the commonly used techniques for measuring economic diversification. The Gini coefficient is used for characterizing inequalities. It is defined as a ratio of the area under the Lorenz curve, and its value ranges from zero (complete equality) to one (complete inequality). Although the Gini coefficient is a popular measure of income and wealth concentrations, it is not additive and decomposable. That is, it is not possible to derive within and between values (see Theil, 1967; Shorrocks, 1980).

The Herfindahl index, also known as the Herfindahl-Hirschman index, is another measure of concentration generally used for measuring industrial concentration (see Hirschman, 1964). It is defined as the sum of the squares of all values of the variable of interest. The maximum value of the Herfindahl index is one (complete concentration) and the minimum is the inverse of all the observations. Unlike the Gini coefficient, the Herfindahl index is multiplicative and decomposable (see Theil, 1967). Since the measure gives heavier weight to the higher values of the variable, this index can often misrepresent the actual concentration, particularly highly variable spatial concentrations.

The Ogive index is a measure of variance of a variable against a uniform distribution. Often, employment is used as the benchmark. However, with the assumption of uniform employment share across the industry sectors and changes arising from the shift in the national economy, it would be difficult to defend the use of ogive as a measure except as a diversity index (see Bahl, Firestine, and Phares, 1971; Wasylenko and Erickson, 1978).

Entropy is a measure of uncertainty and the Shannon's discrete entropy is defined as (Shannon, 1948):

$$H = \sum_{i=1}^{n} p_i \log\left(\frac{1}{p_i}\right) = -\sum_{i=1}^{n} p_i \log p_i \tag{3.1}$$

where $n$ is the number of discrete observations and $\{p_i\}$ are discrete probabilities. $H$ is a continuous and an increasing function of $n$. Its value ranges from 0 to $\log n$ ($H \sim H_{max}$). As $H$ tends to zero, spatial concentration goes toward maximum, and as $H$ tends to $\log n$, spatial concentration goes toward minimum.

Relative entropy is a useful measure for comparing distributions and assumes values between 0 and 1. It is defined as:

$$\text{Relative } H = \frac{H}{\log n} \tag{3.2}$$

Entropy can be decomposed into two components – (1) between-set and (2) within-set entropies (Theil, 1967) – and the decomposed entropy can be written as:

$$H = \sum_j p_j \log\frac{1}{p_j} + \sum_j p_j \left[\sum_{i \in j}\left(\frac{p_i^{(j)}}{p_j}\right)\log\left(\frac{1}{p_i^{(j)} / p_j}\right)\right] = H_j + \sum_j p_j H_i^{(j)} \tag{3.3}$$

where $H_j$ is the between-set entropy and $\sum_j p_j H_i^{(j)}$ is the average within-set entropy for smaller observations $i$ embedded in larger observations $j$. In addition, Equation (3.1) and Equation (3.3) must be equal.

The between-set relative entropy is then calculated as:

$$\text{Relative } H_j = \frac{H_j}{\log j} \tag{3.4}$$

Likewise, the average within-set relative entropy is:

$$\text{Relative } H_{n/j} = \frac{H_{n/j}}{\log\left(\frac{n}{j}\right)} \tag{3.5}$$

Information gain is a measure that can be used for comparing two distributions and the expected information gain is written as:

$$I = \sum_{i=1}^{n} q_i \log\left(\frac{q_i}{p_i}\right) \tag{3.6}$$

where $\{p_i\}$ is the set of prior probabilities and $\{q_i\}$ is the set of posterior probabilities. If the distributions are similar, $I$ tends toward zero.

Entropy measure also has a number of limitations – ambiguous definitions, the assumption of uniform distribution, and spatial boundary definition (see Theil, 1967; Batty, 1974; Haynes, Phillips, and Mohrfeld, 1980). However, since entropy is asymptotically normally distributed, it is commonly used for measuring diversity (see Haynes and Storbeck, 1978; Pannell, 1988; Wheeler, 1990; Kulkarni, Stough, and Haynes, 1999; Feser, Hewings, and Mix, 2013).

According to Acs and Armington (2006) the spirit of entrepreneurship is reflected by the creation of a new firm instead of the expansion of an existing firm. Along this line, entropies of new firm formation will be measured for New England from 1999 to 2009. The data for single-unit new firm births are from the US Census Bureau. Since the relationship between unemployment and entrepreneurial venture is indeterminate, unemployment will be used for measuring the expected information gain. This gives us a measure particularly sensitive to the regional and national fiscal patterns. The unemployment data are from the US Bureau of Labor Statistics.

## 3.4 Results

Figure 3.1 shows the standard deviation of the distribution of total single-unit firm births in New England in 2002.

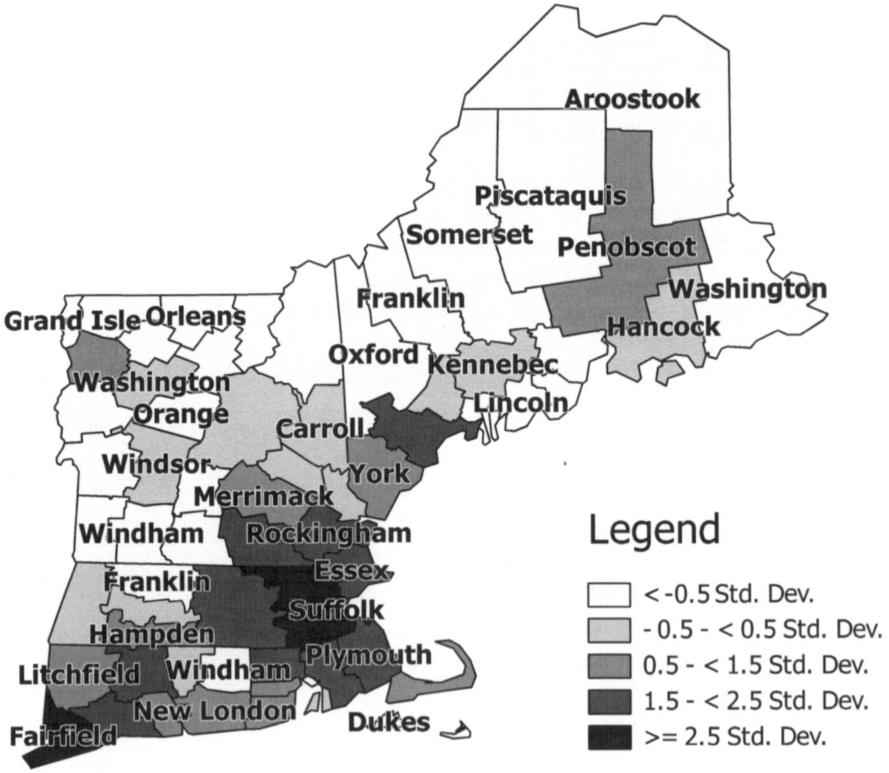


*Figure 3.1* Total single-unit firm births in New England (2002)

Source: US Census Bureau

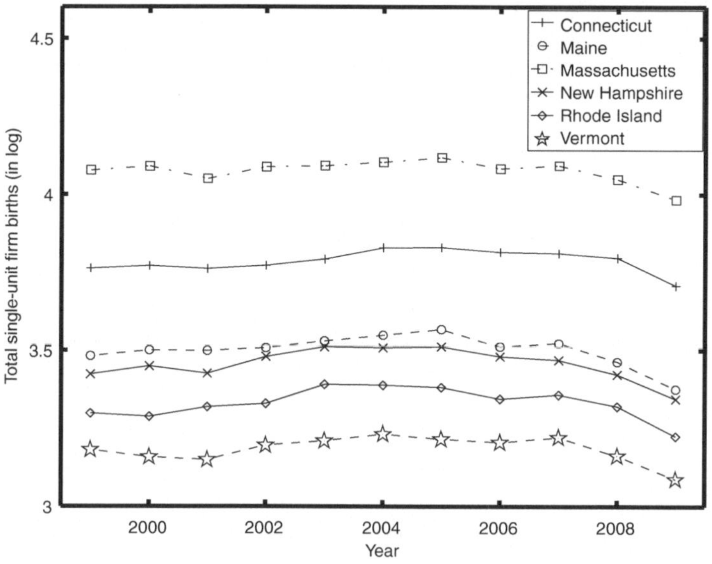


*Figure 3.2* Total single-unit firm births in New England (1999–2009)

In 2002, although Cumberland County in Maine and Fairfield, New Haven, and Hartford counties in Connecticut experienced a more than average level of firm births in New England, Massachusetts had the largest number of single-unit firm births. During the study period, these trends were generally true.

Figure 3.2 shows the total number of new firm formation (in logarithm) from 1999 to 2009. It should be noted that after 2005, and notably from 2007 onwards, the number of new firm formation slowly started to decline as the Great Recession was reflected in the business cycle.

Table 3.1 shows various entropy measures across years. While there was a declining trend in the number of new firms after 2005, all the measures suggest that the distribution patterns of new firm formation were almost consistent across New England. In general, firm formation was dispersed. The between-region entropies suggest that the distribution patterns of new firm births were more concentrated across states. However, the average within-region entropies indicate that, on average, firm formation patterns dispersed more within states than across states. Thus, although the number of new firms was declining, there was a minimal change in the actual distribution pattern of firm births (see also Figure 3.3).

However, it should also be noted that while there is no dramatic change in the relative entropy values, their decline after 2007 suggests that firm births became more concentrated in New England.

The total number of single-unit firm births was disaggregated into the manufacturing and service sectors. The service sector included: (1) Wholesale trade; (2) Retail trade; (3) Transportation and warehousing; (4) Information; (5) Finance and insurance; (6) Real estate and rental and leasing; (7) Professional, scientific,

*Table 3.1* Entropy measures for total single-unit firm births in New England (1999–2009)

| *Aggregate* | | | | | | |
|---|---|---|---|---|---|---|
| *Year* | *Shannon's entropy* | *Relative entropy* | *Between-region entropy* | *Within-region entropy* | *Between-region relative entropy* | *Average within-region relative entropy* |
| 1999 | 3.576 | 0.850 | 1.519 | 2.056 | 0.848 | 0.852 |
| 2000 | 3.568 | 0.848 | 1.512 | 2.055 | 0.844 | 0.851 |
| 2001 | 3.602 | 0.856 | 1.540 | 2.061 | 0.859 | 0.854 |
| 2002 | 3.605 | 0.857 | 1.535 | 2.070 | 0.856 | 0.858 |
| 2003 | 3.620 | 0.860 | 1.555 | 2.064 | 0.867 | 0.855 |
| 2004 | 3.618 | 0.860 | 1.552 | 2.065 | 0.866 | 0.855 |
| 2005 | 3.612 | 0.859 | 1.542 | 2.069 | 0.860 | 0.857 |
| 2006 | 3.618 | 0.860 | 1.543 | 2.074 | 0.861 | 0.859 |
| 2007 | 3.629 | 0.863 | 1.541 | 2.087 | 0.860 | 0.865 |
| 2008 | 3.593 | 0.854 | 1.536 | 2.056 | 0.857 | 0.852 |
| 2009 | 3.582 | 0.852 | 1.524 | 2.058 | 0.850 | 0.853 |

Source: Authors' calculations.

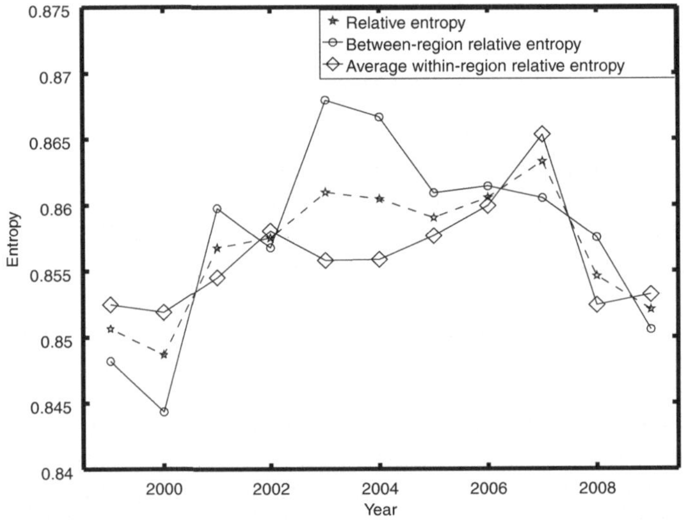


*Figure 3.3* Relative entropies for total single-unit firm births in New England (1999–2009)

and technical services; (8) Management of companies and enterprises; (9) Administrative and support and waste management and remediation services; (10) Educational services; (11) Health care and social assistance; (12) Arts, entertainment, and recreation; (13) Accommodation and food services; (14) Other services (except public administration).

The general trend in the manufacturing and sector sectors was that new firm births were declining, especially after 2007. However, various entropy measures for both these sectors (Table 3.2) indicate that the distribution patterns of new firm formation were not much different from the distribution patterns of total new firm births.

*Table 3.2* Entropy measures for single-unit firm births in the manufacturing and service sectors

| *Year* | *Manufacturing* | | | | | |
|---|---|---|---|---|---|---|
| | *Shannon's entropy* | *Relative entropy* | *Between-region entropy* | *Average within-region entropy* | *Between-region relative entropy* | *Average within-region relative entropy* |
| 1999 | 3.695 | 0.878 | 1.607 | 2.087 | 0.897 | 0.865 |
| 2000 | 3.653 | 0.840 | 1.535 | 2.000 | 0.856 | 0.829 |
| 2001 | 3.619 | 0.860 | 1.582 | 2.037 | 0.883 | 0.844 |
| 2002 | 3.620 | 0.861 | 1.585 | 2.034 | 0.885 | 0.843 |
| 2003 | 3.658 | 0.870 | 1.610 | 2.048 | 0.898 | 0.848 |
| 2004 | 3.595 | 0.855 | 1.604 | 1.990 | 0.895 | 0.825 |
| 2005 | 3.620 | 0.861 | 1.593 | 2.027 | 0.889 | 0.840 |
| 2006 | 3.664 | 0.871 | 1.620 | 2.044 | 0.904 | 0.847 |
| 2007 | 3.673 | 0.873 | 1.607 | 2.066 | 0.896 | 0.856 |
| 2008 | 3.561 | 0.847 | 1.581 | 1.979 | 0.882 | 0.820 |
| 2009 | 3.606 | 0.857 | 1.586 | 2.019 | 0.885 | 0.837 |
| *Year* | *Service* | | | | | |
| | *Shannon's entropy* | *Relative entropy* | *Between-region entropy* | *Average within-region entropy* | *Between-region relative entropy* | *Average within-region relative entropy* |
| 1999 | 3.530 | 0.839 | 1.498 | 2.032 | 0.836 | 0.842 |
| 2000 | 3.524 | 0.838 | 1.496 | 2.028 | 0.835 | 0.840 |
| 2001 | 3.564 | 0.847 | 1.530 | 2.034 | 0.854 | 0.843 |
| 2002 | 3.562 | 0.847 | 1.518 | 2.043 | 0.847 | 0.847 |
| 2003 | 3.579 | 0.851 | 1.544 | 2.035 | 0.861 | 0.843 |
| 2004 | 3.578 | 0.851 | 1.540 | 2.038 | 0.859 | 0.844 |
| 2005 | 3.570 | 0.849 | 1.530 | 2.040 | 0.853 | 0.845 |
| 2006 | 3.573 | 0.849 | 1.531 | 2.042 | 0.854 | 0.846 |
| 2007 | 3.578 | 0.851 | 1.526 | 2.051 | 0.851 | 0.850 |
| 2008 | 3.553 | 0.845 | 1.525 | 2.028 | 0.851 | 0.840 |
| 2009 | 3.544 | 0.842 | 1.510 | 2.034 | 0.842 | 0.843 |

Source: Authors' calculations.

The temporal changes of relative entropies for the manufacturing and service sectors are shown in Figure 3.4 and Figure 3.5. Note that in the aftermath of the financial crisis of 2007, the distribution patterns of new firm formation became more concentrated in New England.

The service sector data were further disaggregated to individual service sectors. Table 3.3 shows entropies of the information (see also Figure 3.6), and finance and information (see also Figure 3.7) sectors. Firm births in both these sectors were less dispersed than at the aggregate or the manufacturing and service sector levels.

All the relative entropies in both the sectors dropped after 2007. However, the continuous decline in the entropy values of the finance and insurance sector,

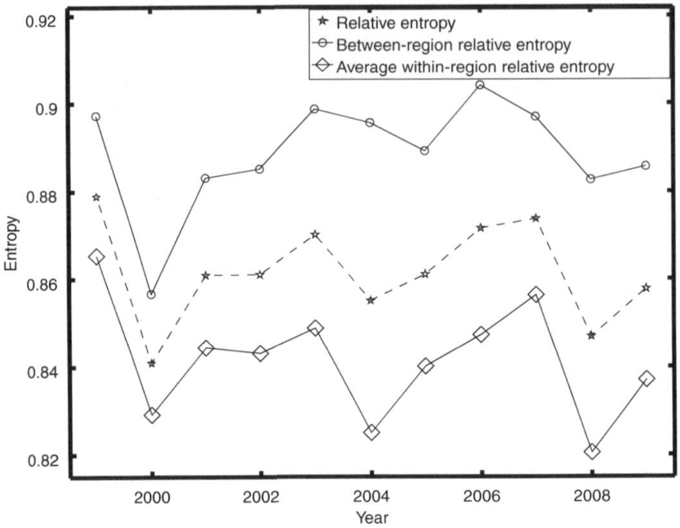


*Figure 3.4* Relative entropies for single-unit firm births (manufacturing)

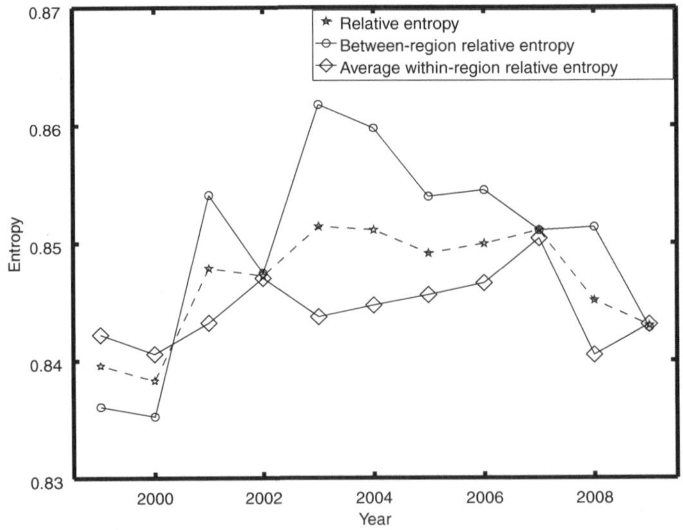


*Figure 3.5* Relative entropies for single-unit firm births (service)

especially since 2006, suggests that firm formation in this sector was more localized than firm formation in other sectors.

Within the service sector, the professional, scientific, and technical services sector also showed similar firm births distribution patterns. However, single-unit new firm formation in the real estate and rental and leasing, health care and

*Table 3.3* Entropy measures for single-unit firm births in the information, and finance and insurance sectors

| *Service* | | | | | | |
|---|---|---|---|---|---|---|
| *Year* | *Information* | | | | | |
| | *Shannon's entropy* | *Relative entropy* | *Between-region entropy* | *Average within-region entropy* | *Between-region relative entropy* | *Average within-region relative entropy* |
| 1999 | 3.145 | 0.748 | 1.319 | 1.825 | 0.736 | 0.756 |
| 2000 | 3.124 | 0.743 | 1.291 | 1.832 | 0.720 | 0.759 |
| 2001 | 3.284 | 0.781 | 1.452 | 1.832 | 0.810 | 0.759 |
| 2002 | 3.211 | 0.763 | 1.434 | 1.777 | 0.800 | 0.736 |
| 2003 | 3.246 | 0.772 | 1.432 | 1.814 | 0.799 | 0.751 |
| 2004 | 3.245 | 0.771 | 1.494 | 1.750 | 0.834 | 0.725 |
| 2005 | 3.294 | 0.783 | 1.436 | 1.857 | 0.801 | 0.769 |
| 2006 | 3.254 | 0.773 | 1.423 | 1.830 | 0.794 | 0.758 |
| 2007 | 3.241 | 0.770 | 1.418 | 1.822 | 0.791 | 0.755 |
| 2008 | 3.153 | 0.750 | 1.432 | 1.720 | 0.799 | 0.713 |
| 2009 | 3.205 | 0.762 | 1.386 | 1.818 | 0.773 | 0.753 |
| *Year* | *Finance and insurance* | | | | | |
| | *Shannon's entropy* | *Relative entropy* | *Between-region entropy* | *Average within-region entropy* | *Between-region relative entropy* | *Average within-region relative entropy* |
| 1999 | 3.211 | 0.763 | 1.450 | 1.760 | 0.809 | 0.729 |
| 2000 | 3.155 | 0.750 | 1.415 | 1.739 | 0.789 | 0.720 |
| 2001 | 3.222 | 0.766 | 1.467 | 1.754 | 0.819 | 0.727 |
| 2002 | 3.176 | 0.755 | 1.409 | 1.766 | 0.786 | 0.732 |
| 2003 | 3.170 | 0.754 | 1.506 | 1.663 | 0.841 | 0.689 |
| 2004 | 3.224 | 0.766 | 1.498 | 1.726 | 0.836 | 0.715 |
| 2005 | 3.168 | 0.753 | 1.481 | 1.686 | 0.826 | 0.699 |
| 2006 | 3.182 | 0.756 | 1.443 | 1.738 | 0.805 | 0.720 |
| 2007 | 3.194 | 0.759 | 1.480 | 1.713 | 0.826 | 0.710 |
| 2008 | 3.121 | 0.742 | 1.434 | 1.687 | 0.800 | 0.699 |
| 2009 | 3.074 | 0.731 | 1.411 | 1.663 | 0.787 | 0.689 |

Source: Authors' calculations.

social assistance, and accommodation and food services sectors were more dispersed.

Figure 3.8 shows the unemployment trends in New England. The expected information gain was calculated to compare the distribution patterns of firm formation and unemployment and is shown in Table 3.4. Unemployment was also lagged by a year to examine the differences in the distribution patterns of firm births. It was found that whether the unemployment was lagged or not, very small values of the expected information gain suggest that firm formation and unemployment patterns more or less follow each other.

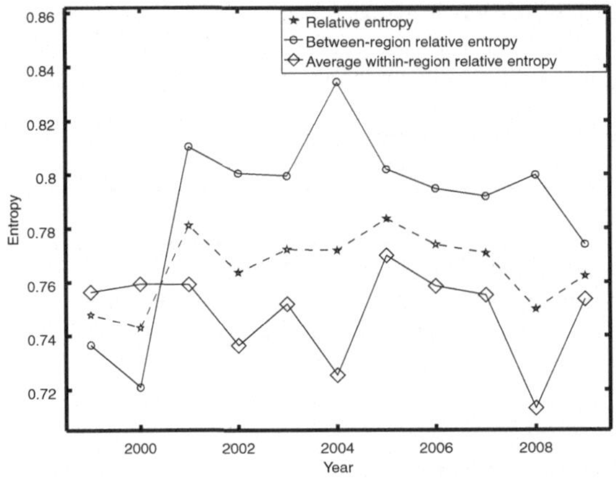


*Figure 3.6* Relative entropies for single-unit firm births (information)

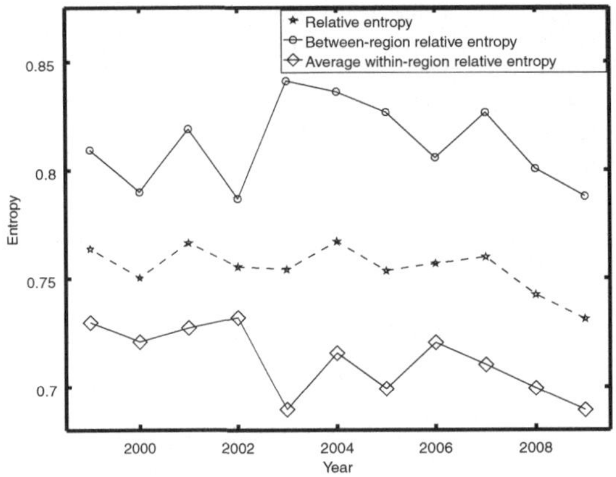


*Figure 3.7* Relative entropies for single-unit firm births (finance and insurance)

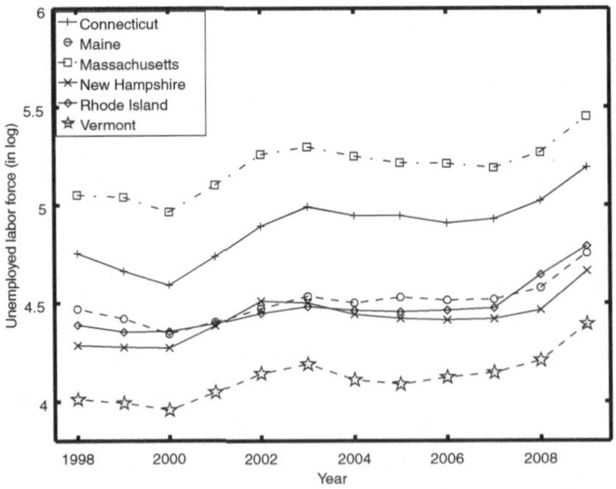


*Figure 3.8* Unemployed labor force

*Table 3.4* Expected information gain

| *Year* | *Unemployment (No lag)* | *Unemployment (One year lag)* |
|---|---|---|
| 1999 | 0.055 | 0.061 |
| 2000 | 0.046 | 0.057 |
| 2001 | 0.042 | 0.045 |
| 2002 | 0.047 | 0.038 |
| 2003 | 0.053 | 0.050 |
| 2004 | 0.052 | 0.051 |
| 2005 | 0.048 | 0.053 |
| 2006 | 0.048 | 0.049 |
| 2007 | 0.045 | 0.047 |
| 2008 | 0.048 | 0.043 |
| 2009 | 0.041 | 0.048 |

Source: Authors' calculations.

## 3.5 Conclusions

The purpose of this chapter was to examine the distribution patterns of new firm formation in New England. It was found that single-unit firm births were generally dispersed across New England between 1999 and 2009. However, distributions of firm births in sectors such as information, and finance and insurance, are more concentrated than sectors such as health care and social assistance, and accommodation and food services. Moreover, within-region entropies suggest that firm births' dynamics could be localized in some counties embedded in their respective states. The values of expected information gains suggest the existence of scale economy and business cycle effects. In other words, regions that have higher unemployment experience a higher number of firm births.

Overall, this study found that economic fluctuation and the distribution of entrepreneurial activities is closely associated in New England. Since the US went through economic disruptions resulting from terrorism to the global financial crisis, it is likely that other regions apart from New England in the US also had a similar experience. It could also mean that New England was an isolated case and the experience of other regions was different in the study period. In any case, government policy to foster economic competitiveness in a region through new firms should take into account the spatio-temporal distributional dynamics of new firm formation, and scale economy and business cycle. For instance, if firm births in some sectors are more concentrated than in others and the distribution patterns follow the scale and cyclical effects, the policy should focus on promoting new firms in a localized space such that new firms can benefit from the sectoral spillover effects. On the contrary, the services from sectors such as health care and social assistance or accommodation and food services are likely to be required more evenly compared to the services of the highly localized sectors. In such situations, the government should create environments to attract new firms beyond some smaller regions embedded within a larger region.

Future research should include spatial entropy extension. In the case of data availability, other economic regions should be included and regional comparison should be made. In addition, econometric analysis of entropy could also be done.

## References


Abarmovitz, M. (1961) The nature and significance of Kuznets cycles, *Economic Development and Cultural Change*, *9*(3), 225–248.

Acs, Z. J., and Armington, C. (2006) *Entrepreneurship, geography, and American economic growth*. New York: Cambridge University Press.

Aghion, P., Howitt, P., and Mayer-Foulkes, D. (2005) The effect of financial development on convergence: theory and evidence, *Quarterly Journal of Economics*, *120*(1), 173–222.

Arbia, G. (2001) Modelling the geography of economic activities on a continuous space, *Papers in Regional Science*, *80*(4), 411–424.

Armington, C., and Acs, Z. J. (2002) The determinants of regional variation in new firm formation, *Regional Studies*, *36*(1), 33–45.

Arthur, W. B. (1990) Positive feedbacks in the economy, *Scientific American*, *262*(2), 92–99.

Audretsch, D. B., and Fritsch, M. (1994) The geography of firm births in Germany, *Regional Studies*, *28*(4), 359–365.

Audretsch, D. B., Dohse, D., and Niebuhr, A. (2010) Cultural diversity and entrepreneurship: a regional analysis for Germany, *Annals of Regional Science*, *45*(1), 55–85.

Bade, F.-J., and Nerlinger, E. A. (2000) The spatial distribution of new technology-based firms: empirical results for West-Germany, *Papers in Regional Science*, *79*(2), 155–176.

Bahl, R. W., Firestine, R., and Phares, D. (1971) Industrial diversity in urban areas: alternative measures and intermetropolitan comparisons, *Economic Geography*, *47*(3), 414–425.

Batty, M. (1974) Spatial entropy, *Geographical Analysis*, *6*(1), 1–31.

Brunnermeier, M. K. (2009) Deciphering the liquidity and credit crunch 2007–2008, *Journal of Economic Perspectives*, *23*(1), 77–100.

Crotty, J. (2009) Structural causes of the global financial crisis: a critical assessment of the "New Financial Architecture", *Cambridge Journal of Economics*, *33*(4), 564–580.

Ellison, G., and Glaeser, E. L. (1997) Geographic concentration in U.S. manufacturing industries: a dashboard approach, *Journal of Political Economy*, *105*(5), 889–927.

Fairlie, R. W. (2013) *Kauffman Index of entrepreneurial activity 1996–2012*. Kansas City, MO: Ewing Marion Kauffman Foundation. Retrieved from www.kauffman.org/uploadedfiles/kiea_2013_report.pdf.

Feser, E., Hewings, G., and Mix, T. (2013) Characterizing local economic diversity in Appalachia: linking industry, workforce skills, function and location. Presented at the 52nd Southern Regional Science Association Meetings, Washington, DC.

Financial Crisis Inquiry Commission (2011) *The financial crisis inquiry report*. Retrieved from www.gpo.gov/fdsys/pkg/GPO-FCIC/pdf/GPO-FCIC.pdf.

Friedman, M. (1970) A theoretical framework for monetary analysis, *Journal of Political Economy*, *78*(2), 193–238.

Fritsch, M., and Falck, O. (2002) *New firm formation by industry over space and time: a multi-level analysis* (Working Paper No. 11). Freiberg, Germany: Freiberg University of Mining and Technology. Retrieved from www.wiwi.uni-jena.de/uiw/publications/pub_1999_2003/fritsch_falck_2002.pdf.

Fujita, M., Krugman, P., and Venables, A. J. (1999) *The spatial economy: cities, regions, and international trade*. Cambridge, MA: MIT Press.

Glaeser, E. L., and Kerr, W. R. (2009) Local industrial conditions and entrepreneurship: how much of the spatial distribution can we explain? *Journal of Economics and Management Strategy*, *18*(3), 623–663.

Goldstein, I., and Razin, A. (2013) *Review of theories of financial crises* (Working Paper No. 18670). Cambridge, MA: National Bureau of Economic Research.

Guesnier, B. (1994) Regional variations in new firm formation in France, *Regional Studies*, *28*(4), 347–358.

Hart, D. M., and Acs, Z. J. (2011) High-tech immigrant entrepreneurship in the United States, *Economic Development Quarterly*, *25*(2), 116–129.

Haynes, K. E., Phillips, F. Y., and Mohrfeld, J. W. (1980) The entropies: some roots of ambiguity, *Socio-Economic Planning Sciences*, *14*(3), 137–145.

Haynes, K. E., and Storbeck, J. S. (1978) The entropy paradox and the distribution of urban population, *Socio-Economic Planning Sciences*, *12*(1), 1–6.

Hirschman, A. O. (1964) The paternity of an index, *American Economic Review*, *54*(5), 761.

Isard, W. (1949) The general theory of location and space-economy, *Quarterly Journal of Economics*, *63*(4), 476–506.

Keynes, J. M. ([1936] 1957) *The general theory of employment, interest and money*. London: Macmillan.

Kindleberger, C. P., and Aliber, R. Z. (2011) *Manias, panics and crashes: a history of financial crises* (6th ed.). New York: Palgrave Macmillan.

Kirchhoff, B. A. (1994) *Entrepreneurship and dynamic capitalism: the economics of business firm formation and growth*. Westport, CT: Greenwood Publishing.

Krugman, P. (1991) Increasing returns and economic geography, *Journal of Political Economy*, *99*(3), 483–499.

Krugman, P. (1998) What's new about the new economic geography? *Oxford Review of Economic Policy*, *14*(2), 7–17.

Krugman, P. (2009) *The return of depression economics and the crisis of 2008*. New York: W.W. Norton and Co.

Kulkarni, R. G., Stough, R. R., and Haynes, K. E. (1999) Towards an information entropy model of job approval rating: the Clinton presidency, *Entropy*, *1*(3), 37–49.

Kydland, F. E., and Prescott, E. C. (1982) Time to build and aggregate fluctuations, *Econometrica*, *50*(6), 1345–1370.

Lee, S. Y., Florida, R., and Acs, Z. J. (2004) Creativity and entrepreneurship: a regional analysis of new firm formation, *Regional Studies*, *38*(8), 879–891.

Levine, R. (1997) Financial development and economic growth: views and agenda, *Journal of Economic Literature*, *35*(2), 688–726.

Lucas, R. E. (1975) An equilibrium model of the business cycle, *Journal of Political Economy*, *83*(6), 1113–1144.

Markusen, A. (1996) Sticky places in slippery space: a typology of industrial districts, *Economic Geography*, *72*(3), 293–313.

Minsky, H. P. (1976) A theory of systemic fragility. Presented at the Conference on Financial Crises, New York.

Pannell, C. W. (1988) Regional shifts in China's industrial output, *Professional Geographer*, *40*(1), 19–32.

Porter, M. E. (2000) Location, competition, and economic development: local clusters in a global economy, *Economic Development Quarterly*, *14*(1), 15–34.

Rajan, R. G., and Zingales, L. (1998) Financial dependence and growth, *American Economic Review*, *88*(3), 559–586.

Reinhart, C. M., and Rogoff, K. S. (2009) *This time is different: eight centuries of financial folly*. Princeton, NJ: Princeton University Press.

Renski, H. (2009) New firm entry, survival, and growth in the United States, *Journal of the American Planning Association*, *75*(1), 60–77.

Reynolds, P. D., Miller, B., and Maki, W. R. (1995) Explaining regional variation in business births and deaths: U.S. 1976–88, *Small Business Economics*, *7*(5), 389–407.

Saxenian, A. (2002) Silicon Valley's new immigrant high-growth entrepreneurs, *Economic Development Quarterly*, *16*(1), 20–31.

Schumpeter, J. A. (1934) *The theory of economic development: an inquiry into profits, capital, credit, interest and the business cycle* (R. Opie, Trans.). Cambridge, MA: Harvard University Press.

Schumpeter, J. A. (1939) *Business cycles: a theoretical, historical, and statistical analysis of the capitalist process* (vol. 2). New York: McGraw-Hill.

Schumpeter, J. A. (1942) *Capitalism, socialism, and democracy*. New York: Harper-Collins.

Shannon, C. E. (1948) A mathematical theory of communication, *Bell System Technical Journal*, *27*(3), 379–423.

Shorrocks, A. F. (1980) The class of additively decomposable inquality measures, *Econometrica*, *48*(3), 613–625.

Storey, D. J. (1991) The birth of new firms: does unemployment matter? A review of the evidence, *Small Business Economics*, *3*(3), 167–178.

Sutaria, V., and Hicks, D. A. (2004) New firm formation: dynamics and determinants, *Annals of Regional Science*, *38*(2), 241–262.

Theil, H. (1967). *Economics and information theory*. Amsterdam: North-Holland Publishing Company.

Tvede, L. (2006) *Business cycles: history, theory and investment reality* (3rd ed.). Chichester: John Wiley and Sons.

US Senate. (2011) *Wall Street and the financial crisis: anatomy of a financial collapse*. Retrieved from www.levin.senate.gov/imo/media/doc/supporting/2011/PSI_WallStreetCrisis_041311.pdf.

Von Mises, L. (2006) *The causes of the economic crisis and other essays before and after the Great Depression* (P. L. Greaves, Ed.). Auburn, AL: Ludwig von Mises Institute.

Wasylenko, M. J., and Erickson, R. A. (1978) On measuring economic diversification: comment, *Land Economics*, *54*(1), 106–109.

Wheeler, J. O. (1990) The new corporate landscape: America's fastest growing private companies, *Professional Geographer*, *42*(4), 433–44.

World Economic Forum (2012) *The global competitiveness report 2011–2012*. Retrieved from www3.weforum.org/docs/WEF_GCR_Report_2011-12.pdf.

# 4 The impacts of policy changes on overseas human capital in Australia

## The implementation of the 485 graduate visa scheme

*Jonathan Corcoran, Francisco Rowe, Alessandra Faggian, and Robert J. Stimson*

### 4.1 Introduction

This chapter evaluates a public policy intervention that resulted in a change in the graduate visa scheme affecting the eligibility of international students studying in universities in Australia to stay-on post-graduation to work. That evaluation focuses on analysing both the employment outcomes and the inter-regional migration patterns.

The phenomenon of skilled and unskilled international migration in the contemporary world is well researched (see, for example, Franklin 2003; Kuptsch and Eng Fong 2006; Özden and Schiff 2007) and is a recurrent issue in policy debates (see Iredale 2002; Lowell 2007). Much of the focus has been on evaluating labour market effects (Borjas 2005) and the impact of selection criteria in the immigrant settlement process (Cobb-Clark 1999). To these ends, a range of methodologies have been used, including: econometric procedures (Dehejia and Wahba 1999), input-output analysis (Glass et al. 2006), simulation techniques (see Bard 1978), spatial and GIS-based analyses (see Courtney 2005), and statistical modelling of data derived from questionnaire-based surveys (Zhang et al. 2006). The analytical approaches within which these methodologies have been applied vary from micro-level analyses using individual-level data to meso-scale evaluations.

Generally, immigration is seen as having positive impacts on a national economy (see, for example, Hoffmann's 1998 study of the US), and it is touted as being an enhancing factor in economic growth (Iredale 2002; Kuptsch and Eng Fong 2006; Lowell 2007). Skilled migration in particular is often seen as being associated with filling gaps in local labour market shortages and in enhancing economic performance. However, studies have also shown that international skilled migration might have negative outcomes for both the individual and the receiving country (Regets 2001, 2007).

In this context, changes in government policy can be a compounding factor influencing the impacts of international migration. This can be especially true when policy changes result in a sudden short-term change, influencing the overall quantum of the immigration flow and its balance of skills across industry and occupation categories. Often short-term policy changes – especially in temporary immigration schemes – seek to address issues, such as skills shortages in specific occupations. In Australia, a working holiday visa scheme operated to address shortages of low-skilled workers in accommodation and agricultural industries (see, for example, Tan and Lester 2011).

This chapter provides an exploratory analysis of a specific aspect of the skilled immigration phenomenon in Australia resulting from a recent policy change. The focus is on investigating the impact of the Australian Government's graduate visa scheme (known as the 485 visa). This visa precipitated a sharp increase in the number of international students who then stayed-on post-graduation and entered the workforce. Surprisingly, there is little research that sheds light on the fields of employment, working conditions and inter-regional migration of patterns of those graduates who stay-on and on their potential impacts on Australian regions. Along with other papers by the authors and their collaborators (Rowe et al. 2013; Tang et al. 2014, 2015, forthcoming; Faggian et al. 2015), this chapter fills some of that void.

## 4.2 Background

Historically, Australia has operated a comprehensive immigration program throughout the post-World War II era. Over this period, the balance of the source countries has changed from the UK, then Europe, and more recently to the Middle East, New Zealand and Asia. That shift in source countries is related to a significant degree to policy shifts. In the 1980s there was a substantial focus in Australia's immigration program on family reunion, but since then that focus has shifted to favour skilled migrants and, more recently, students, with immigration flows increasingly being sourced from Asian countries (Shah and Burke 2005; Hugo 2006a, 2006b; Koleth 2010; Birrell et al. 2011).

The Australian Government's General Skilled Migration Program (GSMP) reflects a growing view that skilled immigration is important for the nation. As a result, immigration policy has shifted to place a greater emphasis on attracting and retaining highly-skilled immigrants (Hawthorne 2005; Birrell et al. 2006; Department of Immigration and Citizenship 2010). This has occurred at the same time as there has been established that Australia's rural and regional urban areas are more disadvantaged than the major metropolitan cities in terms of attracting and retaining population (Sher and Sher 1994). In response, a number of policies have been put in place to address skills shortages in rural regions, with the potential to attract the flow of immigrants that have traditionally concentrated in a few major cities, particularly Sydney and Melbourne.

In recent years there has also been an associated distinct shift in the quantum from long-term (permanent) to short-term (temporary) immigration, with policy initiatives such as the 457 visa scheme (Khoo et al. 2003, 2007a, 2007b, 2008). This scheme was introduced to enable easier entry of temporary immigrants to fill labour market shortages (Kinnaird 2002, 2005; Birrell and Rapson 2005) and to facilitate Australian firms to foster international collaboration and trade with overseas counterparts (Department of Employment and Workplace Relations 2002; Hugo 2006a, 2006b). An important component of that balance-shift has also been a large increase in the number of international students studying in

Australian universities as well as other components of the education and training sector. The number of international students studying in Australia has grown to represent about 11% of the global international student market (Koleth 2010), with the education sector becoming a key new export sector.

On 1 September 2007, the Australian Government implemented a significant change in immigration policy by introducing a graduate visa scheme (the 485 visa sub-class). This scheme provides an opportunity to overseas students who have studied in the nation's universities to stay-on in Australia post-graduation to work for 18 months. That policy initiative reflects a wider view internationally in advanced countries that the retention of international graduates who have studied in a country is a key factor for economic development and represents a means of enhancing a nation's human capital stock (Suter and Jandl 2008). Not surprisingly there has resulted a sharp increase in the number of international students who have stayed-on (Birrell et al. 2011) as the 485 visa scheme was seen as a pathway to achieving permanent residency in Australia. The number of applicants under this scheme increased steadily from 11,807 in 2007–08 to 22,888 in 2008–09, and to 28,126 in 2009–10 (Birrell et al. 2011).

While Corcoran et al. (2010) have analysed the inter-regional migration of domestic university graduates in Australia, little has been done to document the spatial movements and job outcomes of the recipients of the 485 visa scheme and the degree to which the inter-regional migration behaviours and job outcomes of the international graduate cohort might differ from Australian graduate cohort. Only three analyses, by Rowe et al. (2013), Tang et al. (2014) and Faggian et al. (2015), provide a recent view on these issues. This chapter aims to deliver an integrated and cohesive representation of the findings of these studies, and draws a number of key implications for regional development and policy. Fundamental questions the research investigates are:

1. Has the introduction of the 485 visa scheme changed the characteristics of international students graduating from Australia's universities?
2. Has the introduction of the 485 visa scheme altered the spatial dynamics (that is, the location choices) and the occupational choices of international graduates staying-on in Australia post-graduation?
3. How do the post-graduation spatial dynamics (location choices) and the occupation choices of the international graduate cohort differ from those of the Australian graduate cohort, and what are the effects of those choices on salary levels?
4. The data analysed enables a comparison of the inter-regional migration patterns and the working conditions *vis-à-vis* the salary levels of these cohorts of graduates both pre and post the introduction of the 485 visa scheme.

## 4.3 Data

In the studies reported here, individual-level data were used. These data were collected in the Australian Graduate Survey (AGS) conducted by Graduate

Careers Australia (GCA). The micro-data included information describing the following:

- a graduate's personal characteristics (such as gender, ethnicity, age, citizenship);
- the course studied (postcode of university location, subject studied, degree obtained, mode of attendance – that is, full-time versus part-time); and
- the graduate's employment circumstances between six and 18 months post-graduation (including location of employer at postcode level, initial salary, industry and type of occupation).

For the analysis, two waves of AGS data were used:

- the AGS 2006 survey (for the 2005 graduating students); and
- the AGS 2009 survey (for the 2008 graduating students).

Those dates correspond to 'pre' and 'post' the introduction of the skilled graduate 485 visa scheme. Thus, we analyse the survey data for four explicit cohorts of graduates. The analyses are restricted to those graduates remaining in Australia and who are in paid employment. Those engaged in further study are not included in the analyses.

The final dataset used in the analyses contained 55,569 Australian and 3,844 international students that were 'valid' (that is, with no missing information on variables relevant to the analyses), graduate observations for the 2006 survey (that is, the 2005 graduating cohorts), and 64,249 Australian and 7,291 international students for the 2009 survey (the 2008 graduating cohorts).

## 4.4 Methodology

A series of methods were employed to answer the research questions. These methods were intended to:

- identify the characteristics of the international graduates *vis-à-vis* domestic graduates;
- compare the employment outcomes for the international and domestic graduate cohorts both pre and post the introduction of the 485 visa scheme;
- map and analyse the spatial dynamics of internal migration flows of the cohorts in taking up employment post-graduation, and the way in which those patterns might have changed following the introduction of the scheme; and
- identify the influence of demographic, course and employment characteristics on employment location choices of international graduates post-graduation.

The explicit methods that were used to analyse the GCA data are described in summary form in what follows (for a detailed discussion see Rowe et al. 2013; Faggian et al. 2015).

### *4.4.1 Using descriptive statistics and a 'difference-in-difference' methodology*

Descriptive statistics were used to provide a profile of the international and Australia graduate cohort, with respect to their course/subject of study, professions of employment post-graduation, and differences in personal, job and employment location characteristics. In addition, a 'difference-in-difference' methodology, which was employed by Dynarski (2003) in a study of aid for college students in the US, was used to compare the two graduating cohorts – Australian and the international – both pre and post the introduction of the 485 visa scheme. As explained by Faggian et al. (2015), this enables us to '... uncover the real effect of the 485 graduate visa scheme excluding other macro-economic factors that might have affected the employability of the overall student population'.

Z-tests for means and proportions were used to assess whether there was a significant difference before and after the introduction of the 485 visa scheme for each of the graduating cohorts. In addition, *t*-tests were then used to test whether any differences over time for international and Australian graduates were significant.

### *4.4.2 A 'difference-in-difference' regression model*

A 'difference-in-difference' regression model was then used to explicitly examine the effect of the introduction of the 485 visa scheme on the average salaries of the international graduating cohort. As explained by Faggian et al. (2015), the challenge is to '... disentangle the effects of the introduction of the graduate program visa from those emerging from Australian firms to pay higher or lower salaries'.

The Australian cohort was used as a 'control' group as the salaries being received in post-graduation employment are not expected to change due to the introduction of the new visa policy. The 'difference-in-difference' regression was thus used to compare the differences in the salaries between the international and the Australian cohort before and after the introduction of the 485 visa scheme and also to consider a set of 'control variables' (independent variables in the regression), including:

- personal characteristic such as age, gender, disability and level of qualification; and
- job characteristics such as industry sector, occupation, type of contract, number of working hours, and the location of the job;
- dummy variables are used to identify the cohort of international student graduates and the year 2005 graduating cohort. As explained by Faggian et al. (2015), this 'ensures that the regression parameters capture the effects of the visa'.

The dependent variable is the salary level for both cohorts. Three regression models are run:

- a *basic model* where no control variables are used;
- a fully *specified model* using the personal and the job characteristics effects; and
- a *restricted model* run on a sample of graduates working full-time.

### *4.4.3 Migration flow patterns*

Patterns of migration flows for both the international and Australian graduating cohorts were analysed. The location of higher education institution and first employment after graduation were considered, both pre and post the introduction of the 485 visa scheme. Coupled to flow mapping visualisation techniques, a set of migration measures of intensity and connectivity proposed by Bell et al. (2002) were employed.

Migration flows were analysed in two contexts:

1. The patterns of inter-State/Territory migration flows across Australia were mapped.
2. Those flows were analysed within the framework of five broad Remoteness Areas into which Australia may be classified. That framework has been used to compile an Accessibility Remoteness Index for urban centres and localities across Australia. Figure 4.1 shows that spatial classification.[1]

### *4.4.4 A Multinomial Logit Model (MLM)*

Using the Remoteness Areas framework, a MLM was then used to determine the influence of individual characteristics of graduates on the probability of a graduate selecting a particular employment destination post-graduation. As discussed in detail in Rowe et al. (2013), the model is based on a random utility model framework, and it is assumed that a rational graduate can migrate within Australia in order to maximise his/her utility. The utility associated with selecting a particular area to work is assumed to be a function of a deterministic and a random component. The objective is to estimate how the deterministic part of the utility is influenced by selected individual characteristics of a graduate by estimating the beta coefficients in the systematic utility component of the model.

The model allows estimating the effect on location choice for post-graduation employment. Rowe et al. (2013) defined the utility function to be determined by:

- a graduate's personal characteristics (for example, gender and age);
- attributes of the studied course (subject studied, mode of attendance, degree level and intensive research university);

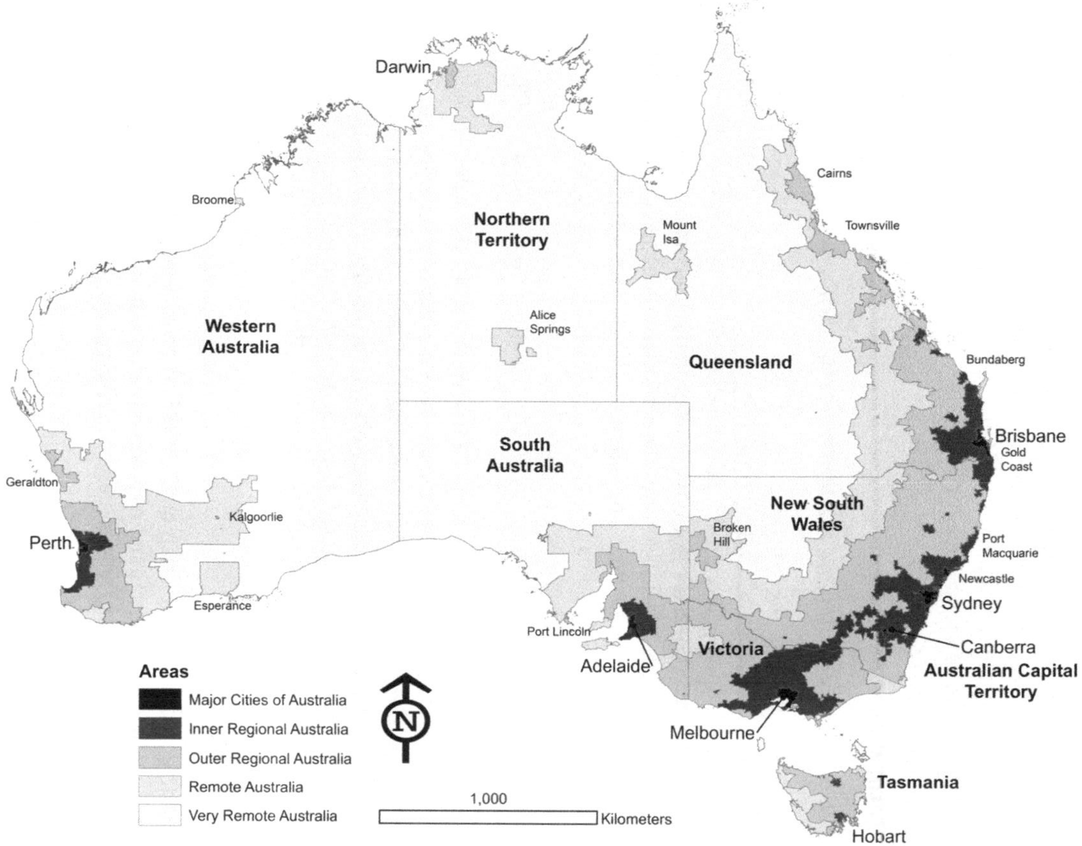


*Figure 4.1* The five Remoteness Areas across Australia

- attributes of the job post-graduation (salary, government affiliation, contractual status and self-employment status); and
- the university location (remoteness classification dummy and state/territory dummy).

The estimated probabilities from the model are interpreted relative to the base category which is taken as being the Major Cities category. Probabilities are expressed as relative risk ratios donating a rise or decline in the chances of being in a particular category, given a one unit change in a particular explanatory variable relative to the reference category.

## 4.5 Results

### *4.5.1 Descriptive statistics and the 'difference-in-difference' tests*

As mentioned earlier, the introduction of the graduate 485 visa scheme in September 2007 in Australia was followed by a sharp increase in the number of international graduates staying-on post-graduation. While Rowe et al. (2013) point out that the AGS does not collect information according to visa type and tends to report low responses among international graduates, they argue that the data effectively capture patterns of international migrants staying in Australia. This is because their response rates tend to be higher than those returning to their home countries, and because even controlling for differences in response rates, there is a 5% growth in the total number of international graduates staying in Australia post-graduation.

Faggian et al. (2015) have conducted a descriptive analysis of the survey data to identify changes pre and post the introduction of the 485 visa scheme. They considered changes with respect to the course/field of study undertaken for both the international and Australian graduating cohorts, the professions in which employment has been gained post-graduation, and in the personal and the job characteristics of the cohorts. The results are summarised in the sections that follow.

#### *Course/subject studied*

The results obtained by Faggian et al. (2015) from an assessment of the courses/subjects studied by international and Australian graduates, both before and after the introduction of the 485 visa scheme, indicate that:

- there were greater changes for the international graduating cohort than for the Australian graduating cohort in regards to educational choices;
- for the international cohort, the popularity of accounting increased substantially to top the list ahead of business studies following the introduction of the scheme; computer science decline from second to third choice, while engineering declined and nursing sciences rose from fifth to fourth place; and

- in contrast, the courses/subjects studied by the Australian cohort remained relatively stable, with business studies, education, accounting and health all in the top five, while, contrary to the international cohort, life sciences became less popular and humanities came into the top five fields of study.

*Professions for employment*

The analysis of the top five professions in Faggian et al. (2015) revealed the following:

- for the Australian graduating cohort, both before and after the introduction of the 485 visa scheme, the top professions were in the education or health (nursing) sectors; and
- for the international graduating cohort there were substantial changes with:
  - an increase in the incidence of employment in the accounting sector, either as an accountant or as a clerk; and
  - an increase in the importance of employment in the retail sector in low-skilled occupations, such as sales assistant and waiters, for which overseas graduates were over-qualified. Their incidence of employment in these occupations was higher than for the Australian cohort.

This last point reflects a greater misalignment between educational qualifications of international graduates staying in Australia and their employment occupations, suggesting that the working conditions of the international graduating cohort 'worsened' following the introduction of the 485 visa scheme.

*Personal characteristics*

Analysing the personal characteristics for both the international and Australian graduating cohorts, following the introduction of the 485 visa scheme, Faggian et al. (2015) found:

- both the Australian and international graduate cohorts were younger, but the difference was not significant;
- there was an increase in the number of female Australian graduates, but an insignificant change in gender for the international cohort;
- the incidence of graduates with a disability remained constant;
- the ratio between undergraduates and post-graduates did change significantly, with a slight increase for the Australian cohort and a slight decrease for the international cohort.

*Job characteristics*

Regarding the characteristics of the jobs in which the graduating cohorts found employment, Faggian et al. (2015) found that:

- in comparison to the international graduating cohort, the Australian graduating cohort was more likely to find employment in managerial jobs and more unlikely to be seeking employment, and also, surprisingly, to be in permanent work or on an open-ended contract;
- the international cohort was less likely to be employed in creative industries or in government employment, but more likely to be working in the education or health sectors, and that might be as a result of the visa requirements;
- for the Australian cohort, the average annual nominal salary rose by $3,000, whereas for the international cohort the average annual nominal salary fell by around $200 post the introduction of the visa. This drop was statistically significant.

*Geographic location of jobs*

Concerning the employment location of international graduates, Faggian et al. (2015) found that:

- for the international cohort both New South Wales and Victoria remained the main states for employment; but
- compared with Australians, international graduates were more likely to undertake jobs in Victoria and Western Australia and less likely to move to South Australia.

### 4.5.2 'Difference-in-difference' regression modelling

Table 4.1 shows the results of the 'difference-in-difference' regression modelling undertaken by Faggian et al. (2015) that was used to investigate the differences in salary levels. They revealed the following:

*Model 1: Basic model using no control variables*

In this model, when no control variables are incorporated, the coefficient on the interaction term between the dummy variable for the international graduate cohort and the year 2005 are identical to the difference in salary.

*Model 2: Fully specified model*

For this model specification, the coefficient on the interaction term between the dummy variable for international and the dummy variable for the year 2005 changes marginally when controlling for personal and job characteristics. The difference between the salary levels for the Australian and international cohort rises just a little from Model 1 to Model 2. The explanatory power of Model 2 also increases from an $R^2 = 0.3225$ to 0.0219, and for Model 2 the adjusted $R^2 = 0.3222$. By taking into account the determinants of the salary differences in Model 2 provide a better bit to the data.

*Table 4.1* The 'difference-in-difference' regression modelling undertaken by Faggian et al. (2015) used to investigate the differences in salary levels

| | **All workers** | | **Only full-time Workers** |
|---|---|---|---|
| | **Diff in Diff** | **Diff in Diff plus Covariates** | |
| | *Model 1* | *Model 2* | *Model 3* |
| **Dependent variable: Salary** | | | |
| Overseas * Yr2006 (Before) | 3,263.72*** | 3,275.02*** | 4,082.10*** |
| | −751.66 | −689 | −1,102.57 |
| Overseas | −20,206.02*** | −14,906.95* | −26,248.41** |
| | −447.47 | −8,863.46 | −12,129.78 |
| Yr2006 | −3,068.36*** | 335.15 | −2,263.42 |
| | −209.78 | −3,045.27 | −4,036.59 |
| **Personal Characteristics** | | | |
| Age | – | 530.00*** | 650.07*** |
| | | −14.36 | −19.61 |
| Male | – | 6,779.14*** | 7,953.28*** |
| | | −254.54 | −340.39 |
| Disability | – | −4,236.92** | −5,314.81*** |
| | | −825.59 | −1,175.57 |
| Undergraduate | – | −9,718.30*** | −11,292.23*** |
| | | −280.76 | −372.66 |
| **Job Characteristics** | | | |
| Manager | – | 16,398.58*** | 16,742.19*** |
| | | −446.12 | −538.75 |
| Self employment | – | 320.51 | 2,023.28** |
| | | −575.26 | −880.33 |
| Permanent/open ended | – | 6,177.46*** | 4,439.83*** |
| | | −269.86 | −370.38 |
| Hours per week | – | 1,010.56*** | 712.81*** |
| | | −9.86 | −18.51 |
| Health sector | – | 94.77 | −1023.05** |
| | | −333.76 | −446.65 |
| Education sector | – | −3,850.35*** | −5,319.08*** |
| | | −371.88 | −507.15 |
| Creative sector | – | −6,402.12*** | −9,346.15*** |
| | | −504.29 | −827.07 |
| Government | – | 5,204.59*** | 3,146.08*** |
| | | −433.01 | −520.29 |
| **Job Location** | | | |
| NSW | – | 367.72 | 704.69 |
| | | −2,059.38 | −2,617.69 |
| ACT | – | 440.62 | 420.29 |
| | | −2,167.24 | −2,757.69 |
| VIC | – | −2,986.12 | −2,946.59 |
| | | −2,060.52 | −2,619.36 |
| QLD | – | −2,264.76 | −2,477.31 |
| | | −2,065.19 | −2,625.16 |
| SA | – | −1,818.11 | −2,218.29 |
| | | −2,088.51 | −2,663.69 |
| WA | – | 2,667.36 | 3,063.04 |
| | | −2,083.62 | −2,652.79 |
| TAS | – | −2,291.90 | −2,387.24 |
| | | −2,223.73 | −2,877.66 |

(*Continued*)

*Table 4.1* (Continued)

| | All workers | | Only full-time Workers |
|---|---|---|---|
| | **Diff in Diff** | **Diff in Diff plus Covariates** | |
| | *Model 1* | *Model 2* | *Model 3* |
| **Interaction terms** | | | |
| All Covariates * Yr06 | – | YES | YES |
| All Covariates*Overseas | – | YES | YES |
| Intercept | 44,185. 63*** | −5,554.05** | 6,871.11** |
| | −142.86 | −2,151.44 | −2,818.62 |
| $R^2$ | 0.0219 | 0.3225 | 0.2048 |
| **Adjusted** $R^2$ | 0.0219 | 0.3222 | 0.2042 |
| **No. observations** | 130,940 | 125,675 | 86,813 |

Source: Faggian et al. (2015), reprinted with permission

The results showed that:

- the personal characteristics variables are affect salaries in the way that human capital theory would predict, with older and male graduates benefiting more;
- job characteristics such as managerial occupations and permanent employment and open-ended contracts generate higher salaries, as do government jobs;
- creative industry jobs and education sector jobs are associated with lesser remuneration; and
- a post-graduate qualification is associated with higher salaries.

Model 2 results also show not significant variations in salaries across Australia's states and territories. That is, job location does not imply significant differences in average salaries.

*Model 3: Restricted model applied to the sample within full-time employment*

The coefficient on the interaction term becomes larger, indicating that the international cohort became even worse-off in terms of salary level in finding full-time employment post visa introduction. However, for the other control variables, only self-employment is associated with increased remuneration, while employment in the health sector has a negative effect.

### *4.5.3 Migration flows*

Turning to the analysis of the migration flows of both the international and Australian graduating cohorts in moving from university to a first job post-graduation, this has been conducted in two contexts by Rowe at al. (2013):

1. To visually show the spatial patterns and spatial concentration of migration flows between Australia's States and Territories.
2. To investigate the propensity of the two graduating cohorts to move across the five broad Remoteness Areas into which Australia may be divided.

*Inter-State/Territory migration flows*

Rowe at al. (2013) mapped the inter-State/Territory flows of graduates and measured the degree of spatial concentration for both international and domestic graduates, pre (the 2005 graduating cohort) and post (the 2008 graduating cohort) the introduction of the 485 visa scheme (Figure 4.2).

It is important to note that most graduates – both international and Australian – tend to remain in the State where they studied after graduation. However, distinctive migration patterns emerge:

- a large proportion of the 2005 graduating international cohort tended to migrate from Queensland to New South Wales;
- after the introduction of the 485 visa scheme, a considerable proportion of the international cohort moved from Queensland to Victoria (up from 10% to 15.5%), and from Victoria to New South Wales (up from 8.8% to 10%), while movements from Queensland to New South Wales remained strong; and
- in contrast, migration flows of both the 2005 and the 2008 Australian graduating cohorts were more spatially dispersed, with a large proportion migrating away from both Victoria and New South Wales to Queensland.

These patterns resemble the picture previously described by Bell and Hugo (2000) for the entire population of recent immigrants in Australia. That is, the great attraction of recent international migrants to New South Wales and Victoria to settle and work. It seems that the capital cities of these States (the largest metropolitan cities in Australia) represent major centres of economic activity employment that attract graduating international students.

*Propensity to move across the broad Remoteness Areas*

Rowe at al. (2013) also investigated the propensity of both the international and Australian graduating cohorts to move post-graduation between the Remoteness Areas of Australia (Figure 4.1). The data indicate that the propensity of the international graduate cohort to move between the Remoteness Areas had increased by about 5% following the introduction of the 485 visa scheme. But this did not occur for the Australian graduate cohort. From a spatial perspective, the analysis revealed that:

- Major Cities are the principal employment centres for both the international and Australian cohorts both pre- and post-485 visa introduction cohorts as they represent the locations for the majority of the university campuses in Australia and thus where the majority of students study.

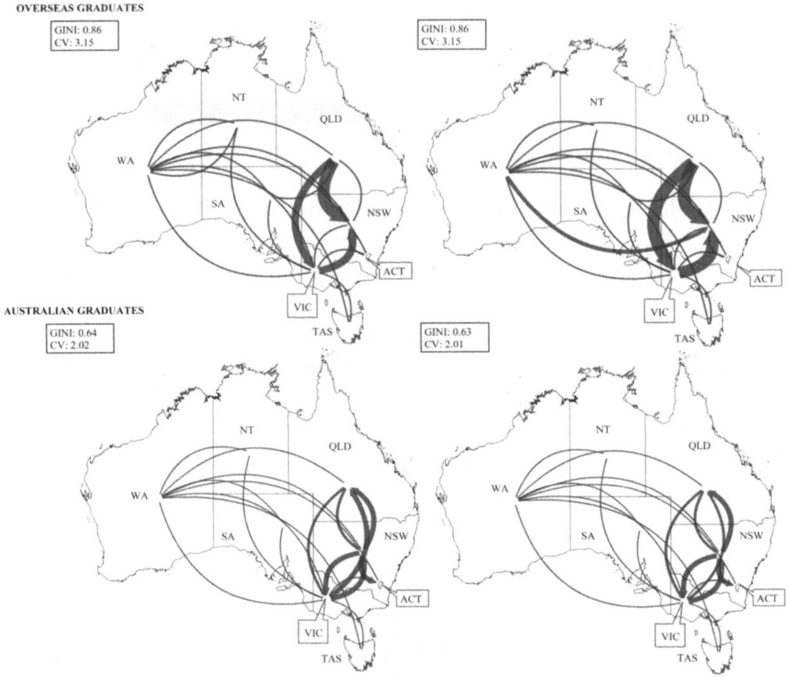


The Australian Capital Territory (ACT), New South Wales (NSW), Northern Territory (NT), Queensland (QLD), South Australia (SA), Tasmania (TAS), Victoria (VIC) and Western Australia (WA). For calculation of the GINI index and Coefficient of Variation (CV), see Bell et al (2002). Only the 20 top flows are mapped. Source: Rowe et al. (2013), reprinted with permission

*Figure 4.2* Inter-State/Territory migration flows of the international and Australian graduating cohorts and measures of migration spatial concentration of the graduates before and after the introduction of the 485 visa scheme

- The retention rates are extremely high for the Major Cities and they drop to low – and increasingly low – levels as remoteness increases. This pattern thus reflects that there are some regional university campuses in towns and cities outside the Major Cities, almost all of them within the Inner Regional and rarely within the Outer Regional areas of Australia.
- For the international cohort staying-on post-graduation, Major City areas are the predominant destination of employment. International graduates tend to stay-on in the same Major City location or move to another Major City location. This pattern has become more pronounced for the 2008 graduating international cohort following the introduction of the 485 visa scheme.
- For the Australian graduate cohorts, the retention rate is also very high in Major Cities. In contrast to the international cohort, for Australian graduates, the retention rates for Major Cities were significantly lower, while the retention rates were significantly higher for the Inner Regional.
- For Australian graduates, there was little change between the 2005 and 2008 graduating cohort in the retention rates across the Inner Regional and the

Outer Regional categories of Remoteness Areas. This pattern contrasts with that observed for international graduates.

- International graduates who graduated from a university in the Inner Regional areas or Outer Regional area of Australia were less likely to find employment in those non-Major City areas than their Australian counterparts. Hence, international graduates educated outside Major Cities in regional areas tend to move to Major Cities for work.

#### *4.5.4 Analysis of the probability of international graduates choosing a location of employment post-graduation*

Rowe et al. (2013) used a MLM model to investigate the determinants of selecting a particular Remoteness Area as employment location post-graduation pre and post the introduction of the 485 visa scheme. The model enables us to investigate what might be the personal, fields of study, and job characteristics that explain those patterns. Table 4.2 shows the results of the modelling.

The findings can be summarised as follows:

1. Focusing first on **personal characteristics** – gender and age – it is evident that:
   - male international graduates are more likely to work in Inner Regional areas, especially after the introduction of the 485 visa scheme; and
   - relatively older international graduates are more likely to work in the more remote areas, but this is significant only pre the visa scheme.[2]
2. Considering the influence of the **characteristics of the course studied**, it is found that:
   - international graduates from educational studies are more likely to select the more remote areas to work, and this increases post the introduction of the visa scheme.
   - international graduates from health studies are more likely to select to work in the more remote areas both pre and post the introduction of the visa scheme, but the probability declines post the scheme, while the likelihood of these graduates taking a job in the Inner Regional areas increases;
   - international graduates having post-graduate qualifications pre the introduction of the visa scheme tend to work in the larger urban areas, while this changes post the visa introduction. The effect of having post-graduate qualification became insignificant; and
   - international graduates from the research intensive Go8 universities tend to find employment in the Major Cities. These graduates are likely to include individuals who completed studies in one of the Major Cities, as Go8 universities are all located in Major Cities. This pattern appears to have persisted following the introduction of the visa scheme.[3]

*Table 4.2* Multinomial Logit Model of the type of Remoteness Areas for the international graduating cohorts (workplace location 6–18 months post-graduation) *Base category: Major cities*

| | Overseas graduates | | | |
|---|---|---|---|---|
| *Cohort* | *Pre-485 visa (2005 graduating) cohort* | | *Post-485 visa (2008 graduating) cohort* | |
| | *Inner regional* | *Distant areas* | *Inner regional* | *Distant areas* |
| Personal characteristics | | | | |
| Male | 1.27 | 0.86 | 1.37** | 0.79 |
| Age | 1.07* | 1.04** | 1.02 | 1.01 |
| Course characteritics | | | | |
| Education studies | 5.93* | 6.48* | 2.87* | 7.50* |
| Health studies | 3.45* | 5.38* | 4.53* | 4.38* |
| Creative studies | 0.30 | 2.83 | 1.40 | 0.53 |
| Full-time | 1.31 | 1.03 | 0.78 | 0.49** |
| Postgraduate | 0.56* | 0.62** | 1.01 | 0.87 |
| Go8 | 0.37* | 1.52 | 0.49* | 0.99 |
| Job experience while studying[a] | 0.76 | 0.70 | 0.87 | 0.69** |
| Job characteristics | | | | |
| Salary | 1.21 | 1.04 | 1.01 | 1.24* |
| Government | 1.89 | 5.22* | 2.02*** | 1.30 |
| Full-time work | 0.83 | 2.49* | 1.45** | 1.60* |
| Self-employment status | 2.01*** | 1.10 | 1.01 | 0.43 |
| HEI location | | | | |
| Distant areas | 0.70 | 11.26* | 0.60*** | 14.20* |
| ACT[b] | 0.03* | 0.02* | 0.01* | 0.01* |
| NSW | 0.01* | 0.02* | 0.01* | 0.01* |
| QLD | 0.01* | 0.02* | 0.01* | 0.01* |
| SA | 0.00* | 0.03* | 0.00* | 0.02* |
| VIC | 0.01* | 0.03* | 0.01* | 0.02* |
| Pseudo-R2 | 0.18 | | 0.19 | |
| N. observations | 3,706 | | 6,963 | |

a: Paid work during graduates' final year
b: Baseline category: NT
Significance: * 1%, **5%, ***10%
Distant areas: Outer Regional, remote and very remote regions
Source: Rowe et al. (2013), reprinted with permission

3. Turning to the influence of **job characteristics**, it is evident that:
   - salary level is not a significant factor for the pre-visa international graduates in working in the more remote areas. However, the probability of these graduates taking up a job in a remote area does increase with an increase in salary for the post-visa graduating cohort; and
   - international graduates working in the government sector are more likely to take up employment in the more remote areas pre-visa change – not so post-visa implementation.
4. To control for **location-fixed effects**, the location of the university from which the international student graduated is included in the modelling. It is evident that:

- both the pre- and post-visa scheme international graduate cohorts from South Australia, Victoria and Western Australia are more likely to select more remote areas over Inner Regional areas to work; and
- the graduates from the Australian Capital Territory, New South Wales and Queensland are indifferent to taking a job in any of these areas;
- international graduates with prior experience in a remote area have a greater likelihood to work in a remote area after graduation.

## 4.6 Conclusion

Certainly, the introduction by the Australian Government of the 485 visa scheme has resulted in a substantial increase in the number of overseas students staying-on and taking up job opportunities in Australia post-graduation. That might be seen as the scheme having been a success story as the objectives of the scheme have been met in terms of it seeming to have retained more international students post-graduation.

However, together with other papers by the authors and their collaborators (Rowe et al. 2013; Faggian et al. 2015), the analyses reported highlight the following:

1. The average working conditions post-graduation of international graduates have worsened *vis-à-vis* their domestic counterparts, worsening that does not appear to be due to the overall state of the Australian economy over the time period under study. International graduates showed a greater propensity for employment in lower skilled occupations, a higher misalignment to their degree, and a widening salary gap to their domestic counterparts. Possible explanations for these findings might be that:
    - on the one hand, international graduates may be willing to sacrifice short-term benefits (that is, accepting less stable and lower remuneration jobs) for potentially achieving a future 'greater good', namely gaining permanent residency;
    - on the other hand, by making it easier for international graduates to stay in Australia, the 485 visa scheme may have allowed the retention of lower quality (poorer-grade) international graduates who prefer to stay in Australia, rather than go back to their original countries where the economy might not be as strong, thus resulting in the scheme having generated a surplus of labour in particular occupations and in general pushing down salary levels.

2. With respect to whether the introduction of the 485 visa scheme has resulted in a dispersal of graduates taking up work outside Major Cities and especially in more remote areas of Australia, there is quite mixed evidence. The following outcomes are evident:

- in general the attraction of Major Cities remains very strong, and has in fact strengthened, as Major Cities are the main locations where international graduates migrated in search for employment. That will probably continue as it reflects where immigrants to Australia in general have shown a strong preference to settle and work;
- however, there is some evidence that international graduates are more likely to be attracted to work outside the Major Cities if:
  - they are older and older males;
  - have studied in education or health;
  - have not studied in a Go8 university;
  - are working in the public sector;
  - have had some experience in a remote area; and
  - if the job has a higher salary;and
- in contrast to the international graduate cohort, the Australian counterpart has shown an increase in propensity to move outside the Major Cities to work, including to the more remote areas, and to be retained in the non-Major Cities to work when they have undertaken their university study at a campus located outside the Major Cities.

It should be emphasised that the research reported here is exploratory in nature. Additional work and more diverse data sources are needed to evaluate the impacts and outcomes of the 485 visa scheme more comprehensively, and that might include a longer-term investigation of employment conditions of the international graduates who stay-on in Australia. That might be achieved through an analysis of data collected in the GCAs longitudinal survey.

## Notes

1 These Remoteness Areas represent a broad geographical regional classification that is used by the Australian Bureau of Statistics. In the modelling reported in this chapter, these Remoteness Areas are used for categorising both the location of the university at which graduating cohorts have studied and the location of the jobs that they have taken up post-graduating:
   - the Major Cities (MC) category includes the large cities that comprise only about 0.2% of Australia's land area in which around 65% of the nation's population live;
   - the Inner Regional (IR) category covering areas that comprise less than 3% of the land area in which about 20% of the population live;
   - the Outer Regional (OR) category covering a large area comprising a little more than 10% of the land area in which about 11% of the population live;
   - the Remote (R) category covering very large areas comprising between 13 and 14% of the land area in which less than 2% of the population live; and
   - the Very Remote (VR) category covering extremely large areas comprising about 73% of the land area in which fewer than 1% of the population live and much of which is uninhabited.

2 This might be a result of the 485 visa scheme which, prior to 2011, required applicants to be under 45 years of age.

3 There may be a considerable take-up of employment by post-graduates within the Go8 universities.

## Acknowledgements

This chapter is based on research conducted on a project funded by the Australian Research Council Linkage program grant LP120100212 with additional support from the industry partners, Graduate Careers Australia and the Victorian State Government. We wish to acknowledge Graduate Careers Australia for their co-operation and the supply of the data on which this chapter is based. Graduate Careers Australia cannot accept responsibility for any inferences or conclusions derived from the data by third parties.

## References

Bard, J. F. (1978) The use of simulation in criminal justice policy analysis, *Journal of Criminal Justice*, 6: 99–116.

Bell, M. and Hugo, G. (2000) *Internal Migration in Australia 1991–1996: Overview and the Overseas-born*, Department of Immigration and Multicultural Affairs, Canberra, Australia.

Bell, M., Blake, M., Boyle, P., Duke-Williams, O., Rees, P., Stillwell, J., and Hugo, G. (2002) Cross-national comparison of internal migration: issues and measures, *Journal of the Royal Statistical Society A*, 165(3): 435–464.

Birrell, B. and Rapson, V. (2005) Migration and the accounting profession in Australia, *Research Reports*, Centre for Population and Urban Research Monash University, Melbourne, Australia.

Birrell, B., Hawthorne, L., and Richardson, S. (2006) *Evaluation of the General Skilled Migration Categories*, Department of Communications, Information Technology and the Arts, Commonwealth, Canberra, Australia.

Birrell, B., Healy, E., Betts, K., and Smith, F. (2011) Immigration and the resources boom mark 2, *Research Reports*, Centre for Population and Urban Research, Monash University, Melbourne, Australia.

Borjas, G. (2005) The labor-market impact of high-skill immigration, *The American Economic Review. Papers and Proceedings of the One Hundred Seventeenth Annual Meeting of the American Economic Association, Philadelphia, PA, January 7–9*, 95(2): 56–60.

Cobb-Clark, D. (1999) Do selection criteria make a difference? Visa category and the labour force status of Australian immigrants, *Discussion paper*, Centre for Economic Policy Research, Australian National University, Canberra, Australia.

Corcoran, J., Faggian, A., and McCann, P. (2010) Human capital in remote and rural Australia: the role of graduate migration, *Growth and Change*, 41(2): 192–220.

Courtney, K. L. (2005) Visualizing nursing workforce distribution: policy evaluation using geographic information systems, *International Journal of Medical Informatics*, 74: 980–988.

Dehejia, R. and Wahba, S. (1999) Causal effects in nonexperimental studies: reevaluating the evaluation of training programs, *Journal of the American Statistical Association*, 94(448): 1053–1062.

Department of Employment and Workplace Relations (DEWR) (2002) *Review of Australia's Skilled Labour Migration and Temporary Entry Programs*, Department of Employment and Workplace Relations, Canberra, Australia.

Department of Immigration and Citizenship (DIAC) (2010) *Introduction of New Points Test*, Canberra, Australia, viewed 28 February 2013, www.immi.gov.au/skilled/general-skilled-migration/pdf/points-fact.pdf.

Dynarski, S. (2003) Does aid matter? Measuring the effect of student on college attendance and competition, *The American Economic Review*, 93(1): 279–288.

Faggian, A., Corcoran, J., and Rowe, F. (2015) Evaluating the effects of Australian policy changes on human capital: the role of a graduate visa scheme, *Environment and Planning C: Government and Policy*, 0263774X15614755.

Franklin, R. S. (2003) Migration of the young, single, and college educated: 1995 to 2000, U.S. Census Bureau, *Census 2000 Special Reports*, CENSR-12, November.

Glass, J. C., McCallion, G., McKillop, D. G., Rasaratnam, S., and Stringer, K. S. (2006) Implications of variant efficiency measures for policy evaluations in UK higher education, *Socio-Economic Planning Sciences*, 40(2): 119–142.

Hawthorne, L. (2005) Picking winners: the recent transformation of Australia's skilled migration policy, *International Migration Review*, 39(3): 663–696.

Hoffmann, R. (1998) Movement of the people, *Science, Essays on Science and Society*, 280 (5362): 386–387.

Hugo, G. (2006a) Australian experience in skilled migration, in C. Kuptsch and P. Eng Fong (eds), *Competing for Global Talent*, International Labor Organization (ILO), Geneva, Switzerland.

Hugo, G. (2006b) Temporary migration and the labour market in Australia, *Australian Geographer*, 37(2): 211–231.

Iredale, R. (2002) The internationalization of professionals and the assessment of skills: Australia, Canada and the US, *Georgetown Immigration Law Journal*, 16: 797–813.

Khoo, S., Voigt-Graf, C., Hugo, G., and McDonald, P. (2003) Temporary skilled migration to Australia: the 457 visa sub-class, *People and Place*, 11(4): 27–40.

Khoo, S., McDonald, P., Voigt-Graf, C., and Hugo, G. (2007a) A global labor market: factors motivating the sponsorship and temporary migration of skilled workers to Australia, *International Migration Review*, 41(2): 480–510.

Khoo, S., Voigt-Graf, C., McDonald, P., and Hugo, G. (2007b) Temporary skilled migration to Australia: Employers' perspectives, *International Migration*, 45(4): 175–201.

Khoo, S., Hugo, G., and McDonald, P. (2008) Which skilled temporary migrants become permanent residents and why?, *International Migration Review*, 42(1): 193–226.

Kinnaird, B. (2002) Australia's migration policy and skilled ICT professional: the case for an overhaul, *People and Place*, 10(2): 55–69.

Kinnaird, B. (2005) The impact of the skilled migration program on the domestic opportunity in information technology, *People and Place*, 13(4): 67–79.

Koleth, E. (2010) *Overseas Students: Immigration Policy Changes 1997–May 2010*, Parliament of Australia, Department of Parliamentary Services, Canberra, Australia.

Kuptsch, C. and Eng Fong, P. (2006) *Competing for Global Talent*, International Labour Organization, International Institute for Labour Studies, Geneva, Switzerland.

Lowell, L. (2007) Trend in international migration flows and stocks, 1975–2005, OECD Social, Employment and Migration, *Working Papers*, 58: 1–23.

Özden, C. and Schiff, M. (2007) *International Migration, Economic Development and Policy*, The World Bank and Palgrave Macmillan, New York.

Regets, M. (2001) Research and policy issues in high-skilled international migration: a perspective with data from the United States, *Discussion Paper Series*, 366, Institute for the Study of Labor (IZA).

Regets, M. (2007) Research issues in the international migration of highly skilled workers: a perspective with data from the United States, *Working Paper SRS 07-203*, Division of Science Resources Statistics, National Science Foundation, Arlington, VA.

Rowe, F., Corcoran, J., and Faggian, A. (2013) Mobility patterns of overseas human capital in Australia: the role of a 'new' graduate visa scheme and rural development policy, *The Australian Geographer*, 44(2): 177–195.

Shah, C. and Burke, G. (2005) Skilled migration: Australia, *Working Paper 63*, Centre for the Economics of Education and Training, Monash University, Melbourne, Australia.

Sher, J. and Sher, K. (1994) Beyond the conventional wisdom: rural development as if Australia's rural people and communities really mattered, *Journal of Research in Rural Education*, 10(1): 2–43.

Suter, B. and Jandl, M. (2008) Train and retain: national and regional policies to promote the settlement of foreign graduates in knowledge economies, *International Migration and Integration* 9(4): 401–418.

Tan, Y. and Lester, L. (2011) Labour market and economic impacts of international working holiday temporary migrants to Australia, *Population, Space and Place*, 18(3): 359–383.

Tang, A., Corcoran, J., and Rowe, F. (2015) Education–job (mis)matches: are overseas graduates getting the right job? Paper presented to International Conference on Population Geographies (IPCG), 31 June–3 July 2015, Brisbane, Australia.

Tang, A. Z. R., Rowe, F., Corcoran, J., and Sigler, T. (2016) Spatial mobility patterns of overseas graduates in Australia, in T. Wilson, E. Charles-Edwards and M. Bell (eds), *Demography for Planning and Policy: Australian Case Studies* (pp. 175–195), Springer International Publishing.

Tang, Z., Rowe, F., Corcoran, J., and Sigler, T. (2014) Where are the overseas graduates staying on? Overseas graduate migration and rural attachment in Australia, *Applied Geography*, 53: 66–76.

Zhang, L., Wang, H., Wang, L., and Hsiao, W. (2006) Social capital and farmer's willingness-to-join a newly established community-based health insurance in rural China, *Health Policy*, 76: 233–242.

# 5 On the three 'laws' of spatial interaction and a string theory finale

## Perspectives from social physics with examples in the digital and retail economy

*Robert G.V. Baker*

### 5.1 Introduction

> Everything is related to everything else, but near things are more related than distant things.
>
> (W. Tobler 1970)

Tobler's 'law' that distance underpins spatial interaction has been a fundamental construct of theoretical geography since the quantitative revolution in the 1950s and 1960s. The gravity model is its most common manifestation. The greatest challenge to this 'law' came with the advent of the internet, where distant things became very near things because information packets could be transferred at speeds approaching the velocity of light. The 'death of distance' hypothesis became the vogue for a short while, until Baker (2005) showed that the gravity model is still relevant, because the rate of information transfer is not infinite, but is limited by the speed of light. Why is physics (again) important to spatial interaction at a geographical scale? An answer is that current telecommunications, GPS determinations, weapon systems and mobile phone networks depend on a very rapid connectivity of a plethora of electronic information, predicated on infinitesimal space and time co-ordinates. The speeds of transmission of information from these origins to destinations are now within the range of the speed of light, hence the juxtaposition between geography and physics. The maxim, however, is still that 'distance always matters' and is very relevant today, despite the technological revolution.

The mixing of physics with the dynamics of the Earth's geography is not new, since it was quite a common interest for seventeenth-century physicists, where, for example in 1650, the terrestrial measurement of longitude, the celestial distance of place and motion and the speed of motion according to the Copernican 'hypothesis', culminated in the *Geographia Generalis Varenii*. The determination of location scientifically was vital to the growth of commerce and trade to the New World and geography was embraced by the Enlightenment. It was no coincidence that Isaac Newton as Lucasian Professor of Mathematics and Natural Philosophy at Cambridge edited two subsequent versions of *Geographia Generalis Varenii* in 1672 and 1681 (preceding his *Principia* in 1687) (Figure 5.1). Newton contributed new figures, corrected a large number of computational errors and supplied better

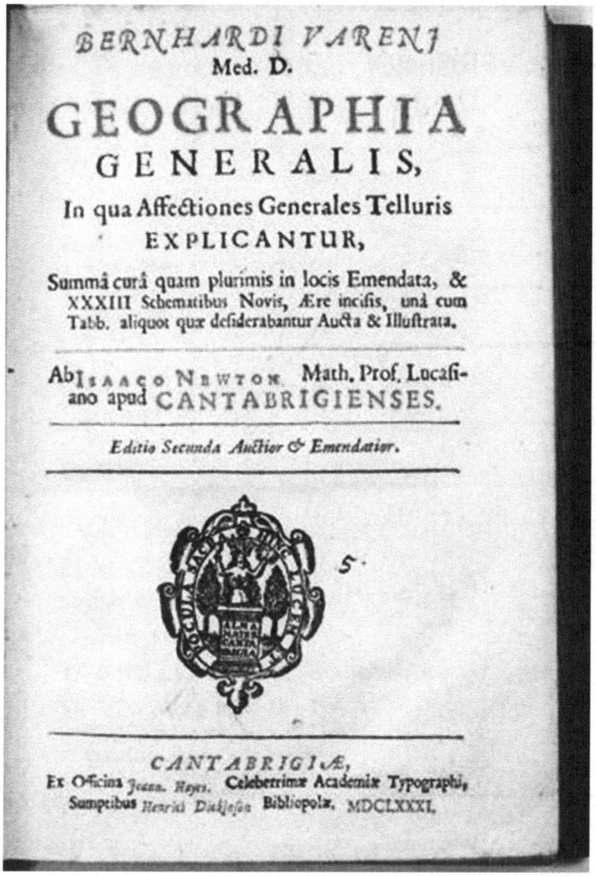
BERNHARDI VARENI
Med. D.
GEOGRAPHIA
GENERALIS,
In qua Affectiones Generales Telluris
EXPLICANTUR,
Summâ curâ quam plurimis in locis Emendata, & XXXIII Schematibus Novis, Ære incisis, unà cum Tabb. aliquot quæ desiderabantur Aucta & Illustrata.
Ab ISAACO NEWTON Math. Prof. Lucasiano apud CANTABRIGIENSES.
*Editio Secunda Auctior & Emendatior.*
CANTABRIGIÆ,
Ex Officina *Joann. Hayes*, Celeberrimæ Academiæ Typographi, Sumptibus *Henrici Dickinson* Bibliopolæ. MDCLXXXI.

*Figure 5.1 Geographia Generalis Varenii*, edited by Isaac Newton in 1672 and 1681

estimates of constants, latitudes and longitudes. He also reviewed a wide range of applications dealing with mathematics (geometrical essentials to the study of geography) and 'mixed' mathematics (applications to such areas as refraction, altitudes, bearings, time, navigation, cartography and map projections). Geography and physics were therefore used collectively to plot location and movement through space using the measurement of time in a revolving sphere.

The movement through space partitioned by time is essence of 'time–space' modelling developed by Baker (1985) and subsequent publications (such as Baker 1994a, 2000, 2006). Technological change in movement economies has seen revolutions in transportation: progressing from sailing ships in a Newtonian world, to steam-powered trains and boats in the nineteenth century, aeroplanes in the twentieth century, to today's emailing and texting via the internet. This technological change has resulted in the convergence of longitude towards a transportation singularity described in Figure 5.2 as the time–space convergence (Baker 2006). The

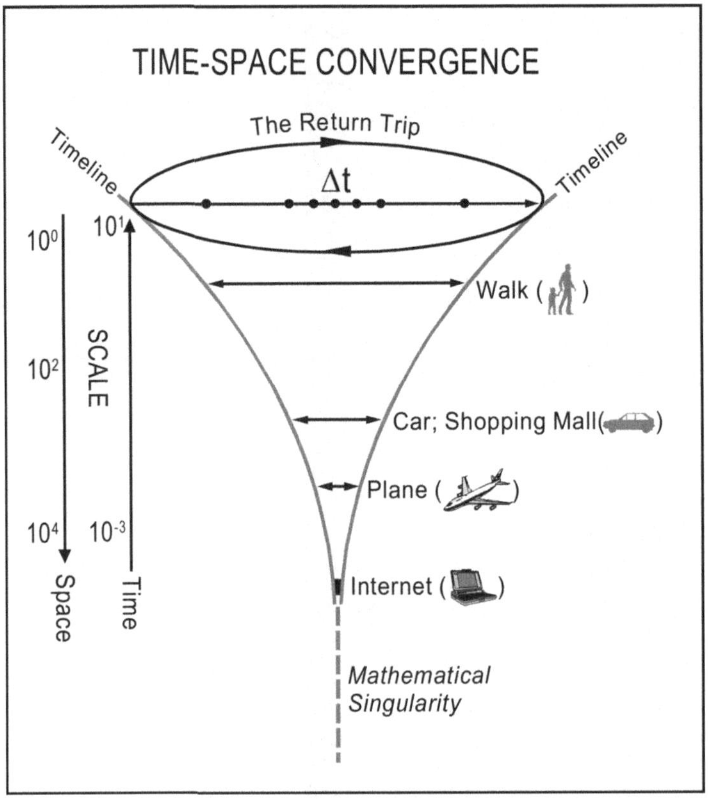


*Figure 5.2* The time–space convergence showing the cone of relative time–space relative to changes to transportation technology (Baker 2005)

convergence of time lines from the internet, where telecommunication speeds can now partition longitude from degrees into nanoseconds, occurs because of technological change in information transmission. This means that a computer origin can capture time lines from destinations in a global network from different time zones and bring them together on a screen. Physically such a process in return pings can result in a 'vortex' where relative time is captured to a point where the convenience of the internet attracts shopping time budgets and trip patronage time away from such entities as planned shopping centres to online shopping. This also means that packets can be lost in this time vortex in information 'black holes', where the network capacity cannot transmit the size of the file. The technology of movement is therefore argued to be a fundamental dimension of spatial interaction.

This convergence of time for origin–destination return trips circumscribes a cone of relative time between the meridians, where the lines never meet because of limits imposed by the speed of light. Why this shape? The physical analogy is that this converging transportation singularity is covered by an event horizon where information transfer at the boundary tends towards a circular line of imaginary time from a rotating Earth. The speed of light is the same for every co-ordinate system on Earth. However, in a geographical sense, the co-ordinate system of origins and destinations are also finite, meaning that the time–space compression is conical. Therefore, geometry can at least show the difference between physical time and relative time. Relative time–space on Earth is bent at high speeds and transcribes a cone of events from the revolution, whereas physical space-time is a trajectory of events, one dimensional and positive. Relative time allows change to appear to flow forwards and backwards, creating a virtual 'time travel' between time zones. For example, it is possible to experience your birthday twice crossing the International Date Line on a flight from Sydney, Australia to Los Angeles. Likewise, electors in Western Australia can change their vote and the outcome of an election from information received from exiting polling after the booths have shut in the eastern Australian states. Physical time is defined as a one-directional positive trajectory of change between two frames of reference, whereas relative time can receive information from the future that can change the present.

There is a technological dimension which can converge the meridians to a fundamental transport event horizon in seconds. Geographical space-time can therefore be constructed for at least five dimensions (two or three spatial, one or two temporal and one technological dimension). An interesting combination of geographic space ($x$, $y$), satellite space ($z$), real and imaginary time ($t$, $it$) and transport technology (@) means we are dealing with six dimensional time–space. The best way to describe this multi-dimensional space-time peculiarity in social physics may be through string theory. An origin–destination pair connected by a closed string of relative time, transmitting information packages in geographic space, is a new worldview of spatial interaction of the internet and possibly the 'everything' in Tobler's 'law'. The visualisation of information as ghostly strings of imaginary time, being converged by real technology into a sea of household singularities at each connected computer, is an extension of the idea of the time-space convergence in Baker (2006). This convergence is where relative time lines are being sucked into the household singularity from a vortex of 'convenience'. Tobler's 'law' therefore raises the question in the context of the internet: what is the nature of the trade-off between Tobler's 'nearness' in space and the technological emergence in the convergence of the 'convenience' in time.

The construction of spatial interaction through what I term 'gravitons' is the first insight from string theory. A graviton is the fundamental construct of interaction of embedded 'nearness' in space independent of the scale of interaction. An example of how to describe a graviton could be through mathematical operators, where the 'nearness' is a function of the process of exchange in space rather than the objects of exchange. Walking to the local shops, driving to a planned shopping centre, air traffic distribution, internet transactions or reception of a wireless signal within a house, are all governed by the same operators, the same 'verbs' of spatial interaction on different spatial interaction subjects (Baker 2006). Implicit nearness to an origin is what matters to a graviton and they can be influenced by time, technology and networks. The operators for a time–space gravitons have the same differentials ($\delta^2/\delta t^2$, $\delta/\delta x$), but it is the gravity coefficient that distinguishes the degree of 'nearness' and the level 'attractiveness' of interaction. A consequence of a graviton is that there could also be fractional dimension independent of the scale and size of interaction, since the operators repeat the same pattern of interaction, even though the rate of movement is significantly different. For example, an around-the-worlds airplane trip around the equator takes 50 hr at 800 km/hr, whereas an internet ping going at 0.6 the speed of light through optic fibre at 200,000 km/hr can circumnavigate the Earth at the equator five times in a second. Both examples can be described by the gravity model, the operators are the same, but the gravity coefficients are orders of magnitude different.

A graviton is therefore defined by a negative exponential function as 'nearness', which we have termed previously the 'gravity model'. It can be acted on by operators and is a solution from the introduction of boundaries of the relevant 'time period' by time–space modelling (Baker 1985). Conversely, the 'nearness' of spatial interaction can also developed through an entropy-maximising method for possible 'networks' (Wilson 1967). Both network and time transactions can approach a singularity within a computer in internet dynamics. Wilson used 'averages' of network configurations from statistical mechanics, whereas Baker initially applied time-based 'movement' from diffusion in thermodynamics and more recently random walks (also from statistical mechanics) in his time–space modelling. Since we are looking at the internet and the transfer of information near the transportation singularity in a time–space co-ordinate mesh, the focus in this chapter is on the time–space worldview. However, both approaches derive the gravity model (that is, the Tobler's 'law') and, whereas the first law is an assertion, both second (networks) and third 'laws' (time) can derive the gravity model and hence the first law (where of course there are still assumptions). This review allows us to further place on a firmer foundation, a social physics in a geography of networks and time, where the further dimensions of time and technology can distinguish it from a classical physics worldview. The end goal of this social physics is a string theory in time rather than space and spatial interaction that is at least physically meaningful, testable and relevant to internet dynamics.

With the introduction of the internet and the emerging possibilities of new dimensions of spatial interaction, a review of Tobler's 'law' is therefore warranted. This chapter suggests that this 'law' can be deconstructed into second and third 'laws' of spatial interaction that the internet revolution has made apparent, involving 'networks' and 'time movement'. This will help us understand and apply spatial interaction modelling (SIM) to some of the new issues facing the complexity of twenty-first-century spatial interaction.

These ideas are unpacked in a number of sections. Section 5.2 reviews the origins of social physics and the first 'law', where geography deserves priority in this area over more recent claimants from physics and computer science. Section 5.3 deconstructs the 'first' law into a second 'law' of spatial interaction based on 'networks', with an example from the stock market. For internet traffic transmitted near the singularity, an inequality, derived from the infinitesimal time–space mesh, is particularly useful to understand the consequences of the second 'law' for internet transactions. Section 5.4 presents a third 'law' of spatial interaction based on time and movement. Section 5.5 applies the 'laws' to problems of satellite lag transmission of TV content, unsolicited call centre calls and the problem of the introduction of a leap second on 30 June 2015. These examples raise broader applications to minimising spatial errors in GPS systems, dealing with computer viruses and financial market security. The potential application areas for the main retail companies are outlined in Section 5.6. Section 5.7 briefly looks beyond the three 'laws' and the path towards a geographical string in time where location are not just points in space but defined by event horizons of closed loops of relative time.

## 5.2 Social physics and the origins of spatial interaction modelling

Spatial interaction modelling in geography has a rich tradition of applied social physics dating back from the beginning of the quantitative revolution. For example, J.Q. Stewart in 1947 stated in the *Geographical Review*:

> There is no longer excuse for anyone to ignore the fact that human beings, on average and at least in certain circumstances obey mathematical rules resembling in general some of the primitive 'laws of physics'.

If there is anything approaching a law in geography, it is the gravity model, where near things are more related than distant things. However, spatial interaction modelling defines a theory of movement rather than a theory of location. It has no rigid underlying assumptions such as in central place theory. The gravity model was originally formulated around a series of mathematical models, based on the gravity concept, analogous to Newton's law in physics. Interaction between two populations was simply proportional to their size and an inverse function of distance separating them. This gravity model has since evolved and now has been applied to many different problems in spatial interaction modelling in space and time (for example, traffic movements, population flows and shopping trips). Therefore, Tobler's 'law' needs to be also inclusive of movement with its implicit connectivity and time connotations, and a review is necessary, particularly with the internet revolution.

There have been many attempts in trying to generate the form and function of gravity behaviour in spatial interaction, such as the entropy-maximising approach from statistical mechanics, the calculus of heat transfer in potential theory or intervening opportunities from psychology. There has always been a strong doubt as to its value, since the complexity of the real world has proved the calibration process problematical and the specification of gravity models are fraught with difficulty. However, the discounting coefficient within the gravity model is a repository of more information than first realised. It is more than just a friction-of-distance measure, where time and socio-economic constructs are imbedded explicitly or implicitly. The effect of distance on shopping is not constant everywhere and can vary according to the time of sampling, socio-economic behaviour of composite sub-populations and statistical constructs such as the size of the aggregation unit and the period used in sampling. Once these limitations are understood, conditional generalisations can be made from gravity modelling and predictions made of future interactions. This is what happened when time-space modelling was applied to shopping hour liberalisation in the 1990s in Australia (Baker 2000, 2006; Baker and Wood 2012). The gravity model and its time derivative have provided key evidence in successful anti-deregulation outcomes in many inquiries in the States of Australia between 1993 and 2005.

However, social physics has been discovered recently by applied physics and computer science publications, claiming priority in a field that has long been occupied by geography. For example, Alex Pentland, in his book *Social Physics: How Good Ideas Spread – The Lessons from a New Science* (2014), aims to use information technology to create a data-driven mathematical model of social behaviour. The central theme of this work is that the overall structure of a given network of interaction, rather than the content that people exchange through that network, determines the 'quality' of the idea flow (Goldberg 2014). An example of Pentland's network theory is that online traders' patterns of interaction give them access to different parts of the information network. The result was that they outperformed isolated traders. In other words, Pentland is arguing that it is better to be globally connected than locally connected in information assimilation. There is a collective intelligence in group social interaction where the sum of parts in the network brings better performance than dominant individuals.

This discovery of the physics of social interaction is not new, where Baker (1982), using work by Weidlich (1971), applied the Ising ferromagnesium model to group social interaction using utility theory. Twenty-five year later, this application has been used in ten papers from 2005 to 2014 (Elsevier Science Publishers 2015) without any reference to previous work in geography. For example, when abstracts from Baker (1982) and Semeshenko et al. (2008) are compared on content, both using the Ising model, the concepts of 'learning', 'social', 'utility' and 'field' are applied to an 'assembly of decision-makers' and 'heterogeneous agents' respectively (Figure 5.3). The adage of 're-inventing the wheel' is appropriate in this case, where a re-awakening of physics occurred decades later without any acknowledgement from work done in other disciplines, such as geography. Likewise, the search using key words in Scopus as gravity model, spatial and distance in the 'Physics and Astronomy' identified a growing number of papers in the applied physics literature, such as Karpiarz et al. (2014) in *Physical Review Letters*, dealing with a gravity model in trade, and Goh et al. (2012) in *Physical Review E: Statistical, Nonlinear, and Soft Matter Physics* on the gravity model and time distance in the Seoul subway system (Figure 5.4). Such applications of the gravity model in geography were undertaken in the 1970s and 1980s, highlighting again physics 're-inventing the wheel' in spatial interaction modelling.

The gravity model can be rather derived than postulated from a number of approaches, beginning with the intervening opportunities model from sociology (Stouffer 1940) where there is a constant probability that a person will be satisfied

The definition of place utility from potential theory shows that a preference and orthogonal indifference field can influence a decision maker, which suggests the operation of behavioural forces. **Learning** energy is defined to be a major behavioural force and a statistical model is proposed for its distribution over **an assembly of decision makers.** An attitude system is derived for a **social network** where the **decision maker** can be in a unitary state of utility or disutility. The resulting partition function shows that the learning energy of the assembly is derived from **nearest-neighbour** interactions and the operation of an external place **utility preference field**. A positive linear relationship is proposed between learning energy and environmental stress. The critical stress in **social interaction** is determined and a variety of **behaviours** in urban sociology are predicted. The model is then applied to a Markov **learning** system and the results discussed.

Source: Baker (1982)

We study the implications of social interactions and individual **learning features** on consumer demand in a simple market model. We consider a **social system** of interacting **heterogeneous agents** with **learning** abilities. Given a fixed price, agents repeatedly decide whether or not to buy a unit of a good, so as **to maximize their expected utilities.** This model is close to Random Field Ising Models, where the **random field** corresponds to the idiosyncratic willingness to pay. We show that the equilibrium reached depends on the nature of the information agents use to estimate their **expected utilities**. It may be different from the systems' Nash equilibria.

Source: Semeshenko et al. (2008)

We consider the problem of enforcing desired behavior in a population of individuals modeled by an Ising model. Although there is a large literature dealing with **social interaction models,** the problem of controlling behavior in a system modeled by the Ising model **seems to be an unexplored field**. First, we provide and analytically characterize an optimal policy that may be used to achieve this objective. Second, we show that complex **neighborhoods** highly influence the **decision making** process. Third, we use Lagrange multipliers associated to some constraints of a related problem to identify the role of individuals in the system.

Source: Cajueiro (2011)

*Figure 5.3* Example of 're-inventing the wheel' in applied physics in the last decade: the Ising model of group social interaction was first published in *Social Science* in 1982

at the next opportunity. This requires an assumption to rank the destinations away from the origins in terms of distance or costs or networking.

Wilson (1967) used the entropy principle from statistical mechanics that requires no a priori assumption of ranking, maximising instead the possible permutations of trips between an origin and destination to derive the gravity model. Wilson's original model was applied to the interaction between the

Globalization is one of the central concepts of our age. The common perception of the process is that, due to declining communication and transport costs, distance becomes less and less important. However, the distance coefficient in the gravity model of trade, which grows in time, indicates that the role of distance increases rather than decreases. This, in essence, captures the notion of the globalization puzzle. Here, we show that the fractality of the international trade system (ITS) provides a simple solution for the puzzle. We argue that the distance coefficient corresponds to the fractal dimension of ITS. We provide two independent methods, the box counting method and spatial choice model, which confirm this statement. Our results allow us to conclude that the previous approaches to solving the puzzle misinterpreted the meaning of the distance coefficient in the gravity model of trade.

Source: Karpiarz et al. (2014)

The Metropolitan Seoul Subway system is examined through the use of the gravity model. Exponents describing the power-law dependence on the time distance between stations are obtained, which reveals a universality for subway lines of the same topology. In the short (time) distance regime the number of passengers between stations does not grow with the decrease in the distance, thus deviating from the power-law behavior. It is found that such reduction in passengers is well described by the Hill function. Further, temporal fluctuations in the passenger flow data, fitted to the gravity model modified by the Hill function, are analyzed to reveal the Yule-type nature inherent in the structure of Seoul.

Source: Goh et al. (2012)

In this paper, we analyze statistical properties of a communication network constructed from the records of a mobile phone company. The network consists of 2.5 million customers that have placed 810 million communications (phone calls and text messages) over a period of 6 months and for whom we have geographical home localization information. It is shown that the degree distribution in this network has a power-law degree distribution k - 5 and that the probability that two customers are connected by a link follows a gravity model, i.e. decreases as d - 2, where d is the distance between the customers. We also consider the geographical extension of communication triangles and we show that communication triangles are not only composed of geographically adjacent nodes but that they may extend over large distances. This last property is not captured by the existing models of geographical networks and in a last section we propose a new model that reproduces the observed property. Our model, which is based on the migration and on the local adaptation of agents, is then studied analytically and the resulting predictions are confirmed by computer simulations.

Source: Lambiotte et al. (2008)

*Figure 5.4* Examples of 're-inventing the wheel' in applied physics in the last decade: the discovery of the gravity model in the 2000s

journey-to-work of a number of workers in a residential zone assigned and the number of jobs in an employment zone. As well, the relationship between potential levels of expenditure in residential and shopping zones could be examined.

The traditional gravity model was generated by calculating the probability of a distribution of trips proportional to the state of the system which satisfies the constraints:

$$\sum_j T_{ij} = O_i \tag{5.1}$$

$$\sum_i T_{ij} = D_j \tag{5.2}$$

$$\sum_i \sum_j T_{ij} c_{ij} = C \tag{5.3}$$

where $T_{ij}$ is the number of work trips; $O_i$ is the total number of work trip origins in $i$; $D_j$ is the total number of work trips; and $C$ the total cost. The solution is of the form:

$$T_{ij} = A_i B_j O_i D_j \exp\left(-\beta\ c_{ij}\right) \tag{5.4}$$

where $A_i$ and $B_j$ are the balancing constants for each origin and destination and $\beta$ is the Lagrangian multiplier associated with Equation 5.3 (that is, the gravity coefficient). The negative exponential is the gravity interaction. The cost of travel $c_{ij}$ is introduced to replace the distance function as a general measure of travel impedance. The reason for the inclusion of total cost $C$ rather than total distance $D$ travelled by the population is that cost structures can be estimated from census data and is meaningful in movement economies. Wilson's smart idea was to substitute 'travel cost' as a proxy for 'distance'. In doing so, it merged the ideas of individual and total household expenditure from economics to origin–destination trips with the journey to work across space from traffic engineering. The spatial economies of trips, through all possible permutations of a potential network, allowed equilibrium conditions to be assumed to solve the equation without any assumption of the ranking.

The weakness in the modelling, in a physical sense, was the arbitrary assignment of values to the balancing constants. The Stouffer and Wilson method followed the worldview of the nineteenth-century social scientist Adolphe Quetelet, who, in 1835, published *Treatise on Man and the Development of His Faculties*, where the attributes of an 'average' man could be grouped under a normal distribution. As Wilson (1974: 393) stated:

> The entropy-maximising method works with individuals, assesses their probability of making a particular journey to work and, essentially, obtains the interaction as a statistical average.

By using 'entropy', Wilson assumes the increasing complexity of the system requires a particular direction of time. As a city increases 'forward' in time, the entropy in a closed system increases, although local entropy can decrease. However, by maximising entropy, there is no condition for any decrease in the entropy of the city and by doing so makes the city a closed system in a classical physical sense of thermodynamics. It is for this reason, it is argued, that the Wilson method cannot deal with the issues of time that global internet transaction presents within 24-hour information cycles produced by the rotating Earth.

The second way of deriving the gravity model from social physics was from time-discounting of space through the imposition of time boundaries on spatial interaction (Baker 1985, 1994a). The timing and limit of spatial opportunities per unit time is what is important. Unlike classical diffusion, time diffusion has the second order operator in time ($\partial^2/\partial t^2$) rather than space ($\partial/\partial x$) and is written for a potential function $\phi$ as:

$$\frac{\partial^2 \phi}{\partial t^2} - M\frac{\partial \phi}{\partial x} = 0 \tag{5.5}$$

This equation is solved by assuming the equivalence of time–space where both the right- and left-hand side is equal to the separation constant (the gravity coefficient) $\beta = k^2/M$ where $k$ is the trip frequency and $M$ the aggregate mobility of the population. The gravity coefficient $\beta$ has imbedded within it both time (frequency) and socio-economics (mobility). Also, Equation 5.5 can be solved equally for an individual or a large population; what is important are the operators, so we can have a gravity model for an individual or for a million internet computers and Equation 5.5 is applicable and empirically verifiable (Baker 2006).

The solution for an individual or a population of the potential function is described as:

$$\phi_o = A\,exp\left(-\frac{k^2}{M}D\right)\begin{bmatrix} sin(kt) \\ cos(kt) \end{bmatrix} \tag{5.6}$$

We have a gravity interaction (*-exp*) constrained by trip frequency, mobility and a time boundary *T*; and a periodic spatial demand wave (*sin*). Equation 5.6 can be dealt with for 'average' trips by squaring the potential function (as in the case of quantum mechanics) to get rid of the positive and negative time function for individual demand, where imaginary time in the 1990s was seen as meaningless before the advent of internet demand. Spatial interaction for aggregate trips were described by squared potential functions that were normally distributed, which is an *ad hoc* probabilistic interpretation of the potential function in relative time.

Following Wilson's (1967) use of statistical mechanics, Baker (2005) derived Equation 5.5 using random walk modelling of movement forwards or backwards between time zones for internet transactions. Imaginary or negative time was no longer meaningless in this new context. There are a number of conditions implicit in Equation 5.5 that this method elucidates, namely, internet traffic within and between time zones can be defined by a set of equations for the exchange rates $E(x,t)$ and continuity conditions for origin–destination pings for bandwidth $w$, the speed of light $c$, the bias in traffic from the Earth's rotation $v$, namely:

$$E(x,t) = -\left(w/c^2\right)\left(\partial\phi/\partial t + v\phi\right) \tag{5.7}$$

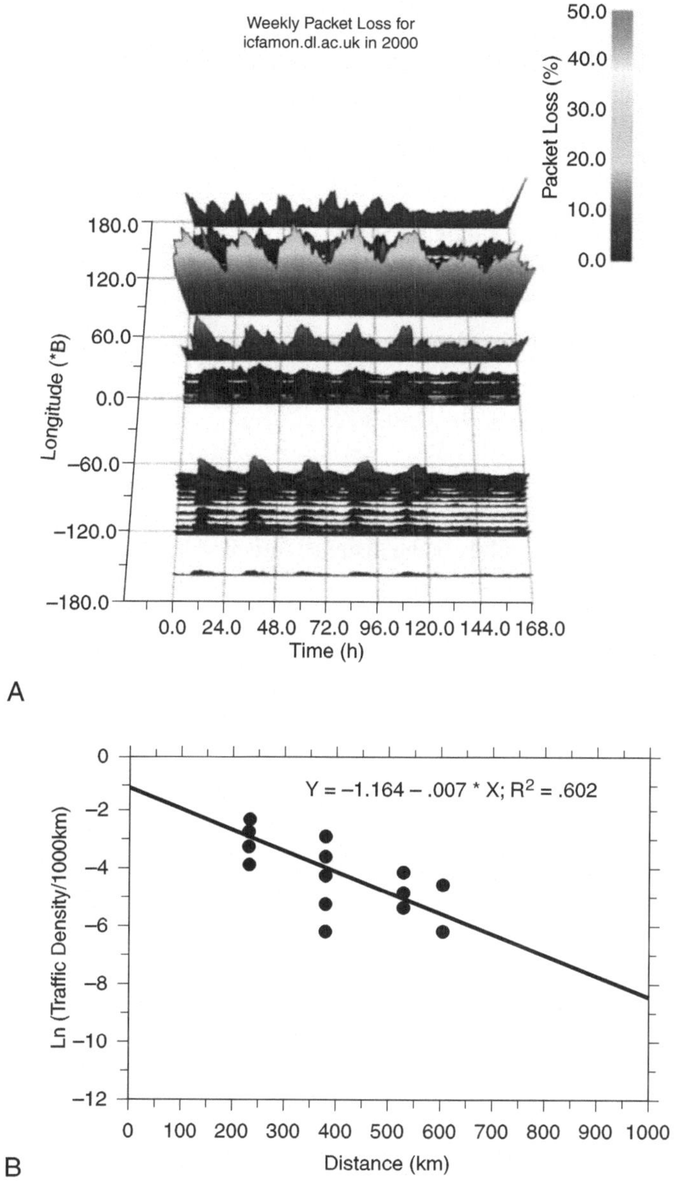


*Figure 5.5* (a) The 2000 global demand wave of congestion for infamon.dl.ac.uk, Liverpool, UK; (b) The gravity model for congestion for infamon.dl.ac.uk to global destination remote host computers

$$\partial\phi/\partial x = -\partial E/\partial t \tag{5.8}$$

Equation 5.5 is therefore expanded into Equation 5.9 below to include the bias of the Earth's rotation as a further time operator and the nature of the transmission constant (bandwidth capacity $w$ and the speed of light $c$), namely:

$$\partial\phi/\partial x = \left(w/c^2\right)\left(\partial^2\phi/\partial t^2 - v\,\partial\phi/\partial t\right) \tag{5.9}$$

where $\phi_o\ (x,t)$ is the site density for a time distance $p$ between pairs of computer remote hosts and the $j^{\text{th}}$ monitoring site. The solution can be tested in form of Equation 5.6 with examples of the global demand wave and gravity model from Baker (2005) (see Figure 5.5).

The Baker (2005) derivation of the gravity model from a random walk problem also provides a condition for the stability of the solutions in the form of the time-space inequality in the co-ordinate mesh, namely:

$$\left(2c^2/w\right)\Delta x \le \Delta t^2 \tag{5.10}$$

where $\Delta t$ is the transmission time and $\Delta x$ is the distance travelled by the computer ping. The inequality means that there will never be instantaneous transfer of information because of the speed of light relative to the limited capacity of information networks. There is always the likelihood of time 'smearing' of information in space and this implication will be explored in section 5.5. This inequality gives the condition for stability related to the capacity of the bandwidth to transfer information packages, namely:

$$w \ge 2c^2\left(\frac{\Delta x}{\Delta t^2}\right) \tag{5.11}$$

If the capacity in the network $w$ is exceeded, then the transfer is no longer stable and information packets in the queue will be lost into an information vortex. This congestion is the proxy that Baker (2005) used to measure the spatial demand wave in Equation 5.6, but with the improvement bandwidth capacity from 2000 to 2004, this congestion wave was significantly damped (Baker 2012).

Spatial interaction modelling has therefore taken three broad philosophical paths (Figure 5.6). The classical origin from the idea of gravity has been picked up from Isaac Newton's *Principia*, where the analogue modelling of gravitational attraction between towns or shopping centres in spatial interaction was argued to be an application from general systems theory. The quantitative revolution in geography in the 1950s and 1960s picked up the mantle of Auguste Comte, who first used the term 'social physics' in 1838 in a series of books (1832–1842) and aspired to explain social reality by a set of universal laws, the same perspective that physicists since Newton sought to present with the goal of a theory of everything. The same aspiration subsequently occurred with the introduction of the gravity model into geography under the guise of general systems theory.

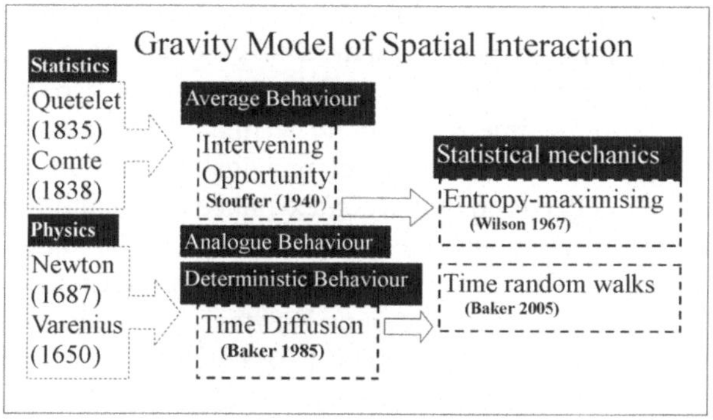


*Figure 5.6* Three paths of spatial interaction modelling to present: Analogue Behaviour, Average Behaviour, and Deterministic Behaviour, and the methods of intervening opportunities, entropy maximising and time diffusion to derive the gravity model of spatial interaction

However, the Comte approach was rejected by natural philosophers such as Adolphe Quetelet, who in 1835 introduced statistical methods into social science and looked at human interaction as 'averages' within a normal distribution. Such an approach inspired Stouffer (1940) in sociology to develop a gravity model of interaction using intervening opportunities where the number of trips between an origin–destination pair is proportion to the number of opportunities in the destination and inversely proportional to the number of destinations. In other words, there is a constant probability that a traveller will be satisfied at the next opportunity (Wilson 1974). Wilson (1967) generalised this type of interaction by applying methods from statistical mechanics to maximise the number of possible permutations available in a way that each state is equally likely. The most probable outcome is the state with the greatest number of individuals in the same transport state; in other words, the statistical average (*à la* Quetelet). By maximising the entropy of the system, we have maximised the information available on the 'average' individual, where, according to Wilson, this individual will most likely take a journey that minimises cost. The best opportunities on average to travel in terms of total household expenditure are the nearest opportunities, so Tobler's law is still appropriate.

A third path comes from Varenius-Newton's benchmark of geography, navigation, longitude and motion applied to the realm of thermodynamics (like entropy-maximising), where classical diffusion of the heat equation is turned into time diffusion. Instead of discounting spatial behaviour movement per unit of time, Baker (1985) discounted time movement and this difference is expressed through the operators described in the introduction. This had no parallel in physics, the idea of 'time diffusion'. The result was that the gravity model was generated by the imposition of time boundaries, such as opening and closing hours of a shopping centres and the gravity coefficient $\beta$ defined by the trip frequency $f$ squared divided by the aggregate mobility $M$ or ($f^2/M$). Baker (2002a) applied time discounting to global internet traffic, and, like Wilson, sourced methods from statistical mechanics by applying random walk dynamics to describe the time diffusion process, where time contributions of demand from destinations can be either positive or negative relative to the time zones of the destinations. The time boundary is now the 24-hour day/night boundary circumnavigating the Earth. The idea of time diffusion is better understood in this context, since physical time is 'smeared' across each time zone, meaning that the time co-ordinate is the same, creating a uniformity in relative time with the partitioning of longitude into segments. This returns to the essence of the Varenius-Newton worldview of spatial interaction through time deconstruction and the role of navigation.

The gravity model has therefore followed three ways of development; firstly, by the 'determining' of spatial behaviour, an analogue from general systems theory to the determinism in Newtonian physics; secondly, by the 'probabilisticism' of averaging spatial behaviour from statistics and statistical mechanics; and thirdly, the 'timing' of spatial behaviour from applied mathematics and statistical mechanics (Figure 5.6). The three-fold way of spatial interaction modelling, has seen new directions and cross-fertilisations between each path.

## 5.3 A second 'law' of spatial interaction?

With the advent of technological change Tobler's law requires a reassessment. The gravity model is the common interpretation, but the law implies something more quintessential about space: namely, the 'nearness' in the relationships. This fundamental 'nearness' is termed a 'graviton' here and can be defined spatially as the fundamental ($\Delta x$) distance in the time–space inequality of the co-ordinate mesh. The graviton is processed repeatedly without any structural change by mathematical operators so it can apply equally to individuals or populations, the microscale or the macroscale or, as Baker (2006) showed empirically, walking to the shops locally or pinging the internet globally (Figure 5.2). Everything is related to everything else implicitly defines relationships between gravitons as having a connectivity that can be formally described as a 'network' or 'a network within a network' with the same characteristics of negative exponentiality. The network is spatially or temporally configured between nodes of exchange and the relationships dependent on the size of the nodes and the distances between them. This distance and size classically defines the attractiveness of the gravitons, but size does not matter near the singularity with internet pings. A hacker, to be successful, does not have to locate in a capital city to bring mayhem to the global network from one computer.

We therefore propose a second 'law' of spatial interaction implicit in Tobler's law, namely:

- Everything can be connected globally.

The internet provides the opportunity for everything to be connected but that does not mean that every person is connected. People wearing a tracking device for detention are globally connected through legal imposition; otherwise there is choice, such as a financial chip inserted in a person's hand to replace a credit card in consumer purchases. Such people can be globally connected with financial information, if they desire. There is no compulsion, but there is choice (at present). The second 'law' conveys that people can be connected through the 'small world problem' where physical and social interaction can be up to six degrees of separation (Watts and Strogatz 1998). Everyone is six or fewer steps away from any other person in the world. However, individuals do not have to be actively connected, but only have the potential to be connected. The second 'law' has obvious implications for internet retailing in terms of global penetration and advertising.

The second 'law' also conjures up connectivity in the context of chaos theory. This also begs the famous question: does one flap of a butterfly's wing change the weather forever? In our context, does once click on a computer change spatial interaction forever. The answer to this question is focused on addressing whether a small change in the initial conditions in time cascades into large-scale implications? For example, do infinitesimal changes in the timing of transmission of global stock markets transactions decides whether a person walks to the shop to buy a newspaper? This question provides the context for the 'stockbroker problem', particularly with the chaos generated by the Global Financial Crisis (GFC) in 2008. A second consequence of global connectivity is the Orwellian thought problem of the inevitability of 'Big Brother' in the informarium of connectivity. This problem will also be reviewed looking at the Stanford Linear Accelerator Experiment reported in Baker (2006, 2012).

## 5.4 Applications of the second 'law' to internet interactions

### *5.4.1 The stock market problem*

The time–space inequality (Baker 2005) for stability in information transmission exchange may be a useful beginning to look at the problem of the stock broker selling shares, namely:

$$\left(2c^2\right)\left(\Delta x\right)\leq\left(w\right)\left(\Delta t\right)^2 \tag{5.12}$$

The term $\Delta t$ could represent the exchange time between two stock brokers ($\Delta x$) apart pinging a stock market in a different time zone with information to sell $M$ stock. This information can travel two directions globally (hence the squared term), but the initial time differences between the brokers is very small. Equation 5.12 summarises the dynamics of the network where the broadband capacity ($w$) is multiplicative and constraining changes in the time co-ordinate, whereas the speed of light ($c$) is multiplicative and constraining on changes in the spatial co-ordinate ($\Delta x$). If there is exponential dependence in the initial conditions in space between the origin and destination (that is, a graviton), then the time lines of information to sell can be simulated depending on the number of times the brokers pushes the computer button to sell. The resulting iterations of a repeated request to sell for both brokers forms a Lorenz attractor at the stock market and a time–space convergence for real (and imaginary) time can be simulated (Figure 5.7a). In our example, the difference ($\Delta t$) in the arrival times of information exchanged is very small. However, the information transfer is not instantaneous and not equivalent as Equation 5.12 indicates. The number of iterations in the trading requests (that is, clicks on the computer) from just small differences in arrival times can have large-scale implications for the changes in number and timing of the shares traded. The simulation in Figure 5.7a shows how the direction that the pings travel can create the conical time–space convergence at the stock market. The time paths to the stock market for both brokers $\Delta x$ apart evolves as the Lorenz attractor, but the amount and time of the sell trade varies significantly between traders over the same period (Figures 5.7b and 5.7c). Such slight changes to the initial conditions of clicking the time to sell produces completely different selling results and underpins the reasons for the chaos generated by global financial shocks, such as the 2008 Global Financial Crisis.

Even when $\Delta x = 0$ and the stockbroker is a dual of himself at the computer, it is always likely that there will be time 'smearing' of information because of the inequality; information in space will always have a potential time lag and the resulting formation of the Lorenz attractor means that there can be significant differences in the sale of the stock from the same stockbroker depending on when the computer is clicked to sell. This result has significant implications in the event of a financial crisis when nanoseconds can produce such different trading outcomes. There will never be instantaneous transfer of information between an origin and destination pair within or outside a time zone, because of the constant speed of light, even if the graviton in $\Delta x$ is very small but never infinitesimally small. Distance always matters but time matters more because of the inequality.

The internet is where physics meets geography with the common factor being a limit to the speed of transmission of information, even if the information is transmitted by light via optic fibre. If $w = c$ hypothetically, the speed of light is still present on the left-hand side of the inequality, because of the squared term. The Lorenz attractor simulations show how the time–space cone can be generated by technology bending the time lines between an origin and a destination. The cone is stable because of the finite co-ordinate system of the internet and so it is an interesting question on how to create instability in the network. All information of spatial interaction is not contained just in space, but is merged within a network embedded in space, where the nodes of the network can be arranged by a certain metric (the graviton) which also contains spatial information. The graviton can be applied across all the network to all computers, meaning that one computer has 'nearness' access to the whole network, hence the hacker's 'joy' of an open door to all computers because of the 'nearness' in the signal.

The translation of a 'relationship' of the first law into a 'network' means that the second 'law' applies, stating that there will never be any instantaneous information within a network transgressing different time zones because of the speed of light and the configuration and capacity of the network.

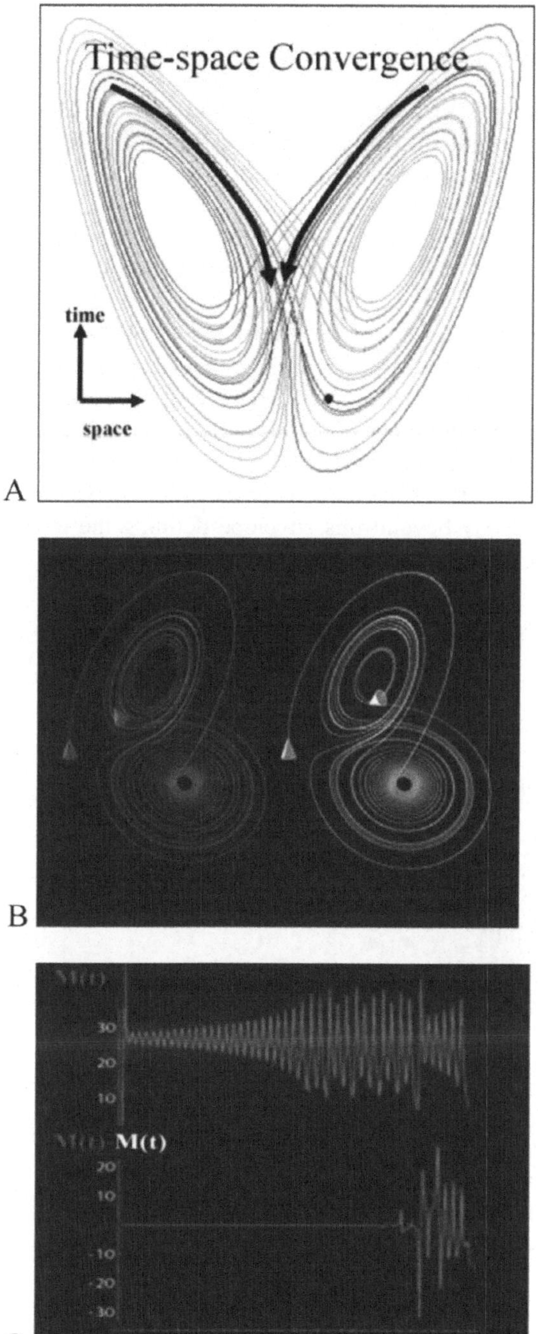


*Figure 5.7* (a) Lorenz attractor forming a time–space convergence of time lines; (b) Two attractors for computers pinging a stock exchange to sell with very small differences in the initial time of clicking; (c) The different trade outcomes of the stock after a set number of clicks of the computer to sell over a time period *t*

### 5.4.2 Global and local connectivity: an Orwellian future?

In the Stanford Linear Accelerator (SLAC) Internet Experiment used by Baker (2012) to test the time-discounting version of spatial interaction (both diffusion and random walk methods), an anomaly appeared in the data where one origin–destination pair was operating within the event horizon; in other words, transmission speeds in the global network were faster than the speed of light and/or the capacity was such that there was some new optic fibre in the network penetrating the event horizon. The result of the SLAC experiment is summarised in Figure 5.8, where the regressions are plotted between ping latencies $\Delta t$ (time taken in pings between an origin and destination computer pairs) and distance $\sqrt{\Delta x}$ between them for all US and European monitoring sites and remote hosts. These origin–destination pairs in the SLAC dataset were regressed for 2000 and 2004 (Figure 5.8). A *J*-curve of time–space interaction relative to a constant bandwidth multiplied by the speed of light has been inserted as a boundary of interaction (dashes). There was one origin–destination pair in 2000 that lay beyond this envelope defining the event

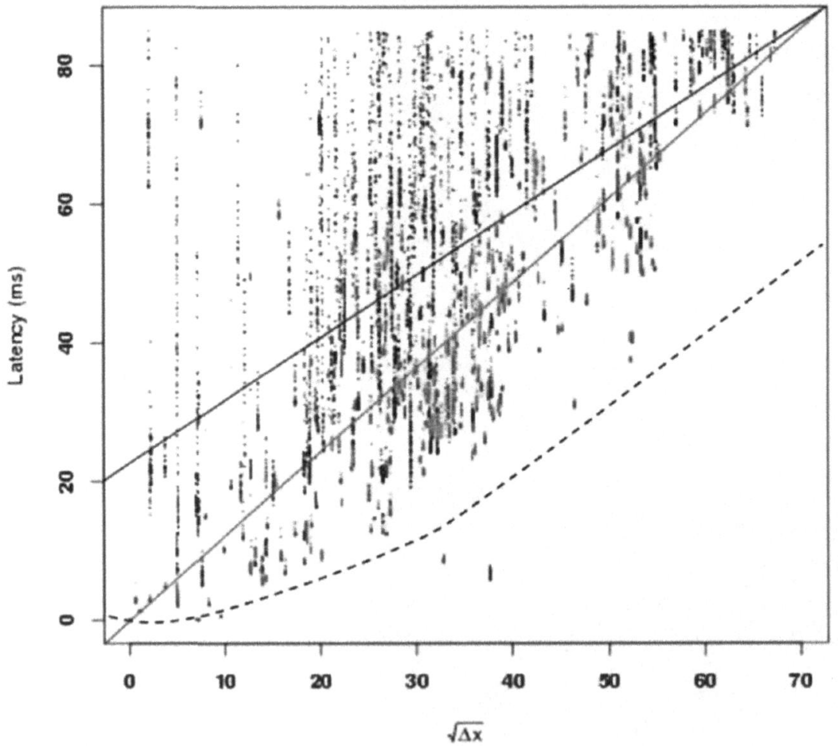


*Figure 5.8* The latency ping relationship with distance approaching the event horizon (dashed) for capacity and the speed of light from 2000 and 2004. Note the shift in increased gradient in the regression line from improved network capacity and the one point (SLAC) beyond the event horizon

horizon for internet traffic. In other words, one information ping appeared to be potentially travelling faster than the speed of light for one origin–destination pair in the US network. This contravenes the gravity inequality of spatial interaction modelling and of course potentially the laws of physics. It was not a random perturbation because when the experiment was repeated in 2004, the same phenomenon was observed even though there was a shift in connective efficiency in the network to the right through improved broadband capacity.

The detailed investigation of the problem revealed that the offending computer pair was actually in the same building at Stanford, that is, the monitoring computer and remote host computer were not globally connected but locally connected within the building, meaning the Stanford site was within the event horizon. The second law of spatial interaction therefore implies all things are connected globally; however, when the embeddedness in space is so small (that is, the distance between the monitoring and remote host computers in the SLAC building), then the time for pings between origin–destination computers is much smaller (because of the squared term), to a point where the capacity $w$ must be *locally connected*, if distance always matters and the transportation speed is limited by the speed of light.

The SLAC building encompassing $\Delta x$ is not 30–40 km long, the pings are not travelling faster than the speed of light; the only explanation in the $(2c^2/w)$ is that the $w$ for that origin–destination pair is not global for the monitoring computer, but rather it is only locally connected within the SLAC server, increasing the $w$ value locally with the resulting observed ping times measured within the $(2c^2/w)$ event horizon. In other words, the building has not been inflated to cover 30–40 km, time is not ticking significantly more quickly in the building; rather, the origin-destination ping is not a global event, but a local event and should be treated as such. This example suggests that the 'second' law needs to be qualified that:

- All people can be, but not necessarily will be, connected globally.

The spatial interaction inequality for stable systems in Equation 5.12 is very important near the event horizon. It defines stability or chaos on the boundary and this is particularly relevant for internet traffic. One way of reaching a tipping point of chaos into the system is to cripple the capacity of the network to deal with the volume of traffic sent, so the inequality is reversed and the system is no longer stable, namely:

$$w \leq 2c^2\left(\frac{\Delta x}{\Delta t^2}\right) \tag{5.13}$$

This difference in network capacity and the inequality in Equation 5.11 (stable) and Equation 5.13 (unstable) in the SLAC data is the periodicity exhibited in the Washington DC–Osaka, Japan origin–destination pair when compared to the poorly-connected Washington, DC–Melbourne, Australia pair (Figure 5.9).

A hacker's aim would be to overload the system from the global connectivity between computers where information packets drop off the resultant queues and

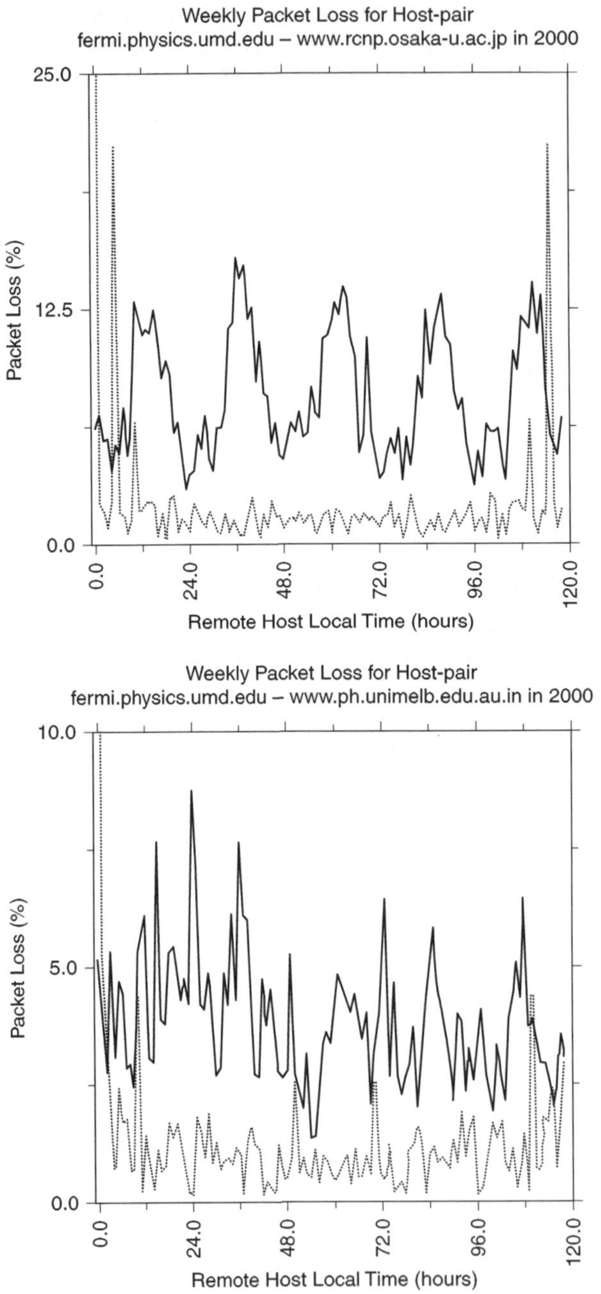


*Figure 5.9* Congestion waves for the origin–destination pairs for Washington, DC–Osaka, Japan and Washington, DC–Melbourne, Australia, showing periodicity and chaos in the transmission over a five-day period (Monday–Friday)

disappear into an information 'vortex' at the singularity. However, to infect the network globally the effective hacker can also use the metric of space to achieve Equation 5.13. This means that we need the possibility of a third 'law' of spatial interaction where the question is addressed: what is the significance that 'distance always matters in internet connectivity'?

## 5.4 A third 'law' of spatial interaction?

The time–space modelling views internet spatial interaction with the potential for technology to bend time lines of information towards computers in space. In the time–space cone (Figure 5.2), the time lines do not intersect in space, because of the limit of the speed of light, and therefore there will never be global instantaneous transfer of information. Even if $\Delta x = 0$, and the origin and destination are the same, then there is still a likelihood of time smearing in the inequality. The result is that a solitary stockbroker operating from one computer to sell shares can still produce the divergent selling outcomes between one click and the next. It also means that there is most likely a singularity for computers with $\Delta x$ very small, producing a vortex in time lines where the second 'law' of global connectivity multiplies data capture. If we define the graviton as lying within the boundary of the event horizon of global connectivity, then the graviton radius $\Delta x$ is defines the boundary as a negative exponential, $e^{-x}$, where $e$ is Euler's number of 2.718281. Therefore, there is not only a fundamental constant $c$ in constructing the space-time mesh, but also a fundamental constant $e$ for 'nearness' in space. A graviton may therefore be Euler's number. A negative graviton is therefore defined as $\Delta x = e\text{–}^{x}$ and the graviton is a function of its own mathematical derivative, meaning that it is always reappearing after successive iterations of proportional change. A negative graviton fluctuates between positive and negative rates of change (contrary to the positive graviton). The exponential (or its logarithmic inverse) are therefore ever present in any spatial interaction and it has what is known in statistical mechanics as a 'memoryless' property. The boundaries of the convergent time lines engender not only 'nearness' but a 'memoryless' property of spatial interaction. This memoryless property is that distance always matters, but we have to look at this statement in the context of the inequality. Therefore, we propose a third 'law' implicit within the first 'law' and the gravity inequality, namely:

- Distance always matters, but time matters more globally.

The gravity inequality in Equation 5.12 summarises this 'law', namely, in the asymmetry between the partitioning between space and time, the standard deviations in time will always be greater or equal to the spatial deviations. This has many applications in internet and telecommunication analysis with some examples described in Section 5.5.

The 'memoryless' property of the graviton means that it can be repeated at every scale from microscale to macroscale spatial interactions. The gravity model can therefore describe origin–destination interactions from walking to a shop, driving to a planned shopping centre, airport traffic configurations and internet pings. Distance always matters in each of these examples. The space and time operators are the same, independent of scale; however, the degree of nearness comes from the gravity coefficient $\beta$ in $e^{-\beta x}$. For planned regional shopping centres, the range of $\beta$ was ~ 0.4 in the Sydney project (Baker 2006), whereas $\beta$ for the *infamon.dl.ac.uk* internet site in Liverpool was ~ 0.007 (Baker 2012). The $\beta$ term, not the operators, is what is critical in distinguishing the scale in spatial interaction modelling. In spatial 'sentences', the spatial operator is the same verb and the difference in context comes from the adverb of the constant coefficient. In other words, the graviton is the verb of the sentence 'to be near' and the gravity coefficient elucidates 'to what degree?'.

The relationship between the graviton and the fundamental relative time partition $\Delta t$ on Earth is simply $\Delta x = \pi \Delta t^2$ and this defines the circular boundary of the event horizon from the rotation of the Earth. Whereas the graviton is fundamentally one-dimensional, relative time is two-dimensional and there is a further fundamental constant of interaction for the time–space mesh, namely $\pi = 3.14159$ for a mesh $\Delta x/\Delta t^2 = \pi$. Therefore, there are universal constants underpinning 'space' ($exp = 2.718$), movement $c = 186{,}000$ m/sec, time–space $\pi = 3.1416$, and a yet to be determined fractal dimension or dimensions.

The determination of the fractal dimension of spatial interaction modelling has been explored in Baker (2012, 2013) and is relevant to this discussion because it will allow us to include the second 'law' on networks into the dynamic set of equations (Equation 5.7–Equation 5.9) for internet traffic. Baker (2013) introduced a fractional dimension Z to the dynamic system (Equation 5.7–Equation 5.9) by adding a further continuity condition which defines the network, namely:

$$\partial^Z \phi^Z / \partial^Z x^Z = \partial^Z E^Z / \partial^Z t^Z \tag{5.14}$$

The use of fractional derivatives of ordinary calculus is an area of growing interest (see Parvate and Gangal 2011 for a review). The $\Phi^z$-derivative is best suited for functions like the Cantor staircase (Figure 5.10), which changes only on the fractal boundary (such as the event horizon). An example of the cumulative distance demand graphs for the *infamon.dl.ac.uk* monitoring machine is equivalent to a gravity model (Figure 5.10) and can be normalised into a Cantor staircase with fractal Z dimension of 0.63 with the general $\Phi^z$-integral and $\Phi^z$-derivative characteristics (Parvate and Gangal 2011).

There is an information exchange function $E^z(\Phi\, x, t)$ for a network that is proportional to the Hausdorff dimension Z and therefore we can write the congestion equation for the internet as composed of a global demand wave plus congestion produced by the network with a fractional dimension of $Z$, namely:

$$\partial\phi/\partial x + \partial^z \phi^z / \partial^z x^z = \left(w/c^2\right)\left\{\partial^2\phi/\partial t^2 - (1-Z)\partial\phi/\partial t\right\} \tag{5.15}$$

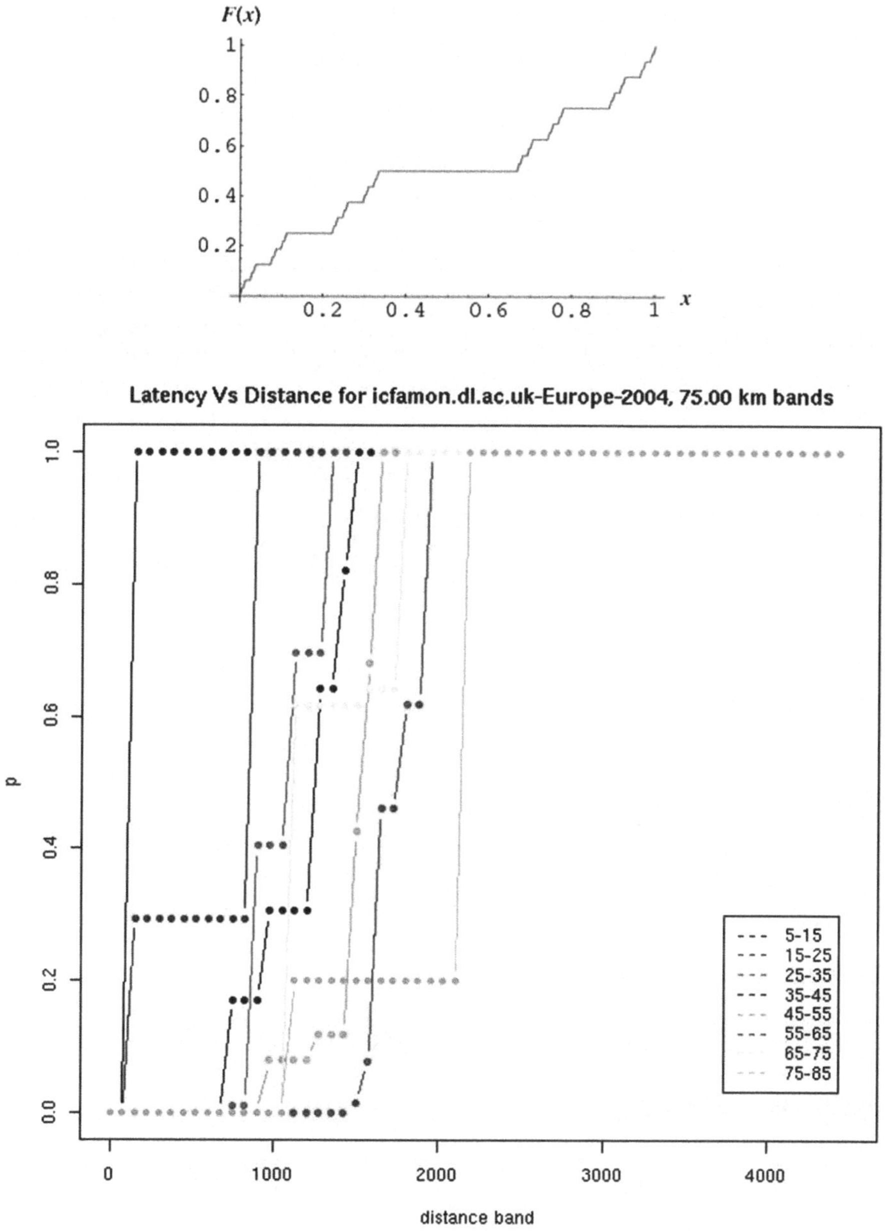


*Figure 5.10* The Cantor staircase from Z = 0.63 fraction (above) and a cumulative patronage distribution for infamon.dl.uk (Liverpool) (below) showing how network can be imbedded into distance decay functions

*Table 5.1* The meaning of the operators and constant in Equation 5.15

| | |
|---|---|
| $\partial\phi/\partial x$ | spatial operator of demand |
| $\partial^z\phi^z/\partial^z x^z$ | network fractional operator |
| $\partial^2\phi/\partial t^2$ | time operator of demand |
| $\{-(1-Z)\partial\phi/\partial t\}$ | bias in traffic from network congestion and the Earth's rotation for ν = 1 |
| $(w/c^2)$ | capacity of the network |

This equation is deconstructed into the different components described in Table 5.1.

Equation 5.15 summarises the dynamics of the internet where distance, networks, time, directional bias and network capacity are summarised in the one equation. It is the most complete description of time–space modelling. When $Z = 2$, Equation 5.15 simplifies to only a bias from the Earth's rotation (Baker 2012) and the global network most likely follows a Lévy random walk of Gaussian behaviour.

$$\partial\phi/\partial x + \partial^2\phi^2/\partial^2 x^2 = (w/c^2)\partial^2\phi/\partial t^2 + \partial\phi/\partial t \qquad (5.16)$$

If the Hausdorff dimension $Z = 1$ (Equation 5.15) simplifies to just the global demand wave, the gravity model and allows us to look at the type of network implicit in the original time–space equation (Equation 5.5).

$$\partial\phi/\partial x = 1/2(w/c^2)\partial^2\phi/\partial t^2 \qquad (5.17)$$

The most likely network configuration for Hausdorff dimension of $Z = 1$ is the Smith-Volterra-Cantor series, where the iteration is defined by:

$$A = \sum_{n=0}^{\infty} \frac{2^{-2n}}{2^{2n+2}} = 1/4 + 1/8 + 1/16 + .... = 1/2 \qquad (5.18)$$

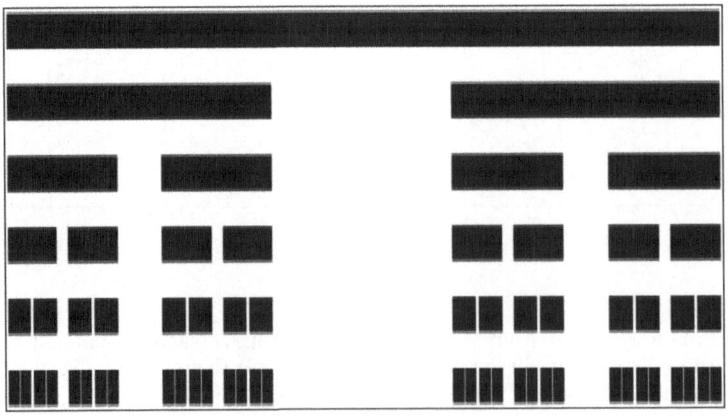

*Figure 5.11* The Smith-Volterra-Cantor network for $Z = 1$ for a closed origin–destination trip

where 1/4 +1/8 +1/16 … = 1/2 is a geometric progression in exponential form for an origin–destination trip [O,D]. The spatial interaction equation is then:

$$\partial\phi/\partial x+\partial^{z}\phi^{z}/\partial^{z}x^{z}=A\left(w/c^{2}\right)\left\{\partial^{2}\phi/\partial t^{2}\right\} \quad (5.19)$$

This means the origin–destination pair [O,D] is a closed interval and Lebesque measurable for an origin–destination trip (excluding the return trip since [A= 1/2]) and this network can be embedded into the spatial differential in Equation 5.5 and that the traffic follows a hierarchical Smith-Volterra-Cantor network of this character (Figure 5.11). There is not only a graviton in the spatial operator and but also in the network operator for Z = 1 so 'nearness' is also fundamental to networks (synthesising 'laws' 1 and 2).

## 5.5 Applications of the third 'law' to telecommunications

'Distance always matters but time matters more globally'. The advent of the telecommunication revolution means that spatial interaction has been thrown into a sea of information ('the informarium') of twenty-first-century living. The third 'law' has practical application to deal with new situations in the 'informarium', particularly through the gravity inequality expressed in Equation 5.12 as:

$$\left(2c^{2}\right)\left(\Delta x\right)\leq\left(w\right)\left(\Delta t\right)^{2} \quad (5.20)$$

The time lag in Equation 5.12 is easily observable in a number of situations described below in the telecommunication of global information.

### *5.5.1 Satellite transmission lags*

The essence of 'time–space' interaction modelling is that spatial behaviour is constrained by time boundaries dependent on the time scale of seconds, minutes, days (much like the scale of a map). We only become aware of the interdependencies over longer periods (days rather than seconds) and over greater distances. We become aware of third 'law' in Australia when a news reporter here interviews a correspondent in the UK and there is a 3–4 second pause from when the question is asked to when the UK correspondent replies. The 'awkward' pause in transmission is just from the imposition that the speed of information transfer is constrained by the network and the physical limit of light speed so we see $\Delta t$ constructed before our eyes in Australia. However, the same phenomenon is observed even when $\Delta y$ is very small between two destinations. For example, suppose there are two TV sets in a house in adjoining rooms in a regional centre receiving the same show on the identical channel from a capital city, except $TV_1$ reception is from free-to-air transmission across geographical space ($\Delta x$), whereas $TV_2$ is by cable TV (via satellite) assumed to be $\Delta z$ from the Earth. On reception, $TV_1$ receives the show approximately a second before the satellite reception of $TV_2$ depending on the relative location of the satellite. The satellite transmission lags and 'smears' the show by $\Delta t$ in adjoining rooms creating the echo in commentary. Distance matters in the transmission, but time matters more in that time in the inequality is squared.

### *5.5.2 The 'undersirable' call centre operator*

The nuisance factor of call operators outside Australia targeting households on private untraceable numbers is a significant problem. The gravity inequality helps households simply filter such operations because a computer first dials the number and transfers, when the call is answered by the household, to the operator who then endeavours to sell their product or service or gain access to your computer. Since the signal comes via satellite to Australia, the $\Delta t$ time lag is discernible before the operator begins their sales pitch. This pause provides the opportunity for the household to terminate call before any engagement. There is no instantaneous transfer of information globally because these call operators are using the network and satellite transmission.

### *5.5.3 The problem of the leap second in 30 June 2015*

This relative time phenomena is no novel academic construct, where the issue of relative time, based on the Earth's rotation, has led to current debate on whether to insert a 'leap second' into Coordinated Universal Time (CUT) on 30 June 2015. The reason for this insertion of a second is to keep the time of day close to the mean solar time (SMT) (from which Greenwich Mean Time is calculated), otherwise time determination from the Earth's rotation will drift away from atomic time because of fluctuations in its rotation. This addition of a second, because we are near to the singularity, means that there are opportunities to shift information through this 'door of relative time'. This has implications for movement economies of information and trading, such as stock prices and currency exchanges at the closing times of stock markets.

The situation is described in Figure 5.12 and the impending problem by the gravity inequality and its interpretation is defined by Equation 5.21, namely:

$$\left(2c^2\right)\left(\Delta x\right)\leq\left(w\right)\left(\Delta t\right)^2 \tag{5.21}$$

The $\Delta t$ drift between CUT and SMT at present is -0.6secs (Figure 5.12) with the proportional effect on the spatial component $\Delta x$ as 0.36 sec/$^2$; but at -.0.8 secs, the spatial variation is 0.64 sec/$^2$ and further, for a drift of -1.0 sec, the $\Delta x$ is of the magnitude of 1.0 sec/$^2$. In other words, the divergence in $\Delta t$ is introducing a positive graviton in space, which means that the errors in time–space co-ordinate systems (such as for GPS) are being geometrically multiplied, decreasing the accuracy of the system. This situation therefore needs to be urgently addressed, particularly for financial markets. The Intercontinental Exchange, the parent body to seven clearing houses and 11 stock exchanges, including the New York

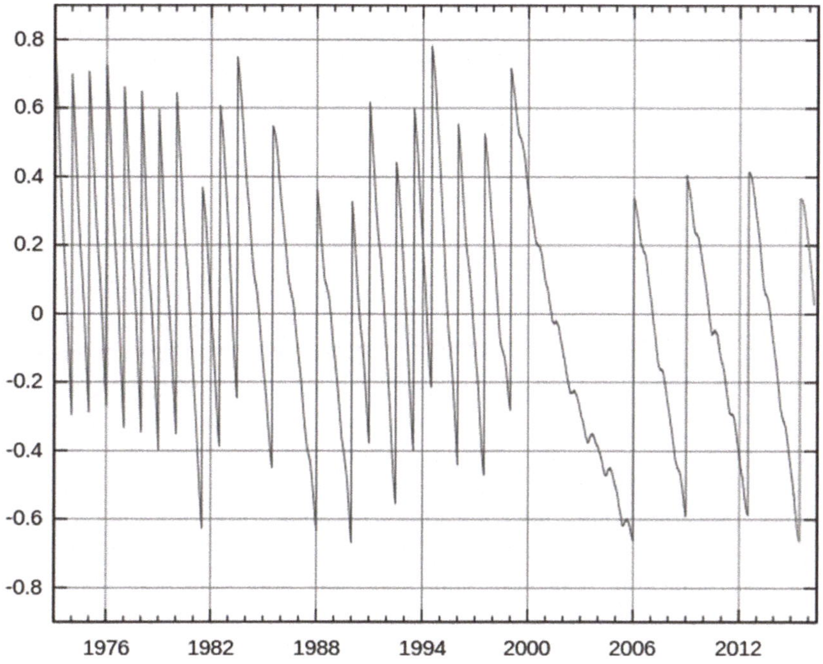


*Figure 5.12* The leap second problem and the time discrepancy between Coordinated Universal Time (CUT) and Greenwich Mean Time (GMT) from changes in the Earth's rotation

Stock Exchange, aims to remedy this problem by ceasing operations for 61 minutes at the time of the 30 June 2015 leap second. Alternatively, instead of inserting a leap second at the end of the day, Google servers will implement a 'leap smear', that is, the idea of smearing time $\Delta t$ in the inequality, by extending seconds slightly over a time period prior to the leap second. Whether a time jump or a time smear, if nothing is done, there will be a multiplier effect in spatial errors that will impede such operatives as Google mapping and GPS co-ordinates.

## 5.6 The multiplicative impact of a time–space compression on high street retailers

There are significant applications of the three 'laws' to retail analysis. Both the second 'law' – that while distance always matters, time matters more globally – and the third 'law' – that all people can be, but not necessarily will be, connected globally – mean that the capture of time shopping budgets occurs from the same operators (processes) for both planned shopping centres and online retailers. The resulting time–space compression means that high street retailers are facing the multiplicative effects of not only planned shopping centres but online retailers (Figure 5.2). The loss of businesses from high streets has been a feature of UK retailing over the past five years, where, for multiple retailers, the desired locations are increasingly in planned shopping centres and retail parks rather than traditional precincts, to combat the growth of online shopping (*Daily Mail* 2014). The shift to mobile technologies and smart phones means that shopping time lines can be captured through time-based connectivity and less by spatial interaction. This is just a manifestation of the second and third 'laws' of spatial interaction modelling, but with online shopping making up to 10% of retail sales, is this the major reason for the current problem?

Regional planned shopping centres with deregulated shopping hours are still argued to be the most potent force behind the collapse of main street retailing in many localities (Baker 2002b). A smart phone offers just a vortex of convenience for individual time budgets whereas a planned shopping centre offers a vortex of convenience for a population of time budgets, combining with it a multiple of spatial choice alternatives within the one centre. Planned regional shopping centres offer the aggregation of integrated temporal and spatial accessibility of convenience and choice. They therefore represent the potency of the time–space compression in a 'sea of deregulation': their scale is analogous to the formation of hurricanes/cyclones over tropical oceans where the consequences are the widespread destruction of adjoining localities.

The above analogy may be useful in describing why high streets are continually under pressure, if the problem is framed within chaos theory (see Dearden and Wilson 2011).

Planned shopping centres, in the language of chaos theory, are chaotic attractors which act as aggregate singularities, distorting travel trajectories by capturing consumption through convenience (time) and choice (space). The travel trajectories converge as vortexes of convenience and choice into their car parks, producing in chaos theory what is termed a 'homoclinic bifurcation', where the planned shopping centre (PSC) attractor sucks in regular shoppers from the market catchment, producing catastrophic consequences for regular shopping in competing main streets (Figure 5.13a). This predator structure of a PSC tends to become stable over time, particularly if there is a floorspace oversupply spatially and/or combined with shopping hour liberalisation. Such floorspace oversupply creates the supercritical conditions, where such events as a Hopf bifurcation occurs and results in the cessation of regular shopping in high streets. In the context of time-space modelling, periodic solutions disappear from the gravity model once the critical condition is reached through either land use or trading hour oversupply. Once consumers regularly and exclusively shop at large PSCs, they create discontinuities of time and space along the boundaries of the PSCs, creating a shock wave of lost demand, invariably affecting the viability and vitality of high streets. Such a shock was illustrated in Baker (2006) and subsequent surveying, where shoppers were tracked in Cessnock and there was little engagement between the new amalgam of shops (a surrogate PSC development) and main street (Figure 5.13b). The consequence of the damage from the shock wave from

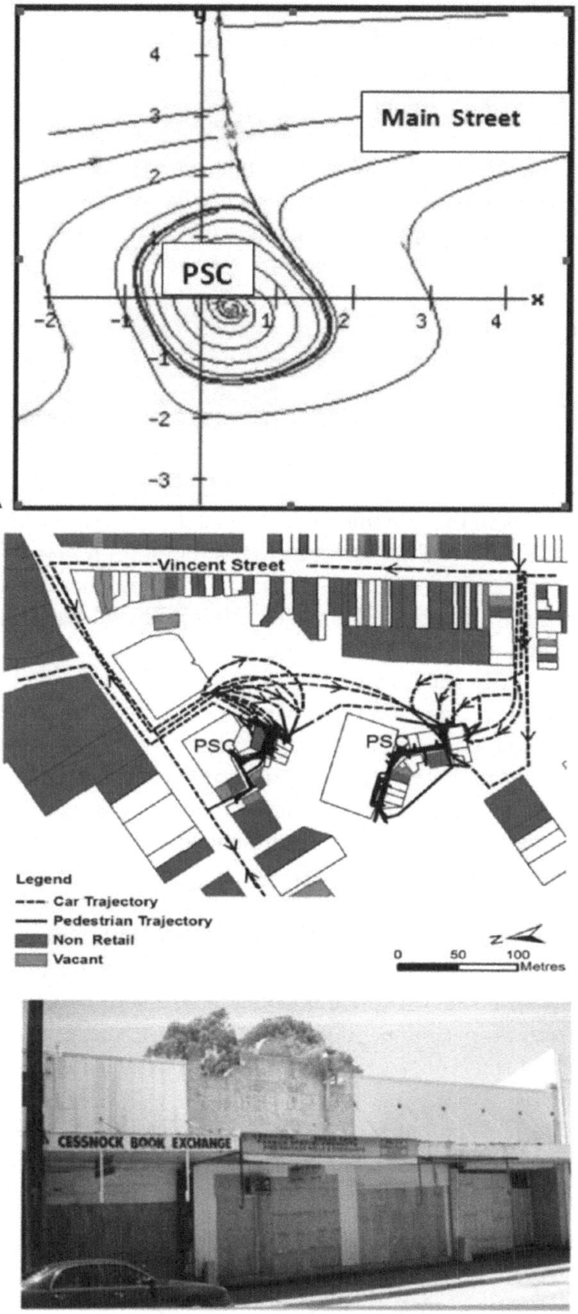


A

B

C

*Figure 5.13* (a) A vortex of time trajectories of trips to and from a PSC away from main (high) street. (b) The tracking of shoppers in 2009 in Cessnock NSW showing their journeys to the edge-of-town PSCs away from main street. The disconnect between both precincts is obvious and the vacant shops noted. (c) A chain of vacant shops in Cessnock in 2004 after the developments opened

Source A: Alian et al. 2015; Source C: Baker 2006

the shift in demand to the new developments was easily observed in the photographic inset (Figure 5.13c). The vacant shops and the invasion of non-retail functions predominated main street Cessnock.

This scenario has been a re-occurring event in the UK and Australia since shopping hour liberalisation in the late 1980s, with added pressure on specific retail functions currently from the mobile communication revolution. The growth of vacancies in main streets in south eastern Australia is well documented (such as Cessnock, see Baker 2006). However, with subsequent shopping hour deregulation in South Australia in 2011, it was no surprise to this author that problems of time–space restructuring soon appeared. For example, on 14 September 2013, it was reported in the financial media that Adelaide's leading high street shopping precincts were facing record vacancies from retail downturn (Tauriello 2013), where vacancies in South Australia's fashion strip had risen from 2% to 12% and shops had lain idle over the past two years. Similar impacts were reported in the high street fashion strip of Oxford Street, Sydney, after Westfield Bondi Junction opened in 2003, where vacancies in this precinct a decade later were surveyed in 2012 to be 15.5% (Alian et al. 2015). Similar long-term structural problems in the UK occurred after the Meadowhall PSC was opened and its impact on the Sheffield high street were observed (discussed in Baker, 2006). The current push in 2015 to extend shopping hours on Sunday in the UK will further exacerbate the problems of high street viability, unless strict land use controls are applied as to where such developments can operate. Out-of-town and edge-of-town developments prosper with trading hour deregulation, particularly if they are anchored by large supermarkets or hypermarkets. The loss of supermarkets from high streets creates the conditions for a Hopf bifurcation (for example, Cessnock). These predictions of such structural changes from the social physics of time–space modelling were made in Baker (1994b: 113), namely:

> The study of the 'when' and 'where' people shop demonstrates that the deregulation of trading hours is going to bring substantial change to the nature of retailing. Such an approach suggests there is going to be a shift in demand towards PSCs away from strip centres (high streets). Furthermore, public policy is increasing market share of large retailers through deregulation.

The same impacts predicted in 1994 continue to occur 20 years later (Baker 2006; Baker and Wood 2012).

## 5.7 The future of spatial interaction modelling: the string theory finale

The time–space compression of time lines through changes in the transportation technology of the internet has introduced a new framework for spatial interaction modelling. The classical space-time dimensions of physics can be expanded in our view of social physics to include two new dimensions of imaginary time (relative time) and technology. We therefore have a six-dimensional construct of space-time, namely: three spatial dimensions, geographic space ($x$, $y$) and satellite space ($z$); two time dimension, real time ($t$) and imaginary time ($it$); and a technological dimension (@). The obvious way of viewing such multi-dimensional constructs of space-time is through string theory, where the strings are in fact the time lines and are closed for return trips. We have introduced the idea of a 'graviton', a fundamental construct of 'nearness', negative or positive exponential in nature and created by the strings of time lines. While there is some overlap with the graviton of physics, it is not a hypothetical elementary particle mediating the force of gravitation within quantum field theory. Rather, it is framed as a result of the operation of time boundaries on $\Delta x$, a fundamental distance of interaction defined by an event horizon, beyond which there is global connectivity and multiplicities of the graviton over space defined by a fractal dimension Z. It is closed by an origin–destination pair. The graviton could therefore be seen to be culmination of the integration of the three laws of spatial interaction.

An origin–destination pair is just a closed time string of transmitted information. With 24-hour rotation of the Earth, the circular event horizon of information from time ($\pi\Delta t^2$) could propagate through a radius of $\Delta t$ around this fundamental event horizon. Alternatively, in string theory, it could wind as a closed string in the form a circle of radius $\Delta x/\Delta t^2$, something akin to a clock being wound up, giving the information an 'energy' to capture more information from the surrounding origins. This winding of time-squared in a $T$-duality closed string may be why near things are more related than distant things because explicit 'nearness' of space is created by the implicit 'convenience' of time in spatial interaction. The end-product from the squared-time term could be the vortex of convenience suggestive in the time–space cone, where information in time can be sucked into a computer from other computers in space. With this view, we can see how destructive computer viruses, worms and hackers can be in this technological landscape because of the potential of time–space vortexes of captured information.

A string theory of spatial interaction is therefore an interesting prospect, since we are not constrained by just the four dimensional paradigm of physics. A formal view of a six-dimensional string is the Calibi-Yau manifold (Figure 5.14). Alternatively, the conception of the spatial interaction string model is of a butterfly diagram of real and imaginary time for return trips reflected around an axis of technology, where the convergence of the time lines is towards a singularity where interactions are no longer globally connected. A corollary to the technological dimension is that beyond the singularity there is convergent imaginary technology or what has been interpreted as artificial intelligence and its relationship to spatial interaction is intriguing but beyond this chapter.

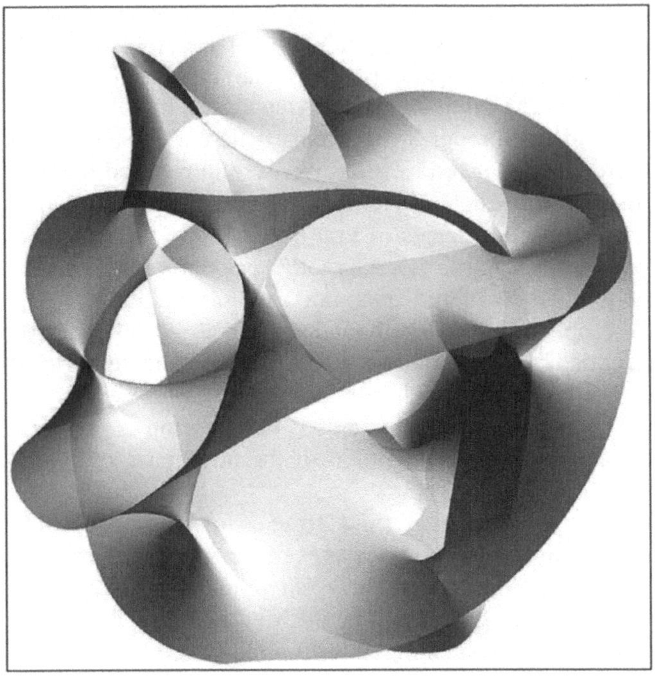

*Figure 5.14* The Calibi-Yau manifold for a six-dimensional string: a representation of the capturing of relative time strings on a three-dimensional rotating sphere

Source: Creative Commons Attribution-Share Alike 2.5 Generic License

## 5.8 Conclusions

This chapter has endeavoured to deconstruct Tobler's 'law' of spatial interaction into two further 'laws' in order to deal with the internet and the revolution of the digital economy in the twenty-first century. The 'laws' are stated as:

- everything is related to everything else, but near things are more related than distant things;
- all people can be, but not necessarily will be, connected globally; and
- distance always matters, but time matters more globally.

Applications of the 'laws' from a time–space modelling perspective for the digital economy are shown from stock market trading, telecommunication reception and the problem of the leap second for 2015. The conception of spatial interaction of a six-dimensional string has attractive elements, such as the 'graviton' and an explanation from a closed string of relative time winding space exponentially so that 'near things are more related than distant things'. The result is the finale of spatial interaction modelling: a butterfly string of relative time where information is being converged by technology towards singularities in networks of home computers in geographical space.

## Acknowledgement

This chapter is written in acknowledgement to my PhD supervisor Professor Barry Garner, an original 'space cadet' and my mentor, without whose encouragement my continuing research into spatial interaction modelling would not have eventuated.

## References

Alian, S., Wood, S. and Baker, R.G.V. (2015) Unfolding Impacts on planned shopping centres upon main street consumption opportunities in advantaged and disadvantaged Sydney. Unpublished Paper, University of New England.

Baker, R.G.V. (1982) A mathematical model of group social interaction, *Environment and Planning A*, 14, 1031–1046.

Baker, R.G.V. (1985) A dynamic model of spatial behaviour to a planned suburban shopping centre, *Geographical Analysis*, 17, 331–338.

Baker, R.G.V. (1994a) An assessment of the space-time differential model for aggregate trip behaviour to planned suburban shopping centers, *Geographical Analysis*, 26, 341–363.

Baker, R. G.V. (1994b) The impact of trading hour deregulation on the retail sector and the Australian community, *Urban Policy and Research*, 12(2), 104–114.

Baker, R.G.V. (2000) Towards a dynamic aggregate shopping model and its application to retail trading hour and market area analysis, *Papers in Regional Science*, 79, 413–434.

Baker, R.G.V. (2002a) Modelling Internet transactions as a time-dependent random walk: an application of the RASTT model, *Geojournal*, 53, 407–418.

Baker, R.G.V. (2002b) The impact of the deregulation of retail hours on shopping trip patterns in a mall hierarchy: an application of the RASTT model to the Sydney project (1980–1998) and the global vacant shop problem, *Journal of Retailing and Consumer Services*, 9, 155–171.

Baker, R.G.V. (2005) Instantaneous global spatial interaction? Exploring the Gaussian inequality, distance and internet pings in a global network, *Journal of Geographical Systems*, 7, 361–379.

Baker, R.G.V. (2006) *Dynamic Trip Modelling: From Shopping Centres to the Internet*, Springer, London.

Baker, R.G.V. (2012) Towards a physics of internet traffic in a geographic network, *Physica A: Statistical Mechanics and Applications*, 391, 1133–1148.

Baker, R.G.V. (2013) The internet differential equation and fractal networks, *Journal of Physics: Conference Series*, 410, 012099, 1–8, doi:10.1088/1742-6596/410/1/012099.

Baker, R.G.V. and Wood, S.N. (2012) Geographical modelling, public policy and informing the 'store wars' sovereignty debate in Australia, in Stimson, R.J and Haynes, K.E. (eds), *Applied Geography: Using Geographic Analysis to Address Real World Problems and Issues*. Edward Edgar, Cheltenham, UK, pp. 160–183.

Cajueiro, D.O. (2011) Enforcing social behavior in an Ising model with complex neighborhoods, *Physica A: Statistical Mechanics and its Applications*, 390(9), 1695–1703.

Dearden, J. and Wilson A.G. (2011) A framework of exploring urban retail discontinuities, *Geographical Analysis*, 43, 172–187.

Daily Mail (2014) High street decline continues, *Daily Mail*, 9 October. Retrieved from: www.dailymail.co.uk/wires/pa/article-2785794/High-street-decline-continues.html (accessed 9 October 2014).

Elsevier Science Publishers (2015) Scopus (Electronic database) Retrieved from www.scopus.com (accessed 28 May 2015).

Goh, S., Lee, K., Park, J.S. and Choi, M.Y. (2012) Modification of the gravity model and application to the metropolitan Seoul subway system, *Physical Review E – Statistical, Nonlinear, and Soft Matter Physics*, 86(2), 026102.

Goldberg, A. (2014) Going with the idea flow, a review of A. Pentland (2014) *Social Physics: How Good Ideas Spread – the Lessons from a New Science*. Penguin Press, New York.

Karpiarz, M., Fronczak, P. and Fronczak, A. (2014) International trade network: fractal properties and globalization puzzle, *Physical Review Letters*, 113(24), 248701.

Lambiotte, R., Blondel, V.D., de Kerchove, C., (…), Smoreda, Z. and Van Dooren, P. (2008) Geographical dispersal of mobile communication networks, *Physica A: Statistical Mechanics and its Applications*, 387(21), 5317–5325.

Pentland, A. (2014) *Social Physics: How Good Ideas Spread – The Lessons from a New Science*. Penguin Press, New York.

Parvate, A. and Gangal, A.D. (2011) Calculus on fractal subsets of real line II: conjugacy with ordinary calculus, *Fractals*, 19(03), 271–290.

Semeshenko, V., Gordon, M.B. and Nadal, J.-P. (2008) Collective states in social systems with interacting learning agents, *Physica A: Statistical Mechanics and its Applications*, 387(19–20), 4903–4916.

Stewart J.Q. (1947) Empirical mathematical rules concerning the distribution and equilibrium of population, *Geographical Review*, 47, 479–491.

Stouffer S.A. (1940) Intervening opportunities: a theory relating mobility and distance, *American Sociology Review*, 5, 845–867.

Tauriello, G. (2013) Adelaide's leading high street shopping precincts face record vacancies from retail downturn. Retrieved from http://www.news.com.au/finance/real-estate/adelaide8217s-leading-high-street-shopping-precincts-facing-record-vacancies-from-retail-downturn/story-fndba8zb-1226719124475 (accessed 3 August 2016).

Tobler, W. (1970) A computer movie simulating urban growth in the Detroit region, *Economic Geography*, 46(2), 234–240.

Watts, D.J. and Strogatz, S.H. (1998) Collective dynamics of 'small-world' networks, *Nature*, 393(6684), 440–442.

Weidlich, W. (1971) The statistical description of polarisation phenomena in society, *British Journal of Mathematical and Statistical Psychology*, 24, 251–266.

Wilson, A.G. (1967) Statistcial theory of spatial distribution models, *Transportation Research*, 1, 1–32.

Wilson, A.G. (1974) *Urban and Regional Models in Geography and Planning*. John Wiley, London.

# Part II
# Housing and settlements

# 6 Spatial typology of the private rental housing market at neighbourhood scale

## The case of South East Queensland, Australia

*Yan Liu, David Wadley and Jonathan Corcoran*

### 6.1 Introduction

Divergences in living standards have been the subject of many studies in Australia and globally (Baum and Gleeson 2010; Kanbur and Venables 2005; Shorrocks and Wan 2005; Beer and Forest 2002; Hunter and Gregory 1996; Troy 1995). In 2002, the World Institute for Development Economics Research of the United Nations University launched a research project to study the geographical economic disparities in developing countries. They found that the spatial inequalities are high and rising between rural and urban areas, and between geographically advantaged and disadvantaged regions (Kanbur and Venables 2005).

Australia's metropolitan areas have undergone significant socio-economic and demographic changes over the past two to three decades, with certain areas accumulating a disproportionate share of disadvantage, resulting in a more complex socio-economic landscape (Vinson 2007, 1999; Larsen 2007; Randolph and Holloway 2007, 2005; Baum 2006; Baum *et al.* 2005; Gleeson and Randolph 2002). While many studies describe and measure inequalities in Australia and overseas, research focusing on the housing market and how spatial divergence relates to, or reinforces, social disadvantage is limited (Gleeson and Randolph 2002). For instance, de Souza Briggs (2005) and Rosenbaum *et al.* (2002) studied the geography of opportunity in the United States and noted that housing choice and location can adversely affect a range of life outcomes. In Australia, Randolph (2000) examined western Sydney and found that concentrations of socially disadvantaged households exist in both public estates and private sector housing. The research showed that many highly disadvantaged households were concentrated in the private rental stock, in poor quality accommodation, with little security and often unaffordable rents (Randolph 2000).

Affordability is thus a keynote of our enquiry, the aim being to shed greater light on the concept by interrelating household incomes and rental outgoings in fine-grain, small-scale local areas. This nexus is important to challenge blanket statements about apparent unaffordability which are made by politicians or the commentariat and take in entire cities or provinces. Factually inaccurate assertions are amplified in the media and can become unnecessarily unsettling for numbers of actual or would-be renters. The concept of affordability has otherwise attracted and troubled scholars. To illustrate, Wulff *et al.* (2009) and Yates *et al.* (2004) studied temporal change in the private rental market in Australia and highlighted the need for low-rent dwellings. More recently, Wulff *et al.* (2011) show that the level of access to affordable rental housing for low-income households in Australian cities reduced between 2001 and 2006 due to the absolute shortage of affordable housing, as well as an increase in housing displacement, despite overall growth of the private rental market by 11% during the same period. The concept of displacement is theoretically important and offers a route to progress in studies of broader affordability. It refers to the extent to which higher income households have accessed low-rent housing stock and thereby reduced such stock for access by low-income households (Wulff *et al.* 2011; HUD 2007). However, apart from general observations on the spatial correlation between housing (and, to a certain extent, private rental housing) and concentration of social disadvantage, there remains insufficient account of the spatial disparity between private rental housing supply and demand at neighbourhood scale, and whether such disparity is related, or has contributed to, social disadvantage and its spatial concentration.

This chapter brings a new perspective to the issue by linking graduated household income and rent outgoings data at neighbourhood scale. We used Australia's South East Queensland (SEQ) region as a case study, given the fine-scale data available from the censuses conducted by the Australian Bureau of Statistics (ABS) as well as the leading role that the private sector plays in Australia and the selected region's rental housing. The study addresses two key research questions: (1) how can the spatial disparity between the supply of, and demand for, private rental housing stocks be quantified at the local neighbourhood level?; and (2) how does the private rental housing market vary among different neighbourhoods in SEQ and is there any clustering of neighbourhoods as distinct sub-markets in the region?

The rest of the chapter is organised as follows. The next section introduces the study area and data used to define the spatial metrics to map and visualise the geographical disparity in the supply of, and demand for, private rental housing stocks using Geographical Information System (GIS) technologies. The methodology used to identify the spatial clustering of the private rental housing sub-market in the case study region is also presented at this point. This account is followed by analytical results which identify four distinctive clusters of the private rental housing sub-markets in SEQ, representing a range of socio-economic and demographic outcomes at the neighbourhood scale. The final section presents discussions about these clusters, including likely practical and policy implications, as well as conclusions and future research directions.

## 6.2 Study area, data and method

### *6.2.1 The study area*

South East Queensland is the most populous region in Queensland, with approximately 3.05 million people, or 66.3% of the State population as at the 2011 Census (ABS 2011). There are ten local government areas (LGAs) in this region,

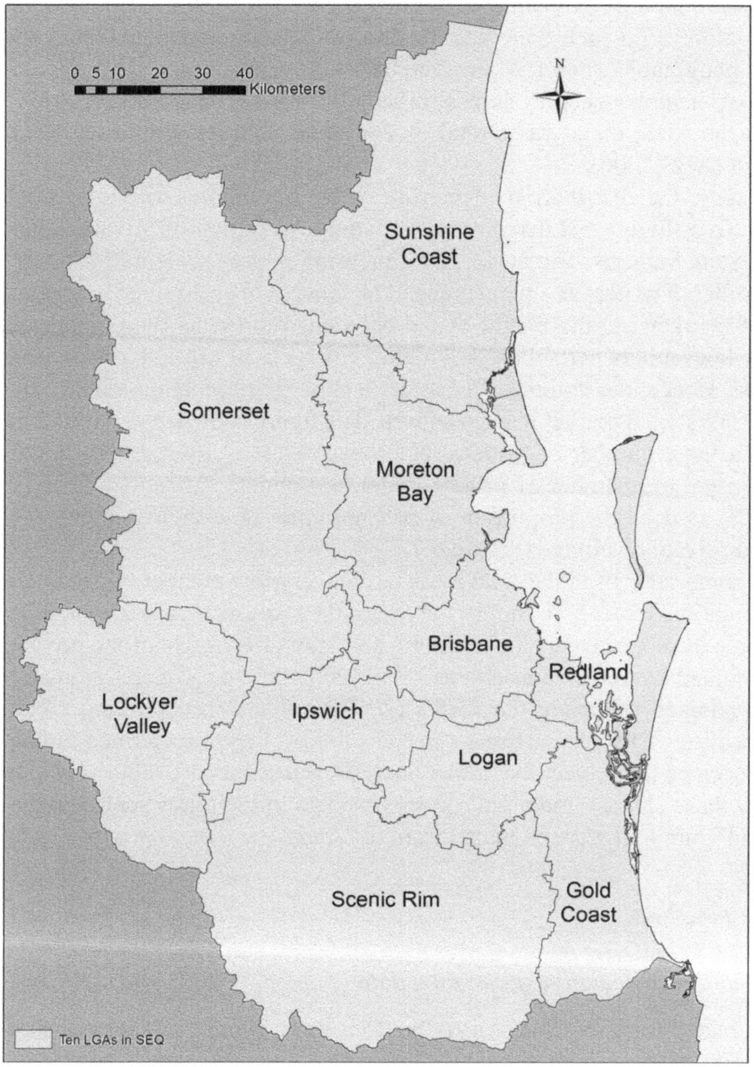


*Figure 6.1* Local government areas in South East Queensland

including Brisbane as the State capital and largest city in Queensland; the Gold Coast and Sunshine Coast areas as major tourist destinations; Ipswich as an outer-suburban city west of Brisbane with industrial and mining heritage; Logan, Redland and Moreton Bay as largely residential municipalities with some industrial and agricultural areas; and Somerset, Lockyer Valley and Scenic Rim as rural areas in the Great Dividing Range (Figure 6.1).

We chose to use the ABS State Suburb as the basic geographic unit in this study. State Suburbs are an approximation of localities gazetted by the Geographical Place Name authority in each State and Territory in Australia since 1996 (ABS 2011). This geographical unit offers the most consistent and appropriate delimitation of neighbourhoods across urban and rural areas (ABS 2011). According to the ABS, there was a total of 744 State Suburbs within the SEQ region in 2006 (ABS 2006).

Geographically, the distribution of private renter households varies significantly across SEQ. Figure 6.2 examines the spatial distribution of private rental dwellings by State Suburbs. The horizontal dimension in grey scale in this figure shows the number of renters as a percentage of all tenures in each suburb, ranging from 0 to 75%. Likewise, the height of the polygons represents the number of private renter households per square kilometre, enlarged 30 times vertically for visual contrast. Hence, the volume of each vertical bar represents the size of the private renter household market in each suburb. The figure shows that, within the SEQ region, a large number of private renters was clustered within the urban suburbs. However, the number of private renter households varies significantly across suburbs, as does the proportion of private renter households among the total occupied private dwellings (Figure 6.2).

Given that some suburbs in the rural areas have a very low number of occupied private dwellings and renter households, the rest of the analysis focuses on suburbs with 200 people or more per square kilometre and with at least 5% of the private dwellings occupied by renter households. The 200-people-per-square-kilometre criterion was adopted following the ABS's (2010) standard in delimiting urban centres and localities. This choice turned up 392 suburbs; they were termed 'urban suburbs' and became the subject for further analysis. Subsequently, suburbs which did not satisfy these criteria (mainly those areas shown in light grey scales on the flat surface in Figure 6.2) were excluded from the study.

### 6.2.2 Data and methods

#### *Scaled household income and rent outgoing data*

Data were extracted from the 2006 Australian Census of Population and Housing (ABS 2006). The primary sources include the number of private renter households categorised by the equivalised household income and dwelling rent. According to the ABS (2006), the equivalised household income is the total household income adjusted by the application of an equivalence scale to facilitate comparison of income levels among households of differing size and composition, reflecting the requirement of a larger household to have a higher level of income to achieve the same standard of living as a smaller household. The measure can be viewed as an indicator of the economic resources available to a standardised household, or to each individual in a household (ABS 2006). Renter households with income not stated, with negative income or income partially stated, were not included in the study. Households which did not state their dwelling rent were also excluded.

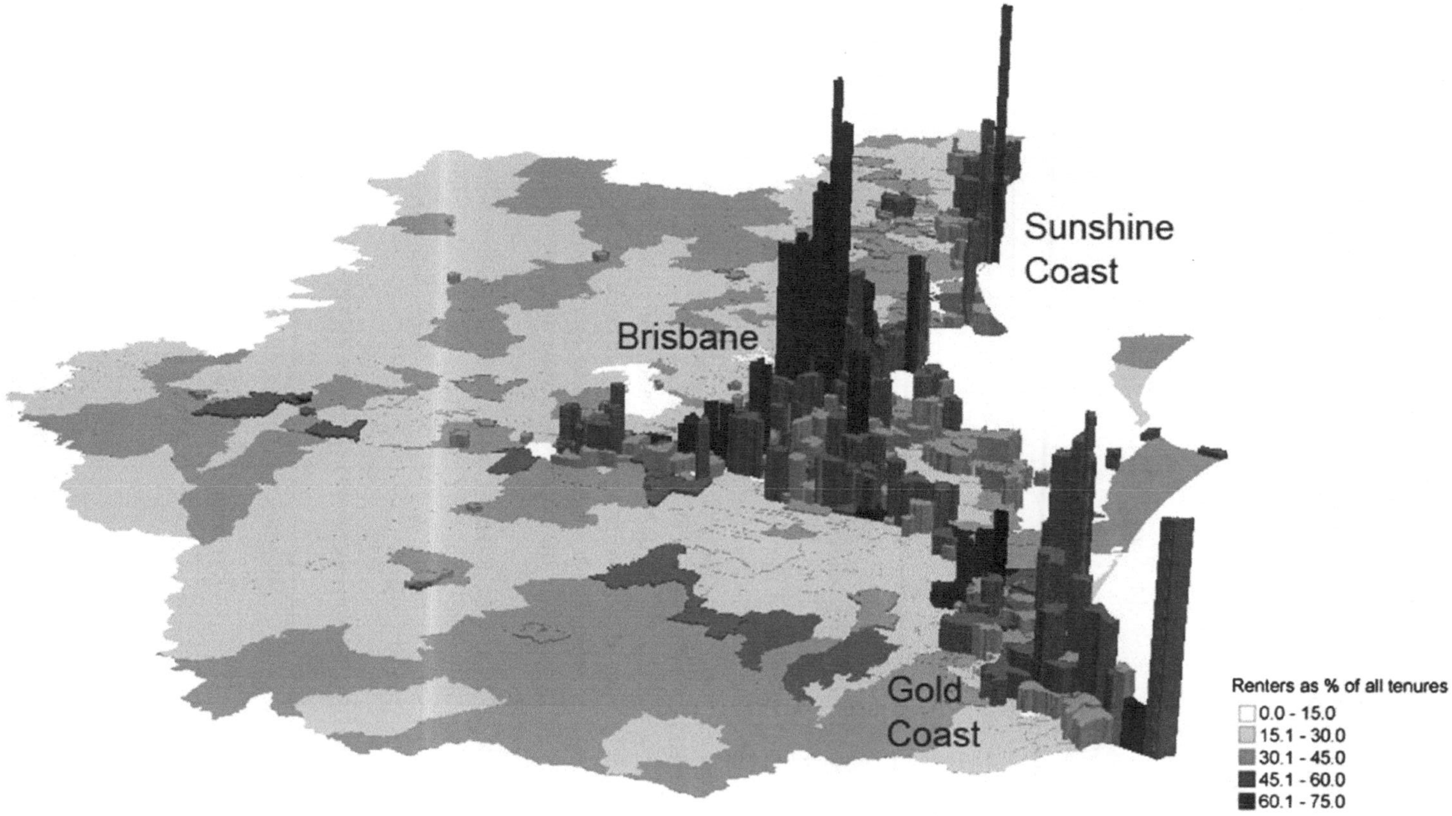


*Figure 6.2* Spatial distribution of the private rental households in South East Queensland. The grey colour scheme indicates the proportions of the private renter households in all tenures, and the vertical bar the number of private renter households per square kilometre in each suburb

*Table 6.1* Nominal household income and rent categories in SEQ

| *Household equivalised gross income per week in 2006 Census (AUD)* | *Household income categories* | *Household rent per week in the 2006 Census (AUD)* | *Household rent categories* |
|---|---|---|---|
| \$0–\$149 | Very Low | \$0–\$74 | Very Low |
| \$150–\$249 | | \$75–\$119 | |
| \$250–\$399 | | \$120–\$159 | |
| \$400–\$599 | Low | \$160–\$199 | Low |
| \$600–\$799 | | \$200–\$249 | |
| \$800–\$999 | Medium | \$250–\$299 | Medium |
| \$1,000–\$1,299 | | \$300–\$349 | |
| \$1,300–\$1,599 | High | \$350–\$449 | High |
| \$1,600–\$1,999 | Very High | \$450–\$549 | Very High |
| \$2,000 or over | | \$550 and over | |

Source: ABS Census TableBuilder: 2006 Australian Census of Population and Housing

While income and rent are the two main factors driving the demand for, and supply of, private rental housing, they rarely match geographically. According to Wulff *et al.* (2009), households are defined as having a financial affordability problem if they pay more than 30% of their gross income on rent. Derived from the initial categorisation of the equivalised household income and their weekly rent as per the 2006 Census, five nominal household income scales and five rent scales were defined. Each rent scale corresponds approximately to 30% of the upper boundary of the income scale (Table 6.1).

By cross-tabulating the scaled income and rent data through the ABS's TableBuilder (2006), a matrix of five rows and five columns was realised for each suburb to describe renter households in terms of their income and rent distributions. A sample cross-tabulation matrix of income and rent scales in one suburb is shown in Table 6.2.

The data depicted in bold down the main diagonal in Table 6.2 are the number of households which spent just 30% of income on rent; they have no housing affordability problem, according to Wulff *et al.* (2009). Cells located in the lower

*Table 6.2* A sample cross-tabulation matrix of income and rent categories in one suburb in SEQ (2006)

| | | *Number of households in each rent category* | | | | | *Total* |
|---|---|---|---|---|---|---|---|
| | | *Very High* | *High* | *Medium* | *Low* | *Very Low* | |
| Number of households in each income category | Very High | **7** | 8 | 37 | 11 | 0 | 63 |
| | High | 0 | **14** | 31 | 41 | 8 | 94 |
| | Medium | 4 | 13 | **168** | 226 | 60 | 471 |
| | Low | 12 | 30 | 188 | **323** | 156 | 709 |
| | Very Low | 10 | 3 | 87 | 185 | **361** | 646 |
| Total | | 33 | 68 | 511 | 786 | 585 | 1,983 |

triangular area beneath the main diagonal represent the number of households which outlay more than 30% of their income on rent; they do have an affordability problem. On the other hand, cells in the upper triangular area above the main diagonal depict households which pay less than 30% of their income on rent; they live in lower rent dwellings than their income permits, potentially causing a housing affordability problem for lower income households in the suburb, other things being equal.

### *6.2.3 Spatial diversity measures*

Four types of variable were developed to measure the spatial diversity between the supply of, and demand for, private rental housing in SEQ's urban suburbs. They include: (1) population and rental variables, i.e., population density (which is the number of people per square kilometre in each suburb) and rental incidence (which is the percentage of renter households in total occupied private dwellings); (2) measures of income and rent, i.e., median renter household income per week, median rent per week by renter household, percentage of households paying more than 30% of their income on rent, and percentage of households paying less than 30% of income on rent; (3) measures of rental housing concentration and diversity, i.e., the Shannon index and Herfindahl index; and (4) a measure of socio-economic disadvantage using the Socio-Economic Indexes for Areas (SEIFA) developed by the ABS (Pink 2006). A summary description of all these variables and their computation is provided in Table 6.3. The Shannon and Herfindahl indexes and the SEIFA disadvantage index are also discussed below.

#### *Shannon index*

The Shannon index was originally developed to measure the richness and evenness of species in ecology (Shannon 1948). It was thereafter extended to measure

*Table 6.3* Variables used in analysis

| *Variables* | *Descriptions* |
|---|---|
| Population density | Number of people per square kilometre |
| Rental rate | % of renter households in total occupied private dwellings |
| Median income | Median renter household income per week |
| Median rent | Median rent per week by renter household |
| Rent over | % of household paying more than 30% of income on rent |
| Rent under | % of household paying less than 30% of income on rent |
| Shannon index | The Shannon index is used to measure the diversity of the private rental housing market across the five income and five rent categories |
| Herfindahl index | The Herfindahl index is used measure the concentration of private rental housing across the five income and five rent categories |
| SEIFA disadvantage index | Index of relative socio-economic disadvantage composed by ABS from the 2006 Census |

diversity in categorical data. It was applied in this study to record the diversity of the private rental housing market across the five income and five rent categories. A maximum of 25 categories of income and rent (or 'species' in ecological term) can be observed in any suburb. The calculation of the Shannon index is as follows:

$$H' = -\sum_{i=1}^{S}\left(\frac{n_i}{N} ln \frac{n_i}{N}\right) \quad (6.1)$$

where $S$ is the number of household types based on their income and rent categories. Here $S$ equals 25; $n_i$ is the number of private renter households in category $i$ ($i$ = 1, … 25); $N$ represents the total number of private renter households in all categories.

The Shannon index ranges from 1.5 (the lowest richness and evenness in the combination of income and rent categories which would occur if there were only one category among all 25 possible combination of income and rent scales) to 3.5 (the highest richness and evenness in the combination of income and rent categories, as when all 25 categories among the income and rent scales are available in a suburb). Suburbs with a mixture of rental housing and household types will receive a larger Shannon index, indicating a diverse range of renter households. Alternatively, a lower Shannon index indicates a lack of diversity of renter households.

### *Herfindahl index*

The Herfindahl index is a commonly accepted measure of market concentration. Here, it is used to measure the concentration of the private rental market across the various income and rent scales, and is calculated as:

$$H = \sum_{i=1}^{S}\left(P_i^2\right) \quad (6.2)$$

where $P_i$ is the market proportion of category $i$ in the market, which is reckoned as the number of renter households in category $i$ divided by the total number of private renter households of the suburb. $S$ denotes the number of categories.

The Herfindahl index takes into account the relative size and distribution of the different categories of households based on their income and rent. The value ranges from $1/S$ to 1 (or 0.04 to 1 since, here, $S$ equals 25) A small index indicates a competitive market with no dominant players, i.e., there is a diverse range of housing stocks for renter households at all income levels. The value increases both as the number of households in the market decreases and as the size among the categories of households increases.

### *SEIFA disadvantage index*

SEIFA is a suite of indexes which ranks all areas in Australia according to their socio-economic advantage and disadvantage (Pink 2006). The Index of Relative Socio-economic Disadvantage or SEIFA disadvantage index is derived from census variables related to disadvantage, such as low income, low educational attainment, unemployment, and dwellings without motor vehicles. The average SEIFA disadvantage score for all statistical suburbs across Australia is 1,000 and the middle two-thirds of index values lie appropriately between 900 and 1,100 nation-wide (Pink 2006). The SEIFA disadvantage index was employed to quantify how one suburb compares with another in terms of socio-economic disadvantage.

### *6.2.4 Data standardisation*

All spatial variables were standardised prior to any analysis to remove the impact of measurement scale on results. The method of standardisation was given by:

$$I_{std} = \frac{I_{\max} - I_{ori}}{I_{\max} - I_{min}} \tag{6.3}$$

where $I_{std}$ is the normalised value of each variable which ranges from 0 to 1; $I_{ori}$ is the original values of each variable; $I_{max}$ and $I_{min}$ are the maximum and minimum value of each variable, respectively.

### *6.2.5 Principal Component Analysis (PCA)*

Among the nine variables used to measure the spatial disparity of the private rental housing market in SEQ, it is highly likely that some of the indicators could correlate with each other due to the nature of the indicators and the spatial proximity of the suburbs. Indeed, a multi-collinearity effect caused by the use of highly correlated spatial variables can result in misleading outcomes in statistical analysis and modelling (Rogerson 2010). Therefore, Principal Component Analysis (PCA) was applied to tackle the problem of correlation among the indicators and extract key components differentiating the rental market of SEQ urban suburbs. Through a set of orthogonal transformations, it converts the datasets into an array of values of linearly uncorrelated variables (Li and Yeh 2002).

Mathematically, for a set of n variables, the covariance matrix with n rows and n columns is defined as:

$$C = \{c_{i,j}, c_{i,j} = \text{cov}(\text{var}_i, \text{var}_j)\} \tag{6.4}$$

where $C$ represents the covariance matrix and $c_{i,j}$ the covariance between variable $i$ ($\text{var}_j$) and variable $j$ ($\text{var}_j$).

In order to obtain the principal components, the eigenvectors and eigenvalues of the covariance matrix are computed mathematically as:

$$v = ECE^T \tag{6.5}$$

where $v$ is the eigenvalue matrix, $E$ is the eigenvector matrix and $E^T$ is the transposition matrix of $E$.

Therefore, the PCA transformation is based on the following equation:

$$y_i = v^T x \tag{6.6}$$

wherein, for all variables of *i*, $v^T$ is the transposition of the eigenvector corresponding to the principal component, and $y_i$ is the $i^{th}$ independent principal component.

Through the PCA transformation, a set of independent, compressed components was selected, and subsequently used for classification and typology of the private rental housing sub-markets in SEQ.

### *6.2.6 Typology of private rental housing sub-markets*

Research literature has highlighted the theoretical significance as well as the policy relevance of identifying sub-markets within the urban housing system (Leishman 2009; Jones *et al.* 2003; Maclennan and Tu 1996). The diverse range of urban suburbs in SEQ in terms of the supply of, and demand for, private rental housing indicates that there could be some distinct rental housing sub-markets.

While numerous researchers have proposed classifications for identifying housing sub-markets, the most common approach is to begin with some prior notion of where sub-markets are likely to exist and then test for the existence of the sub-market. Nevertheless, there is 'a lack of consensus regarding the appropriate methods for identifying (as opposed to testing for) housing sub-markets' (Leishman 2009: 568). Using the principal component results extracted from the PCA, a two-step clustering method was accessed to identify the typology of the private rental housing sub-market in the study region.

The two-step cluster method is a scalable cluster analysis algorithm designed to handle very large datasets (Zhang *et al.* 1996; Chiu *et al.* 2001). The approach thus involves two steps. The first is to pre-cluster the cases into many small sub-clusters based on either the Log-likelihood distance or the Euclidean distance criterion. In this study, we chose to use Log-likelihood distance because it is a probability-based measure in which the distance between two clusters is related to the decrease in log-likelihood when they are combined into one cluster. For the second step, we used the hierarchical clustering method to assess the multiple sub-clusters resulting from the pre-cluster step and classify them into the desired number of clusters. Details about the two-step clustering method can be found in Zhang *et al.* (1996) and Chiu *et al.* (2001). It is a ready-for-use tool in the SPSS statistics program.

## 6.3 Results

Results from the PCA show that the first three principal components account for over 85% of the variance within the original nine variables (Table 6.4).

Using these three principal components, the two-step cluster analysis method was applied to categorise all 392 urban suburbs in SEQ, resulting in four spatial

*Table 6.4* Principal components extracted from PCA and their contributions

| *Principal components* | *Individual contribution (%)* | *Cumulative contribution (%)* |
|---|---|---|
| 1st | 42.6 | 42.6 |
| 2nd | 26.9 | 69.4 |
| 3rd | 15.8 | 85.3 |

*Table 6.5* Number of suburbs and mean values of selected variables for each cluster in SEQ (2006)

| *Selected variables* | *Cluster I: Inner City Urban Lifestyle* | *Cluster II: Outer Suburb Living* | *Cluster III: Coastal City Lifestyle* | *Cluster IV: Low Cost Country Living* |
|---|---|---|---|---|
| Numbers of suburbs | 68 | 130 | 104 | 90 |
| Percentage of renters in all tenure types | 40.6 | 19.1 | 31.0 | 36.4 |
| Population density (no. of people per km$^2$) | 2,436 | 1,028 | 1,345 | 1,037 |
| Median income per week ($) | 1,207 | 1,315 | 981 | 809 |
| Median rent per week ($) | 239 | 251 | 264 | 179 |
| Percentage of households paying more than 30% income on rent | 32.6 | 46.6 | 57.4 | 32.1 |
| Percentage of households paying less than 30% income on rent | 28.5 | 19.7 | 11.9 | 20.6 |
| Shannon index | 2.64 | 2.23 | 2.40 | 1.89 |
| Herfindahl index | 0.09 | 0.14 | 0.12 | 0.20 |
| SEIFA disadvantage index | 1,056 | 1,063 | 1,023 | 925 |

clusters of suburbs, each with distinctive features spatially and structurally which will now be explained (Table 6.5 and Plate 1).

### *Cluster I: Inner-city Urban Lifestyle*

This cluster is generally located around the Brisbane city centre (Plate 1 – purple). A total of 68 suburbs is identified. They feature a very high population density as well as very high proportion of renter households among other tenure types. On average, over 40% of households are renters, with this proportion ranging from 21% to 63%. Due to their close proximity to the CBD, suburbs have good access to workplaces, transportation and other services which generally attracts single persons or young families with medium to high income; the average median household income is AUD$1,207 per week. Subsequently, these central city rental costs are higher compared with the outer suburbs (i.e., Cluster III, see below), with an average median outlay of AUD$239 per week. On average 32.6% of households need to spend more than 30% of their income on rent while another 28.5% spend less than 30% of their income for such purpose. The SEIFA disadvantage index value ranges from 958 to 1,132, with a mean of 1,056, indicating that these suburbs are not disadvantaged in their socio-economic status. A mean Shannon index of 2.64 (range, 2.11 to 2.98) indicates that there are varying types of dwellings available. The average Herfindahl index of 0.09 confirms the range of housing stocks available for renters.

### *Cluster II: Outer-city Suburb Living*

This is a cluster in which renter household have higher income and also pay higher rent. With a total of 130 suburbs, the average median income is the highest among the four clusters, and the average median rent is second only to Cluster III (see *Cluster III* below). However, the proportion of renters in all tenures is very low (averaging 19.1%). With an average Shannon index of 2.23, variation of renter household types is well represented, and the number of households in each type is more evenly spread; hence, the average Herfindahl index is only 0.16. Still, an average 46.6% of households pays higher rent and 19.7% pays lower rent than their income can ideally support. The average SEIFA disadvantage index is slightly higher than that of Cluster I, which indicates no disadvantage in terms of socio-economic status. Suburbs in this cluster are generally located around Cluster I within Brisbane city, and extend to the outer suburbs in the Gold Coast and Sunshine Coast (Plate 1 – yellow).

### *Cluster III: Coastal City Lifestyle*

Most of the suburbs in this cluster are situated along the coastal urban nodes in the Gold and Sunshine Coasts (Plate 1 – blue); only a handful to the south and east of Brisbane are included. A distinctive feature is that this cluster has the highest rent among all others, even though the average median income is lower than Clusters I and II. As such, a very high proportion of households (averaging 57.4%) needs to outlay more than 30% of their income and only 11.9% pay less than 30%. These conditions arise largely because of the proximity of Cluster III suburbs to tourist sites and coastal amenities.

### *Cluster IV: Low-cost Country Living*

This cluster posts a relatively low SEIFA disadvantage index, indicating reduced socio-economic status. With a ratio of renter households in all tenures ranging from 17% to 55%, the grouping records a significantly lower average median income of AUD$807 and a lower average median rent of AUD$179 per week. Overall, about 32.1% of households need to pay higher rent, while another 20.6%

are paying lower rent than their income can support. Geographically, the cluster is located in Logan and Ipswich cities which are on the south and west sides of metropolitan Brisbane, and some suburbs lie in the Moreton Bay Region between Brisbane and the Sunshine Coast. (Plate 1 – green). The low cost of accommodation has attracted renters, especially those in the lower income groups.

## 6.4 Discussion and conclusions

This chapter presents an analysis of the private rental housing market in SEQ by way of a typology of its urban suburbs. Starting with 392 State Suburbs, which correspond to local urban communities, the analysis identified four distinct clusters or groups to represent the broad typology of private rental housing sub-markets across the region. Each cluster possesses distinct features in the supply of, and demand for, private rental dwellings. The city central suburbs feature a high proportion of rental housing and high rent, while suburbs recording high income but a low proportion of rental housing are mostly located in the middle ring around the city centre cluster. Suburbs near the tourist centres in the Gold and Sunshine Coasts feature medium income but high rent, resulting in a significant gap between the supply of, and demand for, low-rent housing. Nevertheless, these three clusters of suburbs are not socially disadvantaged, given that they all score near 1,000 or over of the SEIFA disadvantage index.

The fourth cluster of suburbs exhibits disadvantage as well as low income and low rent, even though rental housing forms a large portion of all the tenure types. Whether renters living in these suburbs are long-term tenants or short-term tenants taking up the advantage of low rent to save in order to advance their housing trajectory should be a topic of further investigation.

The divergence of the private rental market among the different suburbs indicates the pertinence of community and people-based policy responses and solutions to housing problems (Baum 2006). Our study has argued against universalist statements of supposed fact in favour of a more penetrating analysis which can yield an accurate picture of rental practice in an area. The current results dispute a 'one size fits all' commentary in which a lack of affordability would presumably be lamented across an entire population. So much was not the case in SEQ: a more nuanced account emerged. Even so, by means of the concept of displacement, the enquiry does posit that the extent of affordability over many suburbs can be reduced when households of more than sufficient income move down market in their choice of rental housing. Although instances of this process have been demonstrated within the project, displacement will no doubt remain a controversial concept inviting further research. For ideological reasons, some might choose to deny its existence, in that willing participants in a free market should enjoy the liberty to rent wherever their income permits. Others would dispute the practice and outcome of displacement on account of its apparent Pareto deficiency, in that the actions of certain better-off market participants allegedly make poorer players worse off than they might otherwise have been. The focus of such scholars would be to urge public intervention by some form of means testing to restrict wealthier renters to higher priced rental stock. In such a scenario, the market would be manipulated and no longer free in character.

Aside from such substantive debates, it can be asserted that the methodology presented in this chapter is innovative, since it brings a new perspective to the analysis of current, not previously linked, income and rent data at the neighbourhood scale to identify the spatial and structural grouping or clustering of suburbs. However, it is not without its limitations. For instance, even though we excluded suburbs with a small number of rental households, the classification into 25 categories resulted in some suburbs with a small number of households in one or more categories; moreover, such numbers could potentially be randomised by the ABS due to its confidentiality policy. As such, subsequent survey-based enquiry about residences in the region is needed to complement the census data analysis better to understand the causal processes driving the formation of private rental housing sub-markets. On one hand, the physical and historical settings of the cities as well as accessibility to work, the state of the transportation network and access to other services and facilities play a major role in driving spatial disparity and segmentation of the private rental housing market. On the other, the demographic structures of renter households along with their housing and location choice behaviour at various life stages are significant in shaping the market. Clear understanding of these issues is essential to achieve the most effective and efficient policy outcomes.

## Acknowledgements

This research is supported by the University of Queensland early career research grant (project no. 603351) and the University of Queensland new staff research start-up grant (project no. 601871). The authors would like to thank Professor Glen Searle, Dr Thomas Sigler and the anonymous reviewers for their constructive comments on an early version of the manuscript.

## References

ABS Australian Bureau of Statistics. (2006) *Census 2006, 1996*. Canberra: Australian Bureau of Statistics.

ABS Australian Bureau of Statistics. (2010) *1216.0 – Australian Standard Geographical Classification (ASGC)*. Canberra: Australian Bureau of Statistics.

ABS Australian Bureau of Statistics. (2011) *1270.0.55.003 – Australian Statistical Geography Standard (ASGS): Volume 3 – Non ABS Structures*. Canberra: Australian Bureau of Statistics.

Baum, S. (2006) A typology of socio-economic advantage and disadvantage in Australia's large non-metropolitan cities, towns and regions, *Australian Geographer* 37(2), 233–258.

Baum, S. and Gleeson, B. (2010) Space and place: social exclusion in Australia's suburban heartland, *Urban Policy and Research* 28(2), 135–159.

Baum, S., Haynes, M., van Gellecum, Y., and Han, J. H. (2005) Typologies of advantage and disadvantage: socio-economic outcomes in Australian metropolitan cities, *Geographical Research* 43(4), 361–378.

Beer, A. and Forster, C. (2002) Global restructuring, the welfare state and urban programmes: federal policies and inequality with Australian cities, *European Planning Studies* 10(1), 7–15.

Chiu, T., Fang, D., Chen, J., Wang, Y., and Jeris, C. (2001) A robust and scalable clustering algorithm for mixed type attributes in large database environment. In: *Proceedings of the Seventh ACM SIGKDD International Conference on Knowledge Discovery and Data Mining*. San Francisco, CA: ACM.

Gleeson, B. and Randolph, B. (2002) Social disadvantage and planning in the Sydney context, *Urban Policy and Research* 20(1), 101–107.

HUD (US Department of Housing and Urban Development) (2007) *Affordable Housing Needs 2005: Report to Congress, Office of Policy Development and Research (HUD)*. Washington, DC: Government Printing Office.

Hunter, B. and Gregory, R. G. (1996) An exploration of the relationship between changing inequality of individual, household and regional inequality in Australian cities, *Urban Policy and Research* 14(3), 171–186.

Jones, C., Leishman, C., and Watkins, C. (2003) Structural change in local urban housing markets, *Environment and Planning A* 35(7), 1315–1326.

Kanbur, R. and Venables, A. J. (2005) Rising spatial disparities and development, *Policy Brief 3*. Helsinki: United Nations University.

Larsen, K. (2007) *The Health Impacts of Place-based Interventions in Areas of Concentrated Disadvantaged: A Review of the Literature*. Sydney: Sydney South West Area Health Service and UNSW Centre for Health Equity Training, Research & Evaluation (CHETRE).

Leishman, C. (2009) Spatial change and the structure of urban housing sub-markets, *Housing Studies* 24(5), 563–585.

Li, X. and Yeh, A. G.-O. (2002) Urban simulation using principal components analysis and cellular automata for land-use planning, *Photogrammetric Engineering and Remote Sensing* 68(4), 341–351.

Maclennan, D. and Tu, Y. (1996) Economic perspectives on the structure of local housing markets, *Housing Studies* 11(3), 387–406.

Pink, B. (2006) *Information Paper 2039.0: An Introduction to Socio-economic Indexes for Areas (SEIFA) 2006*. Canberra: Australian Bureau of Statistics.

Randolph, B. (2000) Regional disadvantage in Western Sydney. Paper presented to the Responding to Disadvantaged Communities Conference, Liverpool, 23 March.

Randolph, B. and Holloway, D. (2005) Social disadvantage, tenure and location: an analysis of Sydney and Melbourne, *Urban Policy and Research* 23(2), 173–201.

Randolph, B. and Holloway, R. (2007) *Rent Assistance and the Spatial Concentration of Low Income Households in Metropolitan Australia, Final Report*. Melbourne: Australian Housing and Urban Research Institution (AHURI).

Rogerson, P. A. (2010) *Statistical Methods for Geography: A Student's Guide*. London: Sage.

Rosenbaum, J. E., Reynolds, L., and Deluca, S. (2002) How do places matter? The geography of opportunity, self-efficacy and a look inside the black box of residential mobility, *Housing Studies* 17(1), 71–82.

Shannon, C. E. (1948) A mathematical theory of communication, *The Bell System Technical Journal* 27, 379–423, 623–656.

Shorrocks, A. and Wan, G. (2005) Spatial decomposition of inequality, *Journal of Economic Geography* 5(1), 59–81.

de Souza Briggs, X. (ed.) (2005) *The Geography of Opportunity: Race and Housing Choice in Metropolitan America*. Washington, DC: Brookings Institute Press.

Troy, P. (ed.) (1995) *Australian Cities*. Cambridge: Cambridge University Press.

Vinson, T. (1999) *Unequal in Life: The Distribution of Social Disadvantage in Victoria and New South Wales*. Sydney: The Ignatius Centre for Social and Policy Research.

Vinson, T. (2007) *Dropping off the Edge: The Distribution of Disadvantage in Australia*. Richmond, South Australia: Jesuit Social Services and Catholic Social Services Australia.

Wulff, M., Dharmalingam, A., Reynolds, M., and Yates, J. (2009) *Australia's Private Rental Market: Changes (2001–2006) in the Supply of, and Demand for, Low Rent Dwellings*. Position Paper No. 122. Melbourne: Australian Housing and Urban Research Institution (AHURI).

Wulff, M., Reynolds, M., Arunachalam, D., Hulse, K. and Yates, J. (2011) *Australia's Private Rental Market: The Supply of, and Demand for, Affordable Dwellings*. AHURI Final Report No. 168. Melbourne: Australian Housing and Urban Research Institution.

Yates, J., Wulff, M., and Reynolds, R. (2004) *Changes in the Supply of and Need for Low Rent Dwellings in the Private Rental Market, Final report*. Melbourne: Australian Housing and Urban Research Institution (AHURI).

Zhang, T., Ramakrishnon, R., and Livny, M. (1996) BIRCH: An efficient data clustering method for very large databases. In: *Proceedings of the ACM SIGMOD Conference on Management of Data*. Montreal, Canada: ACM.

# 7 Optimal kernel and bandwidth specifications for geographically weighted regression

## An evaluation using automated valuation models (AVMS) for mass real estate appraisal[1]

*Paul E. Bidanset and John R. Lombard*

### 7.1 Introduction

Local governments have the responsibility of fairly and uniformly taxing the properties within their jurisdictions and they must be held accountable for the taxes levied upon property owners. Therefore, it is imperative that residential property assessments be accurate, fair, and defensible. In recent years, there have been large strides in the advancement of mass-appraisal techniques such as automated valuation models (AVMs).

It has long been understood that heterogeneity across geographic stratums results in conventional OLS-based multiple regression analysis models' inability to accurately capture variables' true effects (Ball 1973; Berry and Bednarz 1975; Anselin and Griffith 1988). While spatial consideration in the form of dummy variables and distance coefficients can help improve models, they may fail to fully correct for spatial autocorrelation, and parameter averages may be skewed or averaged out (Berry and Bednarz 1975; Fotheringham *et al.* 2002; McMillen and Redfearn 2010). Inaccuracy in parameter estimation in assessment models can result in the unfair valuation of properties, leading to a host of challenges for the taxing jurisdiction; they face the likelihood of additional costs in time and money in defending their valuations.

Sufficient research has shown locally weighted regression (LWR) methods improve traditional valuation model performance and predictability power (e.g. Brunsdon *et al.* 1996; McMillen 1996; Brunsdon 1998). One such LWR methodology, which more accurately accounts for spatial heterogeneity, is geographically weighted regression (GWR) (Fotheringham *et al.* 2002; LeSage 2004; Huang *et al.* 2010). The use of GWR in property tax modeling has become an area of study as well. GWR has been shown to provide assessment jurisdictions with more accurate valuations than multiple regression analysis (MRA) and other AVM techniques (Borst and McCluskey 2008; Moore 2009; Moore and Myers 2010; Lockwood and Rossini 2011; McCluskey *et al.* 2013). Lockwood and Rossini state that GWR favorably reduces prediction errors that arise from edge effects of boundaries in global models, and that they are more "in-tune with the market." Borst and McCluskey (2008) demonstrate the ability of GWR to detect submarkets within jurisdictions. While GWR improves upon several standard mass valuation techniques, it is a relatively new technique in the appraisal community, and some researchers suggest the need for additional studies to further establish GWR's credibility (Lockwood and Rossini 2011). Therefore, further research aimed at evaluating and understanding GWR performance in valuation is necessary.

The performance of kernel and bandwidth specification within GWR models has been explored in other disciplines, namely forestry and ecology (e.g. Guo *et al.* 2008; Cho *et al.* 2010), but until recently, optimal bandwidth/kernel combinations had not been examined side-by-side with respect to the potential impact on the International Association of Assessing Officers (IAAO) approved statistical measures of equity and fairness. Bidanset and Lombard (2014) examine Gaussian and bisquare kernels with both adaptive and fixed bandwidths and find that the spatial weighting function of GWR models does have significant impacts on coefficient of dispersion (COD) measurements of uniformity. Furthermore, they find that a model which consistently produces superior overall results (i.e. lower COD) can still be outperformed in certain geographic areas by a suboptimal aggregate model – ultimately suggesting that, when calibrating models, results should be extrapolated past the Akaike information criterion (AIC) relative measure of model performance to standard measures of uniformity and equity.

Using residential data provided by the city of Norfolk, VA, this research builds upon Bidanset and Lombard (2014) by examining additional kernel functions (i.e. bisquare, exponential, tricube, and boxcar) in GWR AVM models and their effect on IAAO measures of equity and fairness.

## 7.2 Model descriptions and estimation details

The traditional ordinary least squares regression model within a real estate valuation context is represented by:

$$y_i = \beta_0 + \sum_k \beta_k x_{ik} + \varepsilon_i \tag{7.1}$$

where $y_i$ is the $i^{th}$ home sale, $\beta_0$ is the model intercept, $\beta k$ is the $k^{th}$ coefficient, $x_{ik}$ is the $k^{th}$ variable for the $i^{th}$ home sale, and $\varepsilon_i$ is the error term of the $i^{th}$ home sale. Geographically weighted regression is depicted by the following formula:

$$y_i = \beta_0(x_i, y_i) + \sum \beta_k(x_i, y_i) x_{ik} + \varepsilon_i \tag{7.2}$$

where $(x_i, y_i)$ indicates the xy coordinates of the $i^{th}$ regression point. In this research, the xy coordinates are the longitude and latitude coordinates of each sale. With GWR, a regression equation is generated at each observation.

### *7.2.1 Spatial weighting specifications – kernels and bandwidths*

The respective bandwidth of each observation refers to the encompassing distance of observations to be included in the respective model. A bandwidth value can either be adaptive (including some predetermined value of nearest neighbors), or fixed (including all properties within some predetermined distance). The spatial kernel calculates weights to be assigned to other observations based on their proximity to the regression point. Unlike the prior research of Bidanset and Lombard (2014) that employed Gaussian and bisquare kernels, this chapter examines exponential, tricube and boxcar kernels. Figure 7.1 shows the calculation for each respective function. Each kernel, with the exception of the tricube kernel, incorporates a distance decay function allowing properties closer to a regression point to weigh more than properties further away – exemplified by Figure 7.2. The bisquare, tricube, and boxcar kernels are discontinuous and assign a weight of zero to observations outside of the bandwidth, nullifying their impact on the local regression estimate.

Each kernel will be tested using both fixed and adaptive bandwidths. In GWR the size of the bandwidth is optimized by either distance (fixed kernel) or a

Global Model
$$w_{ij} = 1$$

Gaussian
$$w_{ij} = \exp\left(-\frac{1}{2}\left(\frac{d_{ij}}{b}\right)^2\right)$$

Exponential
$$w_{ij} = \exp\left(-\frac{|d_{ij}|}{b}\right)$$

Box-car
$$w_{ij} = \begin{cases} 1 & \text{if } |d_{ij}| < b, \\ 0 & \text{otherwise} \end{cases}$$

Bi-square
$$w_{ij} = \begin{cases} (1 - (d_{ij}/b)^2)^2 & \text{if } |d_{ij}| < b, \\ 0 & \text{otherwise} \end{cases}$$

Tri-cube
$$w_{ij} = \begin{cases} (1 - (|d_{ij}|/b)^3)^3 & \text{if } |d_{ij}| < b, \\ 0 & \text{otherwise} \end{cases}$$

where: $w_{ij}$ is the weight applied to the $j^{th}$ property at regression point $i$
$b$ is the bandwidth
$d_{ij}$ is the geographic distance between regression point $i$ and property $j$

*Figure 7.1* Kernel functions

Source: Gollini et al. 2013: 5; used with permission

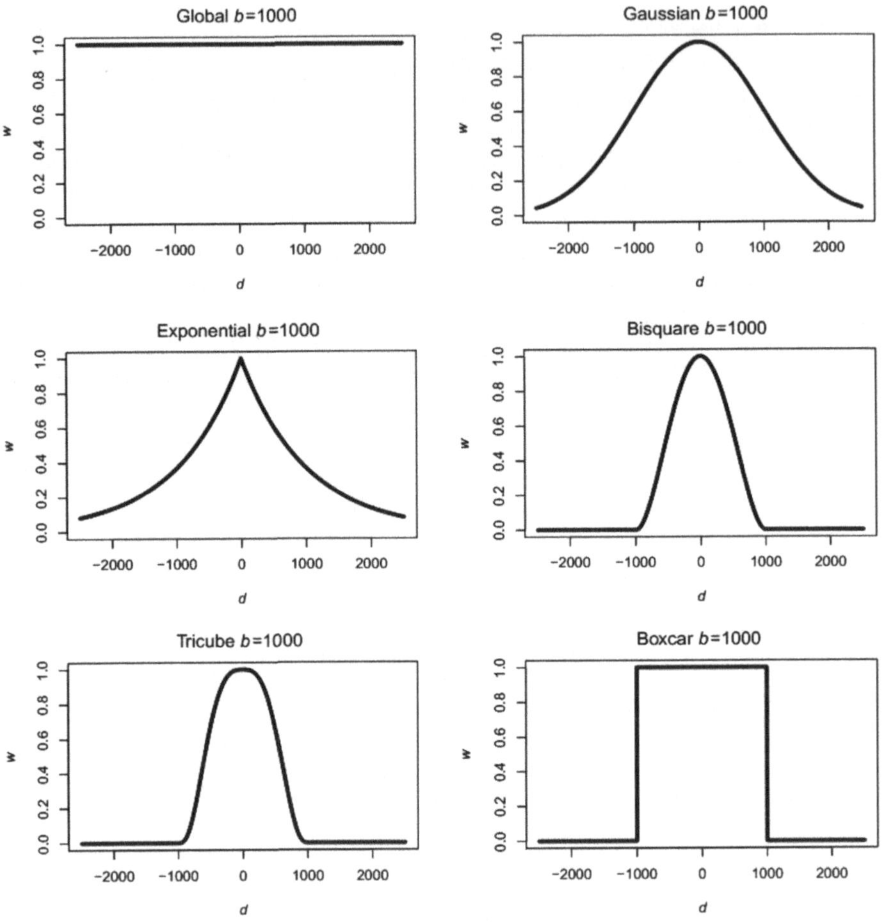


*Figure 7.2* Distribution of kernel functions

Source: Gollini et al. 2013: 6; used with permission

number of neighboring observations (adaptive kernel). During model calibration, bandwidths are tested and assigned AIC corrected (AICc) scores; the bandwidth with the lowest AICc is optimal (Lu *et al.* 2014).

### *7.2.2 Measuring vertical equity and uniformity in valuation models*

As previously mentioned, the International Association of Assessing Officers (IAAO) maintains standards of assessment uniformity and equity to which its member jurisdictions adhere (IAAO 2013). Two indicators known as the coefficient of dispersion (COD) and price-related differential (PRD) are calculated to evaluate assessment uniformity and equity, respectively. The COD measures uniformity of an appraisal stratum and is represented by the following formula:

$$COD = \frac{100}{n} \frac{\sum_{i=1}^{n} \left| \frac{EP_i}{SP_i} - Median\left(\frac{EP_i}{SP_i}\right) \right|}{Median\left(\frac{EP_i}{SP_i}\right)} \tag{7.3}$$

where $EP_i$ is the expected price of the $i^{th}$ property, and $SP_i$ is the sale price of the $i^{th}$ property. The COD is a measure of average percentage deviation of the assessment-to-sale ratios $\left(\frac{EP_i}{SP_i}\right)$ from the median assessment-to-sale ratio (e.g. a COD value of 14.7 means on average, assessment-to-sale ratios deviate from the median assessment-to-sale ratio by 14.7%). The acceptability threshold set forth by the IAAO for single-family homes is a COD value of no more than 15.0, although values of 5.0 or less are suspect of sampling error or sales chasing whereby assessors cherry-pick sales to make assessments appear more uniform than they really are (Gloudemans and Almy 2011).

The price-related differential is a coefficient of vertical equity and is calculated as follows:

$$PRD = \frac{Mean\left(\frac{EP_i}{SP_i}\right)}{\sum_{i=1}^{n} EP_i / \sum_{i=1}^{n} SP_i} \tag{7.4}$$

The acceptability range set forth by the IAAO of a stratum's PRD is between .98 and 1.03, with values above being evidence of regressivity, whereby higher-priced homes are assessed at a lower percentage of market value than lower-priced homes, and values below .98 show evidence of progressivity, whereby higher-priced homes are assessed at a higher percentage of market value than lower-priced homes (IAAO 2013).

The Akaike information criterion (AIC) is a commonly used relative performance measure of models applied to the same sample and has the following calculation:

$$AIC = -2LogL + 2k \tag{7.5}$$

where $L$ is the maximum likelihood of the model and $k$ is the number of free parameters of the model. AIC corrected (AICc) is a similar score that penalizes for parameters that do not offer any explanatory power (Sugiura 1978):

$$AIC_C = -2LogL + 2k + \left(\frac{2k(k+1)}{n-k-1}\right) \tag{7.6}$$

where $n$ is the number of observations in the sample. This chapter utilizes the AICc to compare models.

## 7.3 The data

The sample data consist of 2,450 valid single-family sales in Norfolk, VA, from 2010 to 2012 (three years of sales is a recommended practice in the field of mass appraisal). Valid sales meet several criteria: they must be arm's length transactions where neither party is under duress to buy or sell; the property is listed on the open market; and there is no marital, blood, or previous relationship between the buyer and seller. After a sale is completed and the new deed is registered with the Norfolk real estate assessor's office, a city appraiser will pursue unbiased third-party verification to validate the conditions of the sale and property characteristics. Invalid transfers, such as foreclosures, short sales, and government sales, were omitted from analysis due to the fact that they may not reflect the true market value of a property which conforms to the legal standards assessors are required to follow when validating transactions.

To further reduce data errors, the sample data were further cleaned by meeting the following conditions: a sales price greater than zero, a sales price greater than the land assessment, and properties with a positive net improvement sale price (sale price – assessed land value > 0). For analysis purposes, the sale price was converted to its natural logarithm and outliers were identified using an IQRx3 approach[2] and subsequently removed (approximately 2% of observations).

The dependent variable, *Ln.ImpSalePrice*, is the natural log of the selling price of the house less the land market value. Moore and Myers (2010) and Bidanset and Lombard (2014) utilize this subtraction of the assessed land value, treating it as an offset to help isolate the coefficients' impact on the improved property only. The transformation of the dependent variable into a natural log allows for results to be measured in percentage terms as opposed to dollars. The predicted values are transformed from natural log form and land value is added back in prior to performing ratio tests (COD and PRD). Table 7.1 shows the independent or predictor variables and their descriptions.

*TLA* captures the square feet of livable space within the home; an unfinished attic, for example, is not included in this quantification. *TGA* includes both attached and detached garage space – consistent with Moore and Myers (2010) and Bidanset and Lombard (2014). Dummy variables are used for *bldgcond* and *qualityclass*, and the default for each is 'average'. The effective age is used more often than age because it takes into account the overall state of the depreciation relative to other improvements; homes with depreciation that has been 'cured' (e.g. replacement of aging, rotting exterior wood) receive an 'effective age' that is younger than their actual age to reflect this change (Gloudemans 1999). With *bldgcnd*, *quality*, and *EffAge* accounted for, *Age* is included to capture any potential premium on vintage or historic properties. For the three years of sales data, 11 time-indicator three-year linear spline variables were constructed based on the reverse month of sale (*RM1* through *RM36*), with the earliest month of sale equal to *RM1*, the second earliest month of sale equal to *RM2*, and consecutive months continuing to *RM36*. Only splines *RM12* and *RM21* of our data improved model performance (difference in AIC of at least 2) and were added to the model. Compared to traditional quarterly

*Table 7.1* Independent variables

| *Variable* | *Description* |
|---|---|
| TLA | total living area (sq.ft.) |
| EffAge | effecting age in years |
| TGA | total garage area (sq.ft. detatched + attached) |
| Age | age in years |
| qualityclassFair | fair quality (avg. is default) |
| qualityclassGood | good quality |
| qualityclassVGd | very good quality |
| bldgcondFair | fair condition (avg. is default) |
| bldgcondGood | good condition |
| RM12 | reverse month of sale - spline 12 |
| RM21 | reverse month of sale - spline 21 |

or monthly dummy variables, linear time spline variables remove more residual prediction error of real estate valuation models due to the fact that they transfer knowledge from one time interval to the next, thereby allowing more sensitivity with respect to time-based market fluctuations (Borst 2013).

## 7.4 Results

Table 7.2 shows the model that achieves the most uniform results is Exponential Fixed, with a COD of 8.41, followed closely by Exponential Adaptive (8.95) and Gaussian Adaptive (9.46). The range of CODs fluctuates by 2.70 with the highest, least uniform model being Boxcar Fixed; its COD of 11.11 is equal to that of the global model. The PRD does not change significantly across models – evidenced by a range of only .02. None of the models, including the global model, reaches a COD or PRD level outside of the range of acceptability as outlined by IAAO standards (COD < 15.00 and PRD .98 ≥ PRD ≤ 1.03).

Figure 7.3 demonstrates that a superior AICc score does not necessarily translate to a superior COD score – supporting findings from Bidanset and Lombard (2014).

*Table 7.2* Results by spatial weighting function

| *Weighting function* | *AICc* | *COD* |
|---|---|---|
| Global | −29.69 | 11.11 |
| Gaussian Fixed | −412.50 | 9.86 |
| Gaussian Adaptive | −402.16 | 9.46 |
| Bisquare Fixed | −139.07 | 10.74 |
| Bisquare Adaptive | −250.57 | 10.28 |
| Tricube Fixed | −125.70 | 10.78 |
| Tricube Adaptive | −238.01 | 10.33 |
| Exponential Fixed | −569.69 | 8.41 |
| Exponential Adaptive | −419.01 | 8.95 |
| Boxcar Fixed | −31.46 | 11.11 |
| Boxcar Adaptive | −77.26 | 10.91 |

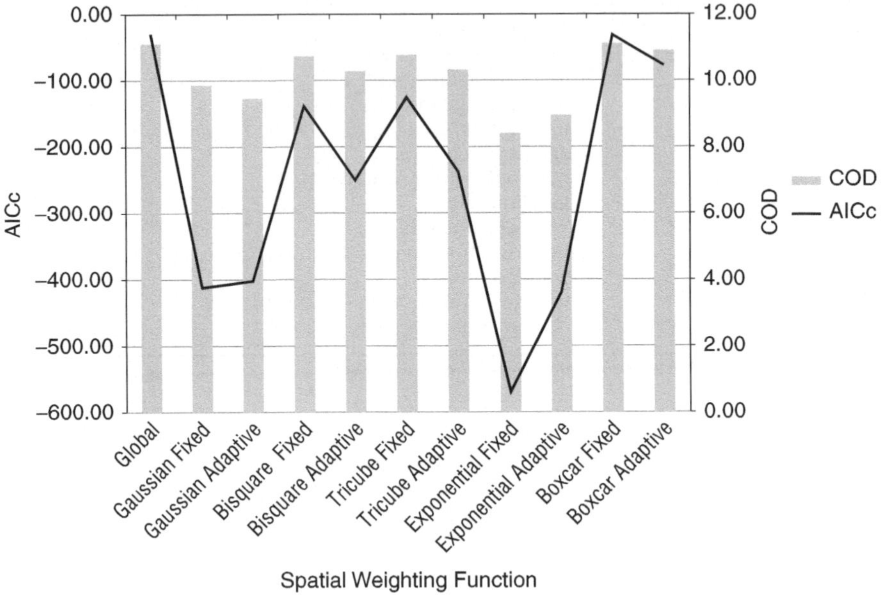


*Figure 7.3* COD and AICc by spatial weighting function

Figure 7.4a shows COD by neighborhood attained by the global model. The darker areas represent a higher COD. The most significant offenders are the Greenhill Farms (29.11), Oak Grove (28.50), Willoughby (25.96), and Huntersville (25.48) neighborhoods. COD maps of other models, such as Tricube Fixed (Figure 7.4f) and Tricubed Adaptive (Figure 7.4g) do not show much improvement from the global model; there are still a number of darker neighborhoods and the COD distribution is less smooth than that of the continuous kernel functions. The Gaussian kernels (Figures 7.4b and 7.4c) show a smoother distribution and the COD levels of the poorest performing neighborhoods have been significantly alleviated. The model using the Gaussian Adaptive kernel has significantly improved aforementioned neighborhoods: Greenhill Farms to 14.83 (improvement of 14.28); Oak Grove to 23.03 (improvement of 5.47); Willoughby to 22.91 (improvement of 3.05); and Huntersville to 24.49 (difference of .99). The maps values of Exponential Fixed (Figure 7.4h) and Exponential Adaptive (Figure 7.4i) are a particularly lighter shade and smoother than the other maps, reflecting a much more uniform COD performance by neighborhood. When compared to the global model, Exponential Fixed brings the COD of Greenhill Farms down to 11.40 (improvement of 17.1); Oak Grove to 20.05 (improvement of 8.45); Willoughby to 8.24 (improvement of 17.72); and Huntersville to 20.73 (improvement of 4.75). Exponential Fixed demonstrates that an appropriate spatial weighting function can bring a neighborhood in-line with IAAO uniformity standards. Tables 7.3 and 7.4 show the number of neighborhoods that do not reach IAAO levels of acceptability

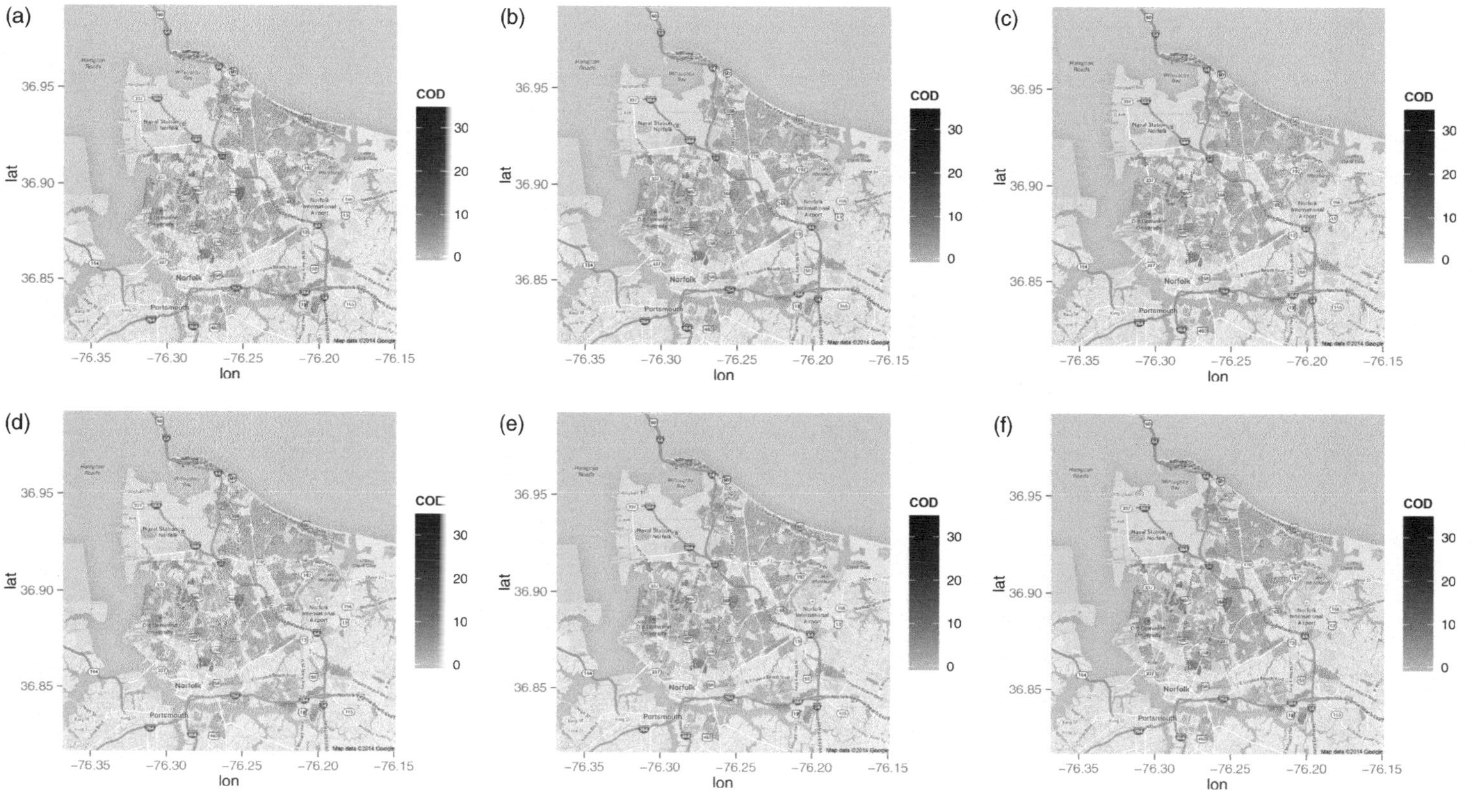


*Figure 7.4* COD maps (a) Global (b) Gaussian Fixed (c) Gaussian Adaptive (d) Bisquare Fixed (e) Bisquare Adaptive (f) Tricube Fixed (g) Tricube Adaptive (h) Exponential Fixed ( ) Exponential Adaptive (j) Boxcar Fixed (k) Boxcar Adaptive

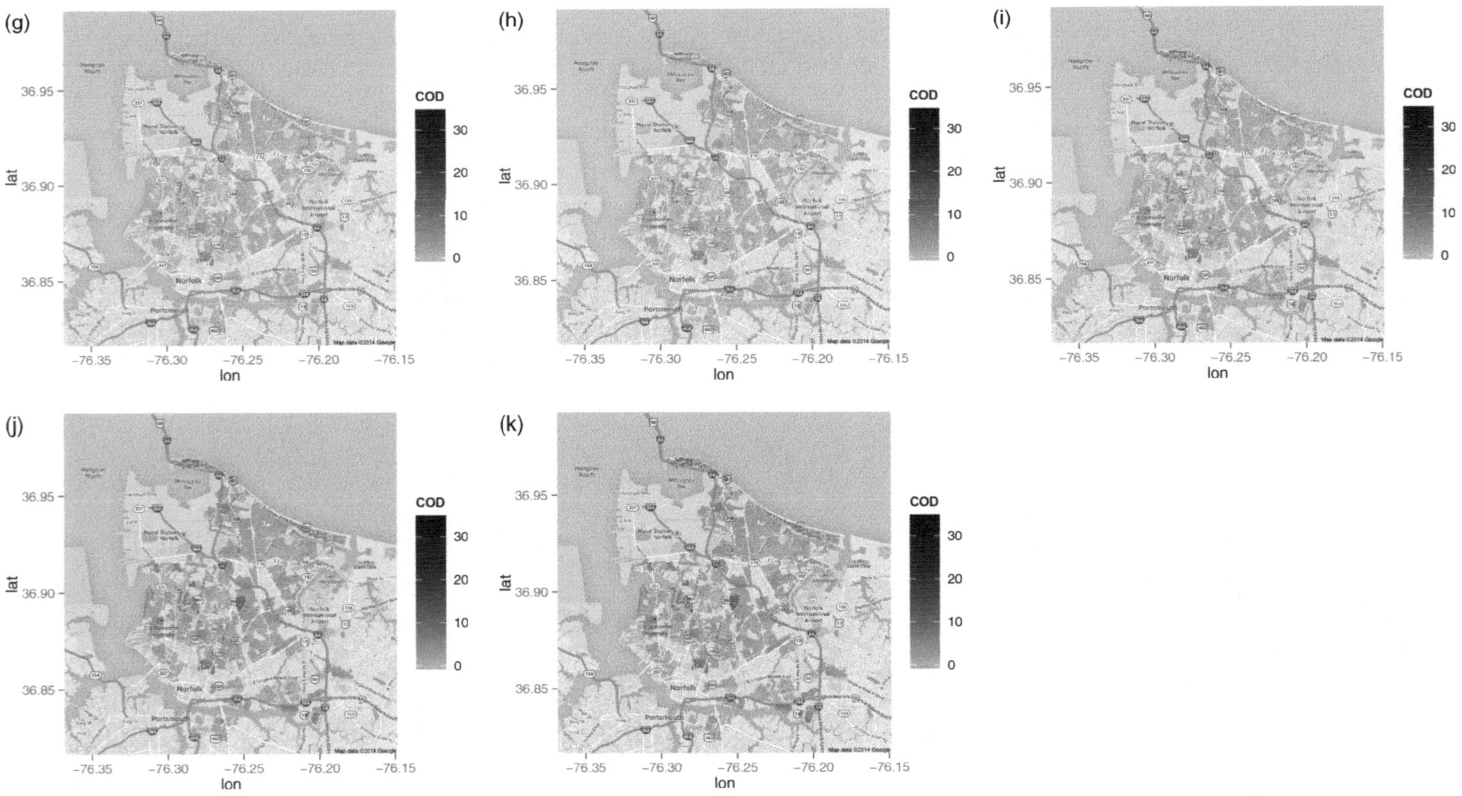


*Figure 7.4* (Continued)

*Table 7.3* Frequency table – neighborhoods out of COD threshold by spatial weighting function

| *n=137* | |
|---|---|
| *Model* | *COD Freq. Count Offenders* |
| Global | 32 |
| Gaussian Fixed | 13 |
| Gaussian Adaptive | 10 |
| Bisquare Fixed | 27 |
| Bisquare Adaptive | 21 |
| Tricube Fixed | 28 |
| Tricube Adaptive | 22 |
| Exponential Fixed | 3 |
| Exponential Adaptive | 8 |
| Box Car Fixed | 32 |
| Box Car Adaptive | 29 |

*Table 7.4* Frequency table – neighborhoods out of PRD threshold by spatial weighting function

| *n=137* | | |
|---|---|---|
| *Model* | *Freq. Count Regressivity* | *Freq. Count Progressivity* |
| Global | 12 | 2 |
| Gaussian Fixed | 10 | 1 |
| Gaussian Adaptive | 8 | 1 |
| Bisquare Fixed | 10 | 3 |
| Bisquare Adaptive | 10 | 1 |
| Tricube Fixed | 10 | 3 |
| Tricube Adaptive | 11 | 2 |
| Exponential Fixed | 5 | 0 |
| Exponential Adaptive | 8 | 1 |
| Box Car Fixed | 12 | 2 |
| Box Car Adaptive | 10 | 3 |

for COD and PRD. The global model produces 12 neighborhoods with evidence of regressivity and two neighborhoods with evidence of progressivity, followed by Tricube Adaptive (11 and 2, respectively), while Exponential Fixed brings these numbers down to five and zero, respectively.[3] The global model results in 32 neighborhoods exceeding the threshold for IAAO COD standards. Exponential Fixed results in just three.

While spatial weighting functions may produce superior results, geographic disaggregation into local areas reveals that optimal overall models can be out performed by models deemed 'inferior' on the aggregate level (Bidanset and Lombard 2014). The neighborhood COD levels support this; Exponential Fixed – while attaining a lower city-wide COD (8.41) than Bisquare Fixed (10.74) and Bisquare Adaptive (10.28) – achieves a COD of 7.73 for the Ingleside Terrace neighborhood, while Bisquare Fixed and Bisquare Adaptive achieve 6.68 and 7.16, respectively.

## 7.5 Conclusions

Using valid sales of single-family homes in Norfolk, VA, from 2010 to 2012, this chapter evaluated the varying predictability power kernel that specifications of geographically weighted regression models lend to mass appraisal of real estate, and the potential improvement each lends to taxing entities in attaining equity, uniformity, and ultimately defensibility in their property assessments, ultimately demonstrating the importance of spatial modeling and analysis within the realm of property taxation and real estate assessment. The weighting specifications that were studied were the Gaussian, bisquare, tricube, exponential, and boxcar kernels – each kernel specification was evaluated with both fixed bandwidth and adaptive bandwidths. The model using an exponential kernel and a fixed bandwidth produced results most uniform by IAAO standards, followed closely by an exponential kernel with an adaptive bandwidth, and both the fixed and adaptive bandwidth models employing a Gaussian kernel. Additionally, on the neighborhood level, the model using an exponential kernel and fixed bandwidth significantly reduced the number of neighborhoods out-of-line with IAAO standards for COD and PRD; COD offenses decreased from 23% to 2%, and PRD offenses (regressivity) decreased from 9% to 4%. This suggests that continuous kernel functions attain more uniform results within the field of mass appraisal modeling than their discontinuous counterparts.

Appraisal uniformity is affected by the spatial weighting scheme chosen by the modeler. While COD was shown to fluctuate with kernel and bandwidth speciation, PRD remained relatively unaffected. For each kernel, the adaptive bandwidth achieved a superior COD score than the respective fixed bandwidth, with the exception of the exponential kernel. None of the models suffered from vertical inequity and each achieved acceptable tax uniformity. Building upon previous research, these findings suggests that careful kernel and bandwidth specifications of GWR models may greatly enhance taxing jurisdictions in more efficiently reaching uniformity in their assessments, and thereby reducing administrative and legal costs associated with inaccurate real estate valuations.

Traditional measures of model performance, such as the Akaike information criterion and adjusted $R^2$, are most often the focus of GWR implications for real estate modeling. Further supporting findings of Bidanset and Lombard (2014), this research suggests modelers should extrapolate beyond such indicators of model fit (e.g. AIC) to include measures of uniformity (e.g. COD) because optimality in one does not necessitate optimality in the other. Therefore, analysts should explore varying kernel and bandwidth combinations during the calibration phase of modeling. Also supporting Bidanset and Lombard (2014), geographic disaggregation into neighborhood COD values reveals that a model which produces superior overall results (lower COD, lower AIC) can still be outperformed within subgeographic areas by a suboptimal aggregate model. Similarly, disaggregating and performing ratio studies at the neighborhood level revealed that suboptimal aggregate models (with respect to COD) can still outperform within certain submarkets. It would behoove taxing entities to evaluate which weighting specifications perform best for each submarket, and subsequently stratify for geographically varying assessment models within a single taxing jurisdiction.

This chapter sets the stage for a wealth of additional research. The implications for optimal weighting specifications can be applied to other locally weighted regression techniques used in real estate modeling, such as temporal or attribute weighting. The superiority of continuous kernels should be tested in other markets of varying size and housing composition. Additional kernel functions (e.g. Epanechnikov, triangular, uniform) can be included in analysis as well.

## Notes

1 Some parts of this paper were published previously in "The Effect of Kernel and Bandwidth Specification in Geographically Weighted Regression Models on the Accuracy and Uniformity of Mass Real Estate Appraisal", *Journal of Property Tax Assessment & Administration*, Vol. 11, Issue 3, 2014. Kansas City, Missouri: International Association of Assessing Officers.

2 IQRx3 Method: Interquartile range (Q3-Q1) is calculated and multiplied by three. Value is then subtracted from first quartile value (Q1) and added to third quartile value (Q3) to create lower and upper bounds, respectively. Values outside of bounds are treated as outliers.

3 Ratio studies were only completed for neighborhoods in the data set with at least five sales, as suggested by IAAO *Standard on Ratio Studies* (2013).

## Acknowledgements

We thank Bill Marchand (Chief Deputy Assessor) and Deborah Bunn (Assessor) of the city of Norfolk, VA, for the opportunity to conduct this research.

## References

Anselin, L., and Griffith, D. A. (1988) Do spatial effects really matter in regression analysis? *Papers in Regional Science*, 65(1), 11–34.

Ball, M. J. (1973) Recent empirical work on the determinants of relative house prices. *Urban Studies*, 10(2), 213–233.

Berry, B. J., and Bednarz, R. S. (1975) A hedonic model of prices and assessments for single-family homes: does the assessor follow the market or the market follow the assessor? *Land Economics*, 51(1), 21–40.

Bidanset, P. E., and Lombard, J. R. (2014) The effect of kernel and bandwidth specification in geographically weighted regression models on the accuracy and uniformity of mass real estate appraisal. *Journal of Property Tax Assessment and Administration*, 11(3). Kansas City, MO: International Association of Assessing Officers.

Borst, R. (2013) *Optimal market segmentation and temporal methods: Spatio-temporal methods in mass appraisal*. International Property Tax Institute, Mason Inn Conference Center, Fairfax, VA.

Borst, R. A., and McCluskey, W. J. (2008) Using geographically weighted regression to detect housing submarkets: modeling large-scale spatial variations in value. *Journal of Property Tax Assessment and Administration*, 5(1), 21–51.

Brunsdon, C. (1998) Geographically weighted regression: a natural evolution of the expansion method for spatial data analysis. *Environment and Planning A*, 30, 1905–1927.

Brunsdon, C., Fotheringham, A. S., and Charlton, M. E. (1996) Geographically weighted regression: a method for exploring spatial nonstationarity. *Geographical Analysis*, 28(4), 281–298.

Cho, S. H., Lambert, D. M., and Chen, Z. (2010) Geographically weighted regression bandwidth selection and spatial autocorrelation: an empirical example using Chinese agriculture data. *Applied Economics Letters*, 17(8), 767–772.

Fotheringham, A. S., Brunsdon, C., and Charlton, M. (2002) *Geographically weighted regression: the analysis of spatially varying relationships*. Chichester: John Wiley and Sons.

Gloudemans, R. J. (1999) *Mass appraisal of real property*. Chicago, IL: International Association of Assessing Officers.

Gloudemans, R., and Almy, R. (2011) *Fundamentals of mass appraisal*. Kansas City, MO: International Association of Assessing Officers.

Gollini, I., Lu, B., Charlton, M., Brunsdon, C., and Harris, P. (2013) GWmodel: an R package for exploring spatial heterogeneity using geographically weighted models. arXiv preprint arXiv:1306.0413.

Guo, L., Ma, Z., and Zhang, L. (2008) Comparison of bandwidth selection in application of geographically weighted regression: a case study. *Canadian Journal of Forest Research*, 38(9), 2526–2534.

Huang, B., Wu, B., and Barry, M. (2010) Geographically and temporally weighted regression for modeling spatio-temporal variation in house prices. *International Journal of Geographical Information Science*, 24(3), 383–401.

IAAO. (2013) *Standard on ratio studies*. Chicago, IL: International Association of Assessing Officers.

LeSage, J. P. (2004) A family of geographically weighted regression models. In *Advances in spatial econometrics*. Berlin and Heidelberg: Springer, 241–264.

Lockwood, T. and Rossini, P. (2011) Efficacy in modelling location within the mass appraisal process. *Pacific Rim Property Research Journal*, 17(3), 418–442.

Lu, Binbin, et al. (2014) The GWmodel R package: further topics for exploring spatial heterogeneity using geographically weighted models. *Geo-spatial Information Science*, 17(2), 85–101.

McCluskey, W. J., McCord, M., Davis, P. T., Haran, M., and McIlhatton, D. (2013) Prediction accuracy in mass appraisal: a comparison of modern approaches. *Journal of Property Research*, 30(4), 239–265.

McMillen, D. P. (1996) One hundred fifty years of land values in Chicago: a nonparametric approach. *Journal of Urban Economics*, 40(1), 100–124.

McMillen, D. P., and Redfearn, C. L. (2010) Estimation and hypothesis testing for nonparametric hedonic house price functions. *Journal of Regional Science*, 50(3), 712–733.

Moore, J. W. (2009) A history of appraisal theory and practice looking back from IAAO's 75th Year, *Journal of Property Tax Assessment and Administration*, 6(3), 23.

Moore, J. W. E., and Myers, J. (2010) Using geographic-attribute weighted regression for CAMA Modeling. *Journal of Property Tax Assessment and Administration*, 7(3), 5–28.

Sugiura, N. (1978) Further analysis of the data by Akaike's information criterion and the finite corrections. *Communications in Statistics-Theory and Methods*, 7(1), 13–26.

# 8 Spatial search

## New settlements for Israel's ultra-orthodox population

*Eliahu Stern*

### 8.1 Introduction

Applied geography is quite an old branch in the wide subject tree of geography (e.g. Stamp, 1960). Although it has been a subject of criticism from Marxist to postmodern theorists, it still enhances the quality of present and future living conditions on earth (e.g. Pacione, 1999). A good example of its continuing vitality is the never-ending practice of spatial search. According to Massam (1980), spatial search is the process of the evaluation of alternate locations and the selection of a particular site for either public or private facilities. Since GIS has become a common tool in spatial analysis, spatial search has become an obvious base for spatial decision making.

This chapter presents a GIS-based search aimed at finding locations for housing the growing ultra-orthodox population in Israel. This sector has an outstandingly high population growth rate, continuously demanding public housing solutions. In order to avoid intra-urban conflicts between the orthodox and the non-orthodox sectors (e.g. Shilhav, 1984, 1993), the Ministry of Housing and Construction has decided to examine the feasibility of building new, purely orthodox, settlements. Bounded by several sectorial restrictions, such as proximity, and/or high accessibility to Jerusalem, and low-rise buildings, the Ministry assigned the author (Enviroplan Ltd, 2009) to find feasible locations to accommodate a minimum of 50,000 dwelling units. Feasibility is defined by several objective and subjective indicators. Both the indicators and the search procedure are presented in this chapter. A paired-comparison analysis of location criteria is also used to examine the rank-orders of feasible locations by the leaders of the ultra-orthodox population and the community of urban planners. An agreeable solution and learned insights are put forward at the end of the chapter.

### 8.2 Background: the ultra-orthodox population

The ultra-orthodox (OU) sector in Israel accounted for 10.7% (598,993 people) of the total Jewish population in 2008 (Central Bureau of Statistics, 2009). It is the fastest growing population sector in the country (Kahaner et al., 2012); 6.1% per year while the national average for the Jewish population is only 1.4% (Table 8.1).

*Table 8.1* Selected characteristics of the ultra-orthodox sector in Israel, 2008

| *Characteristics* | *The Ultra-Orthodox Sector* | *Total Jewish Population* |
| --- | --- | --- |
| Annual growth rate | 6.1 (%) | 1.4 (%) |
| Median age | 16.3 | 31.0 |
| Average No. of children per family | 7.7 | 2.6 |
| Males outside the labour force | 60 (%) | 35 (%) |
| Percent married at ages 25–29 | 93.4 | 47.2 |
| Percent of people over 65 | 4.5 | 11.5 |

Sources: Central Bureau of Statistics, 2009; Gurovich and Cohen-Kastro, 2004

The average number of children per family is 7.7, about three times the number in the total Jewish population, as reflected also by the median age – 16.3 for the ultra-orthodox and almost double for the total population.

The ultra-orthodox people tend to get married relatively early; by the age of 29 almost 94% of them are married, again twice that of the total population. Most of them are poor, obviously resulting from the high percentage of males outside the labour force (60%) and the large families to be supported. Spatially, 80% of them are concentrated in 40 settlements, of which 16 are homogeneous ultra-orthodox (Figure 8.1). These settlements are located in the centre of the country, with two major cities – Jerusalem and Bnei Barak. The Jerusalem area includes 33% of the sector; Bnei Barak and the central region include 39%. The rest are distributed equally in the Northern and the Southern districts.

These data illustrate three processes: (1) an exponential growth rate of the ultra-orthodox population, (2) a spatial concentration in the central region, which is also the area with the highest demand for housing, and (3) a forced diffusion into non-religious settlements, which encourages internal local conflicts. Consequently, the Ministry of Housing and Construction decided: (a) to estimate the current supply and the future demand for ultra-orthodox housing, (b) to examine the planning/physical and environmental feasibility for locating homogeneous urban settlements for the ultra-orthodox population, and (c) to design a consequent spatio-temporal plan. As mentioned above, this chapter presents the methodology used and the insights learned throughout the project.

## 8.3 General methodology

Following the three assignments of the Ministry, the study included three stages (see Figure 8.2), but the current chapter is mainly focused on the second one – the spatial search process. The first stage is a straightforward calculation of the suitable housing supply for the UO population based on the statutory planning inventory held by the planning division of the Ministry of the Interior, and the average rates of the available housing stock in the homogeneous UO settlements. The demand for UO housing is estimated on the basis of the average annual growth rate and the demographic characteristics. The outcome of this stage is an estimation of the annual housing needs for the next 20 years. This estimation is the main

*Figure 8.1* The distribution of homogeneous ultra-orthodox settlements in Israel

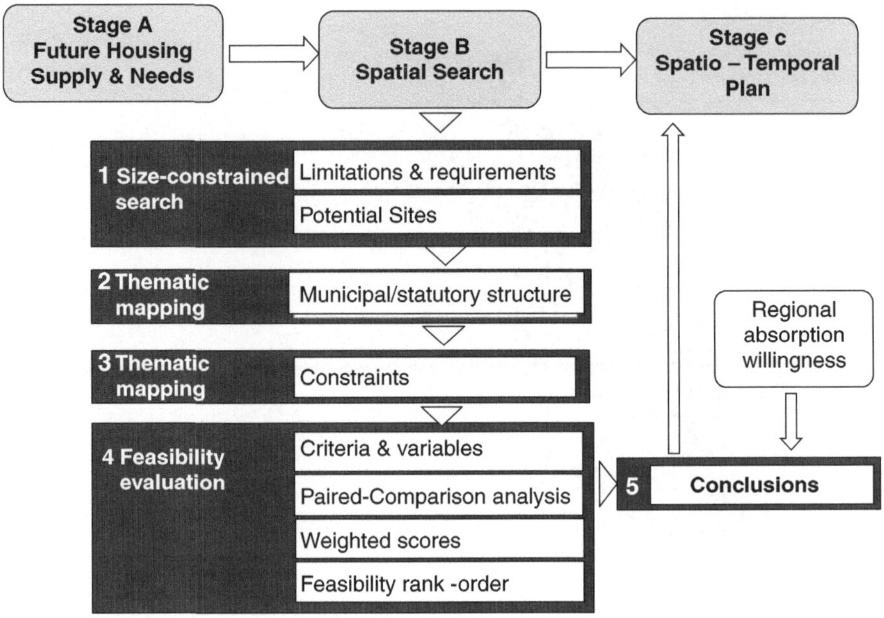


*Figure 8.2* The study working stages

requirement for the first step of the next stage – a size-constrained search (see Figure 8.2). The last, third part of the project is a longitudinal assignment of the estimated demand to the total sources of supply resulting from the spatial search solution, thus providing a spatio-temporal housing plan.

The main, second, stage is the spatial search process, which includes three components: (a) a size-constrained search, (b) thematic mapping of both themes (second step in stage B, Figure 8.2) and constraints (third step in stage B), and (c) feasibility evaluation. The results of each component serve as an input for its consequent step. Further methodological details for each component are presented along with the results in the following sections.

### *8.3.1 A size-constrained search*

The straightforward calculation of the suitable housing supply and demand for the UO population found that the total existing and planned supply of housing until 2025 is 25,000 dwelling units. The respective demand for this period has been estimated to reach 160,000 units, of which 100,000 are new dwellings; that is a demand of 5,000 new dwelling units per year for the next 20 years. Taking into account the cost of erecting new towns and the various characteristics of the target population, one relatively large city was preferred over several small or medium-size settlements. Accordingly, the government and the planning team decided to search for locations suitable for at least 50,000 dwelling units each. Considering the average UO family size (about nine people; see Table 8.1) a city of approximately 450,000 inhabitants should be built every ten years. The area needed for such a city is between 10,000 and 15,000 dunams (one dunam is 1,000 square metres). This area has been taken as the size constraint for the first step of the spatial search process.

In addition to the area-size constraint, the UO representatives (i.e. their chief rabbis) required proximity to Jerusalem and/or a good access to the main national transport arterials. This, however, stood in conflict with another requirement of low land prices (as they are the poorest sector in the Israeli population), which are obviously higher in the national core (i.e. the Tel Aviv–Jerusalem corridor). As a result, the spatial search was stretched to include the northern latitude of Afula in the north and Kyriat Gat in the south (see Figure 8.3) and only (as another sectorial requirement) within the pre-1967 boundary (known as the former 'green line'). This area is crossed by the 'Cross Israel Highway' which provides good access to both the Jerusalem and the Tel Aviv areas.

The search for large open spaces in a small and densely populated country like Israel, and even more so in its central area, cannot avoid the inclusion of agricultural lands. Even so, the actual physical search is also restricted by other statutorily protected open areas like nature reserves, national parks and military zones. These are important constraints in the first step of identifying potential sites that

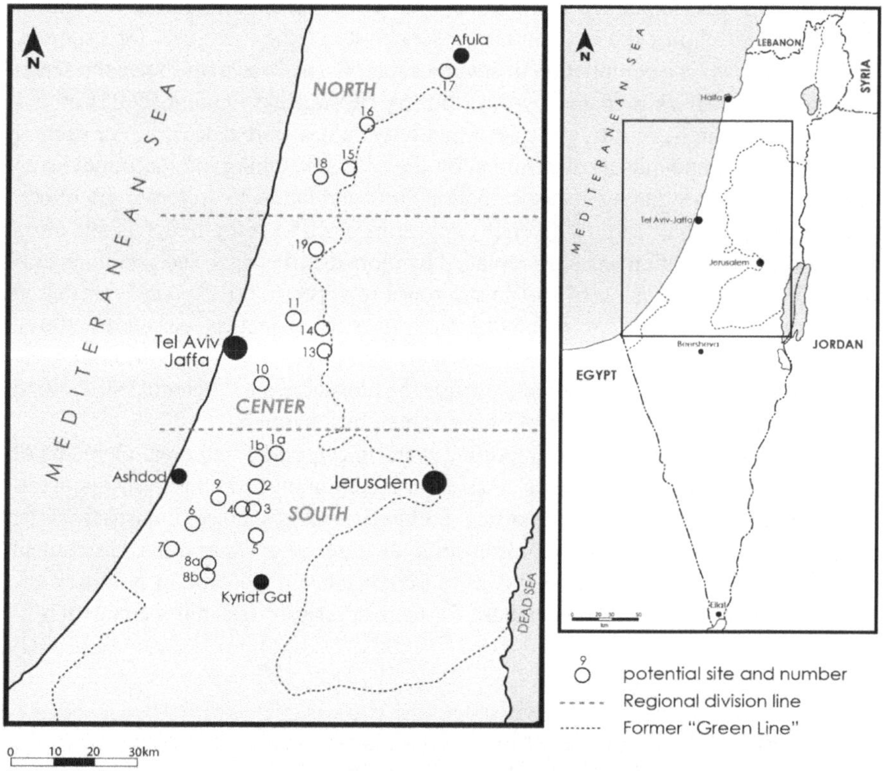


*Figure 8.3* The spatial search area and the 19 identified potential sites

comply with the area-size restriction. The potential sites will be further filtered in the second and the third steps of the spatial search. This first (size-constrained) step has identified 19 potential sites, as shown in Figure 8.3. Five of these are in the central core ('Center' in Figure 8.3) four in the northern-central area, and ten in the southern-central area.

### *8.3.2 Thematic mapping*

The next two steps, 2 and 3, of the spatial search process (stage B in Figure 8.2) include GIS-based mapping of thematic information and further possible constraints which are needed for the feasibility evaluation of the 19 identified potential sites. Step 2 provides the municipal division of the searched area in a three-tier structure: urban area boundaries, regional council boundaries and rural settlement boundaries. In addition, this step also includes the three-tier structure of the physical statutory planning: the district plans, the regional (sub-district) plans, and the local plans. The information can obviously be overlaid in GIS with the layer of the potential sites.

The next step of the thematic mapping provides areal, nodal and linear information on further potential constraints. Apart from nature reserves and national parks, it provides the coverage of other protected open green areas, including forests, biosphere reserves, landscape compounds and river banks. In addition, it builds a comprehensive map of the environmental sensitivity of the area (see, for example, Plate 2). This map is a compilation of several sources. The base map shows the sensitivity of open areas as prepared by the Ministry of the Environment (2005) on a 1 (low) to 7 (very high) sensitivity scale. Sensitivity values were calculated for each of the landscape compounds as determined by the Ministry, taking into account eleven topical layers: physiography, rocks formation, flora and fauna, hydrology, agriculture, heritage, preservation, scarcity, diversity, appearance and tourism. In applicable areas the base map information has been replaced by more detailed data, such as, for example, a five-category division of biospheric zones (e.g. Stern, 2012), which are based on a larger number of environmental indicators (for details see Stern, 2004). Furthermore, antiquity zones and archaeological sites were overlaid on the composite map, both in the biospheric zones and outside the biosphere area. Potential sites found in the biosphere areas were omitted from further consideration.

Other constraints found to be important were linear, nodal and areal elements of infrastructure, such as gas, water, electricity and communication lines, sewage treatment facilities, electrical plants, quarries, polluted zones (by noise, industrial waste, smell, and the like) and areas around sensitive facilities (e.g. water drills, enrichment zones, etc.). Finally, land ownership data was collected and mapped. In summary, the thematic mapping provides the database for the next step of feasibility evaluation.

### *8.3.3 Feasibility evaluation*

The feasibility of each of the 19 potential sites was estimated according to the procedure shown in Figure 8.2 above. Basically, it is a multi-criteria method which involves a subjective weighting of each criteria by two groups of stakeholders: the UO population, represented by their rabbi, and a group of planners. Another factor worth mentioning is the absorption preparedness of the surrounding population as an external effect on the operative conclusions from the search process. Each of the steps is described below.

### *Criteria and variables*

Five criteria involving ten variables were used to evaluate site feasibility (Table 8.2). The criteria include the sectorial subjective preferences and two types of constraints: property and planning constraints. The criteria are:

a. *The subjective preference* of the UO population as represented by their rabbi. The site's preference was examined in several pre-designed interviews in which the participants were asked to rank their subjective preference. The distribution statistics of the preferences for each site were later used to define the subjective preference rank-order of the potential sites.
b. *Planning constraints – compliance with the national planning policy*. Since 2006, planning in Israel is carried out with full compliance to the National Structure Plan 35 (Asif and Schachar, 2006) regarding the spatial nature of the site and its proximity to existing built-up or planned areas. Both became part of the nation-wide policy and they are strictly applied, especially in cases of adding new settlements. A third variable in this criterion is whether there are statutory conflicts between the potential site and the existing land-use plan. A conflict of this type may prolong the implementation period.
c. *Planning constraints – infrastructure/physical*, which include the disturbance of upper or buried infrastructure lines (e.g. the National Water Carrier, electrical high voltage lines, gas pipelines, etc.) and nodal constraints (e.g. quarries and sewage treatment ponds) or other physical elements imposing limitations on local construction activity.
d. *Planning constraints – environmental sensitivity*, as defined by the existence of protected areas (nature reserves, forests, etc.), and the level of environmental sensitivity as defined by the Ministry of Environmental Protection, the National Structure Plan 35 and, where relevant, biosphere zoning.
e. *Property constraints*, as based on three variables: the amount of penetration into the municipal areas of existing settlements, land ownership (private, public, etc.) and actual land use. Each of them affects the implementation availability of the sites.

### *Importance profiles and subjective weighting*

As actual implementation heavily depends on the UO leaders, their preferences must be examined and compared with those of 'objective' urban planners. To that end, a Paired-Comparison technique was employed to define the 'criteria importance' profiles of the two stakeholder groups (on the method,

*Table 8.2* Evaluation criteria and variables for site feasibility

| *Site* | *UO Preference* | *Property Constraints* | | | *Planning Constraints* | | | | | | |
|---|---|---|---|---|---|---|---|---|---|---|---|
| | | | | | *Infrastructure/Physical* | | *National Planning Policy* | | | *Environmental sensitivity* | |
| | | *Actual Land use* | *Land Ownership* | *Local Boundary* | *Infrats. Lines* | *Nodal Constr.* | *Statutory Conflicts* | *N.S.P 35* | | *Environ. Value* | *Protected Areas* |
| | | | | | | | | *Border proximity* | *Texture compliance* | | |
| | | | | | | | | | | | |
| | | | | | | | | | | | |
| | | | | | | | | | | | |
| | | | | | | | | | | | |
| | | | | | | | | | | | |
| | | | | | | | | | | | |
| | | | | | | | | | | | |

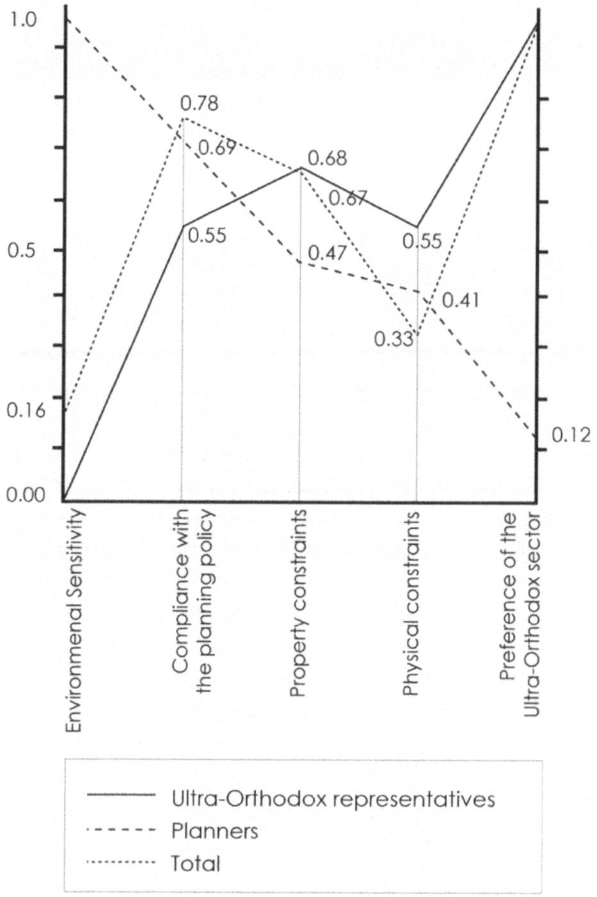


*Figure 8.4* The preference profiles of the ultra-orthodox representatives and the planners as resulted from a paired-comparison analysis

see Penner et al., 1968). The results of the analyses were transformed to a 0.0-to-1.0 scale, as shown in Figure 8.4.

The results show two different profiles. While 'environmental sensitivity' is the most important criterion among the planners, it has no importance at all among the UO leaders. On the contrary, 'the UO (own) preference' is the most important criterion among the UO leaders, while it is the least important (although with some weight) among the planners. Also interesting to examine is the second ranked criterion in each group. While the planners consider the 'compliance with the planning policy' as second in importance, the UO leaders prefer 'property constraints', i.e. price and size. Important to note is also the fact that the preferences among the UO leaders were totally homogeneous, while some variance was naturally found among the planners. These results reflect the relatively ego-centred, 'closed bubble'-type opinions of the orthodox world, in which only themselves and financial considerations are important. The planners are more sensitive to the current emphasis in the planning arena, environmental protection and compliance with national planning policy. The scores of each of the criteria were used as weights in the multi-criteria framework for determining the preference rank-orders of the location sites.

*Preference rank-orders*

Despite the differences in the importance profiles, the same potential locations were found in the first preferred quintet of both the UO rabies and the planners, although in a slightly different order (Figure 8.5). This is explained by the relatively higher criteria values of these five locations, which have masked the difference in the preferences of the two groups.

The location named Harish is, however, the most preferred location of both groups. It is a developed site which has not been populated for over 10 years for various reasons, located near the Cross-Israel highway and relatively isolated from surrounding secular settlements. As it has already been approved in planning terms, with an existing un-used infrastructure, it can provide an immediate solution for 13,000 dwelling units, with a potential to expand to 50,000, as required. A difference exists between the second and third preferred locations. The UO group prefers a limited capacity site called Yesodot (Figure 8.5), located in a religious regional council area on a main transport corridor to Jerusalem, while the planners prefer a site with a larger capacity but slightly farther from Jerusalem. The first quintet of potential locations can provide the estimated demand for the seeable future and it was used to prepare an integrated plan accepted by both parties. This plan was not finalized without an additional enquiry – the preparedness of the surrounding population to absorb the UO new residents.

### *8.3.4 Regional and local preparedness for absorption*

The UO community is known, among others, by its strict beliefs. As such, it is also characterized by geographical segregation, a preference to live in separate neighborhoods and towns. This segregation affects the whole social structure in Israel as well as the UO–secular relationship in society. Its implications, however, are not unambiguous. The mutual, continuing, segregation of the secular and the UO populations may lead to a complete disconnection and to growing negative mutual images. On the other hand, living together creates friction which fans their principal differences. Either way the social threat hiding in the exposure to the other's culture, especially between majority and minority groups, is threatening and consequently leads to a desired spatial segregation of the UO population. A study by Degani and Degani (1999) found, for example, that 60% of the Israeli UO people showed a preference for spatially separated living.

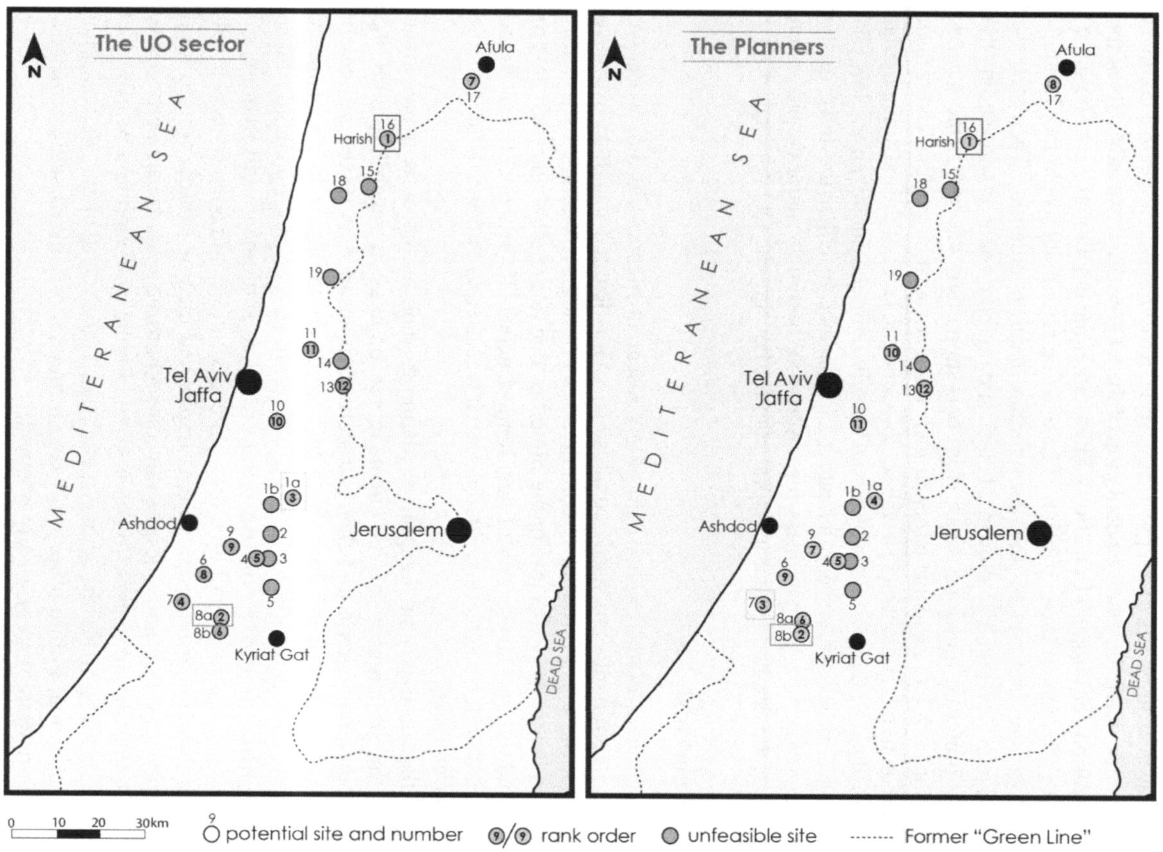


*Figure 8.5* Rank orders of the potential sites

The integration of an UO community in a secular city requires several thousands of UO residents as a threshold for public service needs. Coupled with their high growth rate, many city planners and mayors feel threatened by such anticipated expansion. Leaders of many secular municipalities perceive the UO as an economic burden as well as imposers of a repulsive lifestyle. Research has shown that conflicts begin to merge when the relative population of the UO residents exceeds 20% of the total population (Enviroplan Ltd, 2009). Thus, either the erection of a new homogeneous UO settlement or the expansion of an existing settlement would face difficulties without the consent of the surrounding local population. Therefore we conducted interviews with 15 mayors of cities and regional councils surrounding the three most preferred locations. Apart from two supporters, the rest of the mayors refused to absorb UO residents either in their towns or in their region. The refusal was based on two arguments: (1) as the socio-economic level of the existing population is already low; thus, (2) new OU residents will drain the public purse. New OU residents will break the current delicate social balance in either the town or the region. Therefore, the 'preparedness for absorption' became a major consideration in designing the detailed spatial-temporal plan for housing the OU population.

## 8.4 Learned insights

Being an applied-oriented rather than purely research-oriented piece of work, we replace the commonly practised 'conclusions' with 'learned insights'. We have learned that the feasibility of OU communities to integrate within small secular settlements is low. It is also low in the big cities due to the relatively high cost of housing and living. The erection of new homogeneous OU settlements was found to be the best preferred solution. Spatial search was used to find and define feasible and preferred locations.

We further realized that integrative planning, such as the current one, must consider the preferences of its clients/target population regarding both the location objective criteria and their subjective importance. A mixed settlement solution, which was not accepted by the OU sector, could cause tension, conflicts, and economic harm mainly to economically weak settlements, apparently those which actually absorb them. An ultra-orthodox settlement will require an exceptionally large amount of area for public institutions and open spaces. It will also require local/near-by employment supply, and income sources for the local municipality as well. Finally, despite the comprehensive spatial search procedure, at the end, one should expect, to some extent, public objection to a near-by UO settlement. As regard to the above-mentioned Harish settlement, that is what actually happened. Public objection led to a much more mixed settlement, now under construction, than originally planned.

## References

Asif, S. and Schachar, A. (2006) *The National Structure Plan 35 for Construction, Development, and Conservation*. The Ministry of the Interior, Jerusalem.

Central Bureau of Statistics (2009) *The Annual Statistical Yearbook*. Central Bureau of Statistics, Jerusalem.

Degani, A. and Degani, R. (1999) *The Demand for Housing in the Ultra Orthodox Sector*. A research report submitted to the Ministry of Housing and Construction, Geo-Cartography, Tel Aviv (in Hebrew).

Enviroplan Ltd (2009) *The Housing Needs of the Ultra Orthodox Sector in Israel*. A report submitted to the Ministry of Housing and Construction, Tel Aviv (in Hebrew).

Gurovich, N. and Cohen-Kastro, E. (2004) *Geographic Distribution, Demographic, Social and Economic Characteristics of the Ultra-Orthodox Jewish Population in Israel 1996–2001*. Working Papers Series No. 5. Central Bureau of Statistics, Jerusalem.

Kahaner, L., Yosgof, N., and Sofer, A. (2012) *The Ultra Orthodox in Israel: Space, Society, Community*. The Heikin Chair for Geo-Strategy, University of Haifa, Haifa.

Massam, H.B. (1980) *Spatial Search: Applications to Planning Problems in the Public Sector*. Pergamon Press, Oxford.

Ministry of the Environment (2005) reported in www.sviva.gov.il/infoservices/tools/pages/GIS.aspx (accessed 3 August 2016).

Pacione, M. (1999) Applied geography: in pursuit of useful knowledge. *Applied Geography*, 19(1), 1–12.

Penner, L., Homant, R., and Rokeach, M. (1968) Comparison of rank-order and paired-comparison for searching value systems. *Perceptual and Motor Skills*, 27, 417–418.

Shilhav, Y. (1984) Spatial strategies of the 'haredi' population in Jerusalem. *Socio-Economic Planning Sciences*, 18(6) 411–418.

Shilhav, Y. (1993) The emergence of ultra orthodox neighborhoods in Israeli urban centers. In E. Ben-Zadok (ed.), *Local Communities and the Israeli Polity*. State University of New York Press, New York, pp. 157–187.

Stamp, L.D. (1960) *Applied Geography*. Penguin, Harmondsworth.

Stern, E. (2004) Delineation and planning of biosphere reserves in Israel. *The Geographical Network*, 1(1), 2–15. Available at: www.geo-network.bgu.ac.il.

Stern, E. (2012) A local master plan for biospheric conservation and development: concept, methodology, and application. In Robert J. Stimson and Kingsley E. Haynes (eds), *Studies in Applied Geography and Spatial Analysis: Addressing Real World Issues*. Edward Elgar, Cheltenham, UK, pp. 295–310.

# 9 A tale of two earthquakes:

## Dynamic agent-based simulation of urban resilience

*A. Yair Grinberger and Daniel Felsenstein*

### 9.1 Introduction

As cities increase in size and complexity they also become increasing vulnerable to unanticipated events, both natural and anthropogenic (Godschalk, 2003; Deppisch & Schaerffer, 2011). Large-scale disasters such as the 1995 Kobe earthquake, hurricane Katrina, the Tohoku earthquake and tsunami and Superstorm Sandy have elicited research interest in the way cities cope with such shocks. This work tends to highlight either mitigation measures (Godschalk, 2003; Fleischauer, 2008) or 'bouncing back' strategies (Campanella, 2008; Chang, 2010; Chang & Rose, 2012; Olshansky, Hopkins & Johnson, 2012). It also tends to imply that urban recovery should be directly related to the magnitude of the disaster, with larger shocks to the urban system requiring more drastic mediation or rejuvenation measures. However, as this chapter shows, an exogenous shock does not have any predetermined outcome and multiple (unstable) equilibria may exist. The same shock may elicit wildly diverging urban responses in different environments. This has implications for the notion of urban resilience. It undermines much of the popular literature promoting a 'one size fits all' approach to both urban mitigation and rejuvenation and neutralizes the standard checklist approach to disaster management mechanisms, which, while well-intentioned, may be misleading (Prasad et al., 2009; UNISDR, 2012).

As the urban environment is fashioned by the interaction of many agents, such as residents, workers, local governments, developers, and by sub-systems, such as housing markets and transportation networks (Cruz, Costa, de Sousa & Pinho, 2013), unraveling the key to urban resilience becomes extremely difficult (Müller, 2011). Local shocks may have global effects and innocuous, short-term perturbations may cause long-term change. The result can be a shift of the entire system to one of a few possible unstable equilibria states. This situation plays havoc with attempts to formulate generic post-disaster urban resilience solutions without consideration of context (Kartez, 1984; Kartez & Lindel, 1987).

To illustrate this position, we use dynamic agent-based (AB) simulation of a hypothetical earthquake in the downtown area of Israel's two largest cities, Jerusalem and Tel Aviv. The former is the national capital and seat of government.

The latter is the business and economic center of the country. In the AB world, the complexities of the urban system are decomposed into the operation of 'agents'. These can be both individual entities such as citizens or aggregate institutions such as markets. Each of these operates according to certain (programmable) behavioral rules grounded in classic behavioral foundations such as maximizing utility in terms of residence, minimizing risk and participating in activities such as work, leisure and commercial activities. In so doing, agents affect the behavior of other agents and, in the aggregate, the operation of urban institutions such as land and housing markets and the planning system. We simulate the long-run impacts on the urban system with a view to highlighting the complexity of restoring the urban equilibrium and rejuvenating city life. The rest of the chapter proceeds as follows. We first review current knowledge regarding urban resilience in the wake of a disaster, in light of the multiple possible equlibria states that can emerge. Then we present the AB simulation and the principles guiding its design. In the following section, the different urban contexts of Tel Aviv and Jerusalem are described. The simulation outcomes are then discussed. These are measured by time to recovery, land-use rejuvenation and CBD shifting. Special attention is given to the effectiveness of urban policies aimed at restoring the urban equilibrium. These relate to land-use regulation, public provision of shelter and the restoration of damaged urban services and run the gamut from status quo market-led initiatives to heavy-handed regulation. Our results show very different outcomes from a similar shock and the implications of this with respect to urban resilience are discussed.

## 9.2 Literature review

The concept of resilience emerged in the study of ecology in the early 1960s and the 1970s (Folke, 2006). One of the first definitions of the term sees resilience as a property of a system that has high probability of persistence in form and structure, embodied in an ability to absorb changes to its variables and parameters (Holling, 1973). This definition has been further elaborated to include the self-organizing ability of a system, as well as the ability to adapt and learn (Folke et al., 2002). This dynamic conceptualization of resilient systems extends the previous focus on the ability to restore equilibrium after a temporary disturbance (Holling, 1973; Folke, 2006). The latter, sometimes referred to as 'engineering resilience', is criticized as static and deterministic, ignoring the possibility that the pre-shock state is only one of several states the system could present (Holling, 1973).

The notion of resilience has been imported by other fields of research, including urban planning and disaster management. However, the ideal of a 'resilient city' is still a concept lacking universal definition and acceptance. Some authors follow Holling's definition and regard the resilience of a city as the degree to which it can sustain a shock before shifting to a new state (e.g. Alberti et al., 2003; Alberti & Marzluff, 2004). Others ascribe to the notion of the city's ability to reorganize (e.g. Cruz et al., 2013). And others adopt the 'engineering resilience' conception of rebounding, bouncing-back and restoration (e.g. Godschalk, 2003; Campanella, 2008; Müller, 2011).

All views can be justified. On the one hand, as market mechanisms of supply and demand are involved in the behavior of many of the urban sub-systems (such as the housing and employment markets), the equilibrium-stability view seems to be valid. Yet, cities are complex systems whose state depends on many decisions by a wide assortment of agents and entities (Godschalk, 2003; Müller, 2011; Cruz et al., 2013). The stability approach is thus criticized for its reductionist and deterministic character (Davoudi, 2012; Martin, 2012). In this chapter we frame recovery and resilience within the concept of equilibrium. However, we accept that the pre-shock state is just one of many possible unstable equilbiria states. We therefore explore the feasibility of the bounce-back scenario and also the possibility of reorganization under a new state (i.e. 'bouncing-forward', see Grinberger and Felsenstein, 2014).

When operationalizing resilience and designing recovery strategies, both of these concepts present difficulties. The rigid policy options associated with rebounding and derived from the equilibrium view may paradoxically tilt the system away from stabilization by not allowing the freedom needed to achieve steady state (Folke et al., 2002). Viewing cities as complex systems, on the other hand, leads to confusion regarding the processes and factors promoting urban resilience and to great difficulty in formulating absolute resilience strategies (Müller, 2011; Allan, Bryant, Wirsching, Garcia & Teresa Rodriguez, 2013). As systems differ in terms of inputs, outputs, agents, and parameters, no two urban areas are alike and even the same urban space can change character over time.

The tendency of the discussion on resilient cities and urban recovery to 'focus on process rather than place and form' (Allan et al., 2013: 244) only aggravates the situation. Portraying a picture of a general process, for example a shock leading to loss of lives, damage to property and infrastructure, diminishing accessibility and provision of services, may promote generic perceptions of the recovery process. These are expressed in the common conception that recovery is proportionate to the magnitude of the effect (Chang & Rose, 2012) and in the typical knee-jerk reaction to disaster that involves time-compressing rebuilding and rejuvenation measures (Olshansky et al., 2012). These well-intentioned activities do not consider the existence of multiple and unstable equilibria resulting from different activities recovering at different rates. Neither do they consider the possibility of incongruence between the location of the event and the point of recovery.

## 9.3 Methodology

To deal with the complexities of the urban system, an AB model requires the specification of three elements: the agents and their characteristics, simple behavioral rules driving their actions, and their situation within an environment (Macal & North, 2005). These are discussed in turn below. Uniquely, we move beyond the traditional demand-oriented representations of the environment by including a dynamic element in the form of dynamic house pricing. This system mediates between agents' behavior and the land-use system and reflects supply-side dynamics. It is based on the conceptualization of buildings as semi-autonomous, quasi-agents which lack mobility and initiative but still react to changes in their environment. This system is depicted in Plate 3. Citizen agents (colored yellow in Plate 3) make locational decisions (colored in green) regarding residence and everyday activities. These decisions, along with the routes along which agents move, create an effect on the size and value of both housing and non-residential capital stock (colored purple in Plate 3). A large enough effect may accrue, resulting in morphological change and even a shift in the Central Business District (CBD) location. These changes inform the next round of agent decisions and the migration of new agents. In this manner, the response of the urban system to an earthquake becomes a consequence of the way a shock alters the behavior of city inhabitants, as depicted in Plate 3. Policy interventions are also considered within this model and are treated as exogenous inputs that impact the behavior of citizen agents and the functionality of buildings (that are treated as quasi-agents). While the specification of such a model is complex and requires various assumptions, as summarized in Table 9.1, these are generally logical and intuitive. Model development is done using Repast Simphony 2.0 (North et al., 2013), a popular agent-based development environment, programmed in Java.

### *9.3.1 The urban environment*

The environment of the city reflects the fixed results of previous rounds of investment, in the form of infrastructure, buildings and the land-use system. All of these elements exert powers of attraction and repulsion within the decision process of the individual agent, as detailed in the next section. Therefore, we move beyond

*Table 9.1* Model assumptions

| *Entity* | *Behavior* | *Assumption* |
|---|---|---|
| Citizen agent | Migration | Migration probabilities, both inter and intra-urban, are dependent on previous trends of migration. |
| | Choice of place of residence | Willingness to pay for housing up to one-third of monthly income.<br>Wages reflect social class.<br>Aspiration for residence amongst equal or better. |
| | Choice of activity location | Push and pull factors.<br>Satisfying behavior<br>Risk evasiveness. |
| | Movement path | Satisfying behavior. |
| Building quasi-agent | Land-use change | Traffic load as a proxy for revenue.<br>Floor-space size as a proxy for operating costs. |
| | Housing prices | Spatial trickle-down effect of housing prices.<br>Sensitivity of prices to competition and amenities. |

the grid representation common to agent-based modeling (Brown et al., 2005), which relates one specific value to a unit of space, to a more detailed representation, which characterizes individual buildings and the road network connecting them. This is achieved by importing three GIS-based data layers into the model. These are, first, a statistical areas (SA) layer. SAs are small, homogenous areas defined by the Israeli Central Bureau of Statistics (CBS) that include data on population size, income and migration trends. Second, we utilize a GIS buildings layer, provided by the Israel Land Survey which includes data on buildings by height, number of floors, land use and floor-space. Finally, we use a streets layer from the Hebrew University GIS database.

While these layers contain some information at a sufficiently disaggregated spatial resolution (e.g. use and floor-space for buildings), much of the data is available only at the coarser level of the SA. Therefore, we generate building-level data by using data fusion and proportional allocation, as described in Lichter and Felsenstein (2012). This disaggregates coarse area-level data to individual buildings according to their share of the area or city floor-space, as follows:

a. Initial resident population in a building:

$$Res_i = Pop_{SA_i} * \frac{FS_i}{\sum_{j \in SA_i} FS_j} \tag{9.1}$$

where $Res_i$ is the number of residents of building *i*, $Pop_{SA_i}$ is the population size of the statistical area *SA* in which building *i* is located, and *FS* is floor-space volume (area times number of floors).

b. Initial residential building value:

$$V_i = \frac{HP_{SA_i} * FS_i * {SL_i}/{SL_{SA_i}}}{HHS} \tag{9.2}$$

where $V_i$ is the value of a residential building *i*, $HP_{SA_i}$ is the average housing price per meter (in New Israeli Shekels – NIS) in statistical area *SA* in which building *i* is located, *FS* is floor-space volume, *SL* is the service level – non-residential buildings to residential buildings ratio (*i* indicating within a vicinity of 100 meters from building *i*), and *HHS* indicates citywide average for household size.

c. Initial non-residential building value:

$$V_i = CS * \frac{FS_i}{\sum_{j=1}^{J} FS_j} \tag{9.3}$$

where $V_i$ is the value of non-residential building *i*, *CS* is the citywide capital stock value, *FS* is floor-space volume, *J* is the global number of non-residential buildings.

### 9.3.2 Citizen agents

Agents are generated according to initial population size. Each agent has only two characteristics: place of residence and level of income (in NIS). While residence is determined in accordance with the results of Equation (9.1), income is randomly drawn for each agent from a normal distribution, the average of which is the average income per month in the building's SA, and the standard deviation is 0.1 of this value.

The goals of each agent are simple – attaining an adequate place of residence and participating in daily activities (see Plate 3). In each iteration (representing one day), the agent first makes a decision regarding current place of residence, depending on the citywide probability of out-migration and the probability of intra-urban migration in its SA of residence. These probabilities are calculated based on the assumptions in Table 9.1, as follows:

$$OutP_t = \frac{OC_{t-1} / Pop_{t-1}}{365}$$

$$IUoutP_{SA_t} = \begin{cases} OC_{SA_{t-1}} > 0 & \dfrac{OC_{SA_{t-1}} / Pop_{SA_{t-1}}}{365} \\ OC_{SA_{t-1}} \leq 0 & 0.00001 \end{cases} \tag{9.4}$$

where $OutP_t$ is the global out-migration probability at time $t$, $IUoutP_{SA_t}$ is intra-urban out-migration probability from SA at time $t$, $OC_{t-1}/OC_{SA_{t-1}}$ is the number of citizens leaving the city/SA at time $t-1$, $Pop_{t-1}/Pop_{SA_{t-1}}$ is population size at time $t-1$. Values of $t-1$ elements stay constant during the simulation.

A random number is drawn out of the range [0,1]. If the number fails to exceed $OutP_t$, the agent will leave the city and be deleted from the simulation. Otherwise, if the number fails to exceed $IUoutP_{SA_t}$, the agent will enter a new process of residential choice. The criteria for this process are based solely on the potential residential location price and agents income and are defined according to three assumptions (Table 9.1). These are, first, that the agent will not spend more than one-third of monthly income on housing (see the next section for derivation of house prices); second, similar people earn similar wages; and, third, agents strive to live among agents of similar social class or higher, i.e. earning similar or higher wages. According to this, a potential location will be within a monthly cost range of one-sixth to one-third of an agent's income. The agent searches randomly chosen locations until these conditions are satisfied or until more than 100 locations are searched. In the latter case, the search process fails and the agent leaves the city and is deleted from the simulation.

When the relocation process succeeds, the agent turns to its second goal of daily activities. This is expressed by visitors locations within the simulation area. In each iteration the agent visits three locations, of which at least one is non-residential. The other two have an equal probability of being either residential or non-residential. The location visited is determined according to simple behavioral principles (Plate 3 and Table 9.1). Each building is given an attractiveness score. This is based on the nature of its surroundings. The share of empty buildings nearby is taken to represent risk evasiveness, distance from current location is considered a push factor and, in the case of non-residential uses, the amount of floor-space represents a pull factor:

$$Attract_{in} = \frac{1 - \frac{Empty_i}{Buildings_i} + \frac{D_{in}}{\max D_n} + 1\{LU_i = non\,\mathrm{Res}\}\frac{FS_i}{\max FS}}{2 + 1\{LU_i = non\,\mathrm{Res}\}} \tag{9.5}$$

where $Attract_{in}$ is the attractiveness score of building $i$ for agent $n$, $Empty_i$ is the number of unoccupied buildings in the vicinity of (100 meters from) building $i$, $Buildings_i$ is the number of buildings in the vicinity of (100 meters from) building $i$, $D_{in}$ is the distance of building $i$ from the current position of agent $n$, $\max D_n$ is the distance of the building farthest away from the current position of agent $n$, $1\{LU_i = non\,\mathrm{Res}\}$ is an indicator function receiving the value of 1 if building $i$ is of non-residential land use and 0 otherwise, $FS_i$ is floor-space volume of building $i$, $\max FS$ is floor-space volume of the largest building in the city.

The agent does not search for the building showing the optimal score, but instead looks for the first building whose score exceeds a utility level, randomly drawn from the range [0,1]. When failing to find a building which satisfies this condition (after considering 20 buildings), the agent updates its preferences by drawing a new utility level.

After completing participation in these three activities, the agent returns home. The paths chosen are based on the principle of satisficing behavior (Simon, 1952). The agent moves from current position to the next junction which is closest to the destination measured in aerial distance and chooses the first path that leads to the destination. While this assumption can be questioned, it is needed for decreasing computational load, as the model needs to simultaneously generate paths for thousands of agents.

To balance out-migration trends, the model also generates immigrants in the form of new citizen agents. The number of new citizens is proportional to the volume of current out-migration and is dependent on previous trends of inter-urban migration:

$$InMig_t = outMig_t * P \tag{9.6}$$

where $InMig_t$ is the volume of in-migration at time $t$, $outMig_t$ is the volume of out-migration at time $t$, $P$ is a random number drawn from a normal distribution whose mean is the ratio between in-migration volume to out-migration volume at time *t−1* (this ratio stays constant during the simulation), and whose standard deviation is the absolute value of 1-migration ratio.

Each agent is assigned an income value based on a random draw from a normal distribution whose mean is the global average income, and whose standard deviation is 0.25 of that value. The agent attempts to find a residential location suitable to its preferences, in the same manner detailed above. In the case of failure, it does not move to the city and is deleted from the simulation.

### 9.3.3 Land-use and housing prices dynamics

The residence and activity choices of agents impact the land-use system. This impact is straightforward: residential buildings can become unoccupied and unoccupied buildings can become residences when populated and subject to land-use regulation policy. However, a full articulation of the supply side needs to consider the dynamics of non-residential property and house prices. We consider buildings as quasi-agents, i.e. semi-autonomous entities that are immobile and unable to initiate action, on the one hand, but are sensitive to their environment and respond to changes within it, on the other hand. This implies that direct actions of agents, such as residence or visits, are not necessarily required for a change in land use or land price. This change may occur indirectly through changes in buildings' environment.

This quasi-agent nature of buildings is embodied in the sensitivity of the non-residential stock to traffic loads. This sensitivity induces land-use change. We assume that the number of visits to a building is proportional to traffic load on the nearest road, thus making traffic load a proxy for revenue and floor-space a proxy for operating costs. Two conditions for land-use change can now be formulated: from non-residential to unoccupied (Equation 9.7) and from residential/unoccupied to non-residential (Equation 9.8):

$$\frac{a * e^{b*\frac{t_i}{\max T}}}{1+e^{b*\frac{t_i}{\max T}}} > \frac{\left(\frac{FS_i}{\max FS}\right)^q}{p+\left(\frac{FS_i}{\max FS}\right)^q} \tag{9.7}$$

$$c * \frac{a * e^{b*\frac{t_i}{\max T}}}{1+e^{b*\frac{t_i}{\max T}}} \leq \frac{\left(\frac{FS_i}{\max FS}\right)^q}{p+\left(\frac{FS_i}{\max FS}\right)^q} \tag{9.8}$$

where a,b,c,p,q are constants, $t_i$ is the traffic volume on the road nearest to building *i*, $\max T$ is the maximal traffic volume, $FS_i$ is the floor-space volume for building *i*, $\max FS$ is the floor-space volume for the largest building in the city.

The logic underlying these functions is that the scores on each side of the equations reflect a location in a distribution so that the volume of traffic needed to sustain non-residential use will be proportional to the distribution of both traffic load and floor-space. The constants are used to create proportions between the distributions of the two variables so that the simulation achieves an acceptable rate of land-use change. The logit-like function is chosen to increase the probabilities that large-scale land use will be difficult to sustain as they demand greatest revenue while decreasing the probability that very small-scale land use will become non-residential.

The second feature of the agent-like nature of buildings is reflected in the dynamic housing-price system. This presents a spatial trickle-down process that is sensitive to local supply of housing and amenities. As seen above (Equation 9.2), the value of a residential building is a function of the average house price in its SA and of the level of services in the immediate vicinity. Here we add change in house prices at the SA level. The actions of agents affect the demand, supply and service level within each SA, which in turn affects house prices. They rise when demand or service level increases and fall when supply increases:

$$HP_{SA,t} = HP_{SA,t-1} * \left[ 1 + \log \left( \frac{Pop_{SA,t} / Pop_{sa,t-1} + Res_{t-1} / Res_t + Comm_t / Comm_{t-1}}{3} \right) \right] \quad (9.9)$$

where $HP$ is average housing price per meter in NIS, $Pop$ is population size, $Res$ is the number of residential buildings, $Comm$ is the number of commercial buildings, $t$ is the current simulation iteration, $t\text{-}1$ is the previous simulation iteration.

This effect of overall change induced by the behavior of agents also affects the values of individual buildings. This is achieved by making Equation 9.2 time-dependent. This effect can further trickle down to the level of the individual apartment, by assuming a constant dwelling unit size (Equation 9.10) from which the monthly cost of housing can be derived (Equation 9.11):

$$V_{du} = \frac{V_b}{FS_b / 90} \quad (9.10)$$

$$P_{du} = \bar{Y} * \left[ 1 + \frac{V_{du} - \sigma_{V_{du}}}{\bar{V}_{du}} \right] \quad (9.11)$$

where $V_{du}$ is the value of dwelling unit $du$, $V_b$ is the value of building $b$, $FS_b$ is floor-space volume for building $b$, $P_{du}$ is the monthly cost of living in dwelling unit *du*, $\bar{Y}$ is citywide average income, $\sigma_{V_{du}}$ is the citywide standard deviation of dwelling unit values, $\bar{V}_{du}$ is the citywide average of dwelling unit values.

These prices and the changing market affect the behavior of agents and are carried over to the next iteration, as detailed above (see previous section).

The changes to residential and non-residential stocks may also change urban morphology (Plate 3). This is reflected in change in the location of the central business district (CBD). While not of direct importance to the behavior of agents, a shift in CBD location can indicate the level of disruption wrought by an earthquake. The center of the CBD is identified as the location of the single building with the highest average non-residential floor-space of all the buildings in its vicinity (within 250 meters).

### *9.3.4 Exogenous interventions*

In the sections above all actions within the urban system are determined endogenously based on pre-determined initial values for variables. We simulate the earthquake as an exogenous shock whose epicenter is located randomly in space and with an impact that decays exponentially with distance. This impact makes no attempt to capture the seismic details of such an event but rather focuses on the probability of a building suffering damage and collapse. This probability, along with distance decay, is proportional to building height:

$$I_b = \frac{a * 10^p}{D_b * \left|\log\left(D_b\right)\right| * F_b} \tag{9.12}$$

where $I_b$ is the impact building b suffers, $a$ is a constant, $p$ is the earthquake magnitude (similar to Richter scale), $D_b$ is distance of building $b$ from the earthquake epicenter, $F_b$ is number of floors in building $b$.

Whether or not a building collapses is determined by drawing a random number from the range [0,1]. If the impact exceeds this number, the building is demolished by the earthquake. In such a case, all the streets within a 50 meter radius from the structure become unusable until the building is restored. The duration of restoration is proportional to building floor-space. In the case of collapse all residents have an equal probability of leaving the city or relocating. Relocation will be to a new home via the search process detailed in the *Citizen agents* section (above) or to shelter in accordance with policy intervention (see below).

We specify three stylized policy options, which do not correspond to actual planned responses but span the continuum ranging from passive-liberal through to regulative-rigid scenarios:

a. *Land-use regulation* – this aims at containing impacts by preventing any change to the land-use system. Unoccupied or demolished buildings can only recover to their initial use. When this policy is not exercised, structures can switch uses freely, in accordance with market forces.

b. *Sheltering* – this option outlines the way agents affected by the earthquake are treated. In order to prevent population depletion, agents whose residence has collapsed are clustered into one randomly, pre-selected residential building where they are sheltered until their home is restored. Otherwise, the agents are left to find a new home or move away, according to their ability. This could be thought of as giving the affected citizens an income level-based housing voucher.
c. *Service substitution* – many public structures that offer services to citizens may become unavailable after the earthquake. This policy option 'nationalizes' buildings of commercial use and similar size to the damaged structures and uses them for the provision of public services until the original building structure is restored. When not activated, the service will remain unavailable until restoration. Since public uses are stable and do not depend on market dynamics, this policy creates more stability in the non-residential stock, thus indirectly affecting housing prices.

Activating all three of these binary policy states represents the stability-equilibrium view. Policy attempts to direct the city towards the pre-shock state by minimizing the effect on population, non-residential stock, and the land-use system. The opposite, no policy scenario, leaves the city entirely subject to market forces.

## 9.4 Case studies: one earthquake – two cities

We choose Jerusalem and Tel Aviv as case study locations in order to compare the long-term impacts of a similar event in different urban contexts. An earthquake is a probable hazard in both places due to their proximity to the Dead Sea Fault, a geologically active fissure that has activated a number of earthquakes in the past (Salamon, Katz & Crouvi, 2010). Jerusalem is located on top of the Judean ridge, 30 kilometers southwest of the fault and Tel Aviv is located further northwest on the shores of the Mediterranean Sea and is 90 kilometers from the fault.

To limit computing overload, we define the case study area in both cities as the vicinity of the CBD (Plate 4). Both locations are roughly similar in population size, both contain mixed land uses with residential properties alongside commercial and public sector buildings. Both encompass major traffic arteries (the Ayalon Freeway and Dizengoff Street in Tel Aviv and the triangle of King George, Jaffa and Agripas Streets in Jerusalem) and both have focal commercial concentrations that compete with the CBD, such as the Mahane Yehuda Market in Jerusalem and the Dizengoff Center in Tel Aviv. The cost of housing in Tel Aviv is almost twice as high as the cost in Jerusalem but Tel Aviv's population is characterized by higher incomes and smaller households (see Table 9.2).

An important distinction between the two locations is that while Tel Aviv's CBD is larger in area than that of Jerusalem, the area of Governmental public buildings is greater in the latter (Table 9.2). Moreover, in Jerusalem, public buildings exceed commercial buildings (Table 9.2). By contrast, in Tel Aviv commercial density is higher and commercial buildings have more floors (average = 3.9)

*Table 9.2* Case studies characteristics

| *Variable* | *Tel Aviv* | *Jerusalem* |
|---|---|---|
| Area (square meters) | 5,574,110 | 1,433,277 |
| Population | 2,550 | 2,681 |
| Average income (NIS per month) | 8,378 | 6,003 |
| Average household size | 2.6 | 3.4 |
| Residential buildings | 2,550 | 717 |
| Residential floor space (square meters) | 852,060 | 243,075 |
| Commercial buildings | 482 | 119 |
| Commercial floor space (square meters) | 2,496,457 | 504,347 |
| Governmental (public use) buildings | 139 | 179 |
| Governmental floor space (square meters) | 481,527 | 419,932 |
| Average housing price by SA range (NIS per meter) | 18,400–34,500 | 10,052–20,024 |

than their counterparts in Jerusalem (2.6). These features indicate Tel Aviv as a business-led CBD with a smaller public sector presence than in Jerusalem. This correlates with the public perception of Jerusalem as a national center heavily regulated by administrative functions in contrast to the image of Tel Aviv as a business center with global aspirations (Alfasi & Fenster, 2005).

Despite some similarities, these two cases represent very different urban contexts. A similar shock may evoke very different responses in each case study location and their ability to cope with disaster is not pre-ordained or symmetrical. To test this discord empirically, we simulate two polar scenarios for each city. In the first, none of the three policy interventions is activated (no-policy scenario), while in the other all are initiated (policy scenario). Each scenario is simulated 35 times in each city (140 simulations in total) and while the epicenter of the earthquake is located randomly, in order to avoid possible location bias, the timing of the event is set to the fifth iteration (day). This allows a 'run-in' period for the urban system. Each simulation comprises 1,000 iterations (days). This somewhat arbitrary number is chosen as it allows for a reasonable level of convergence while still being computationally manageable. The results that follow relate to the average (homogenized) values from the simulations by city and policy scenario.

## 9.5 Results

As outlined in the above sections, the simulation model generates initial values for variables at a high level of spatial resolution, such as the individual building or agent. The mechanics detailed above allow these values to vary over time in response to changes in the environment. While these changes can be re-aggregated at various spatial scales, the results below present averages for the case study areas in order to present an aggregate picture of overall trends. This allows for comparing across policies in both urban areas.

Figure 9.1 presents population dynamics over time for all scenarios. As expected, the earthquake causes an immediate loss of population. This is due to either lack of supply of physical stock due to damage to structures or due to the

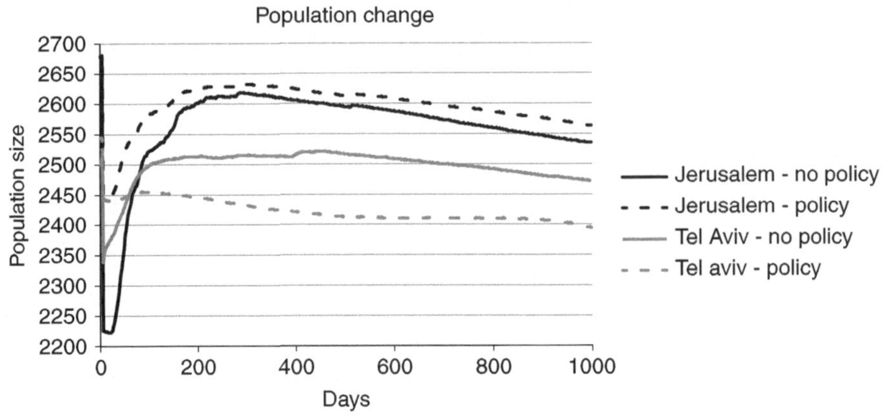


*Figure 9.1* Population dynamics over time

indirect effect of rising prices as demand increases with no commensurate reaction on the supply side. The size of this impact varies over scenarios. In both cities, policy is effective for the short term when the shock is mitigated. However, in the long run, even if population recovers, it is below former levels. In Tel Aviv the picture is even more severe, as policy intervention leads to a sharp decrease with almost no recovery to begin with. This is surprising, since the sheltering policy option strives to contain the initial shock and retain as much population as possible within the city, facilitating a faster recovery to the pre-shock state. This result could be due to insufficient recovery of the housing market, a time delay in response, rising prices, or a combination of all these factors.

Plate 5(a) through 5(d) visualizes the frequency of land-use change by building in the study areas. The policy scenario enforces strict land-use regulation resulting in vacant buildings at the end of the simulation. All of the figures tell a similar story with those buildings that change land use most frequently characterized by commercial use and large floor space. When no policy is exercised, the residential stock seems unstable as many residential structures with limited floor space change use, creating an outward dispersal (sprawl) of commercial activity. Since conversion from residential to commercial use is a function of traffic volume, this trend can be attributed to agents changing movement patterns as they face the damage caused to the traffic network. Over the long run, this sprawl can become self-reinforcing as agglomeration effects start to lock-in. The cluster of buildings changing their character close to the boundaries of the Tel Aviv study area may serve as an example.

Figures 9.2a–9.2b, describe changes to the amount and value of non-residential stock over time. Under the no-policy scenario, the number of non-residential buildings steadily increases over time, in relation to the policy scenario. In the aftermath of the shock, average values of the non-residential stock decreases. This indicates the conversion of small residential buildings into commercial uses

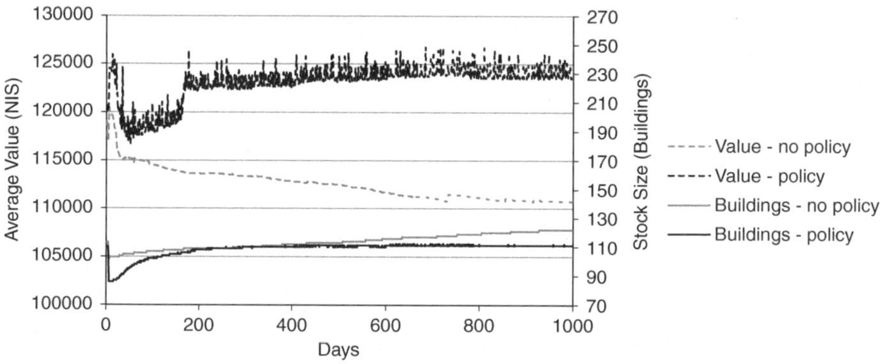


*Figure 9.2a* Changes to non-residential stock over time, for Jerusalem

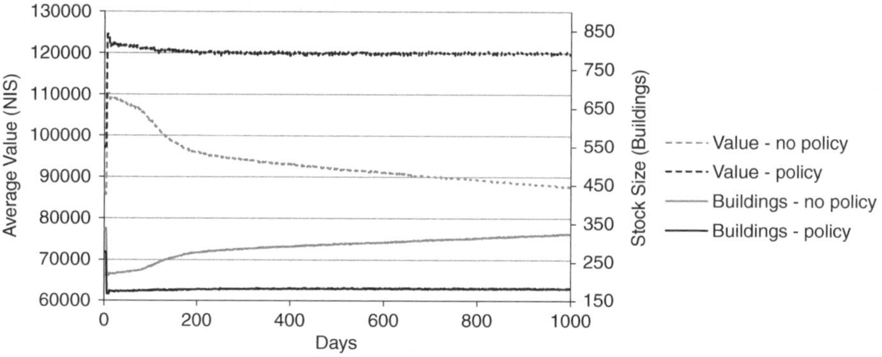


*Figure 9.2b* Changes to non-residential stock over time, for Tel Aviv

('less malls, more convenience stores') since non-residential value is closely related to amount of floor space. While this trend is generally true in both CBDs, Jerusalem displays sharper reactions in the no-policy scenario. Tel Aviv's values rebound to a much lower equilibrium when policy is exercised, suggesting that the urban system is more entrenched.

The dissimilarities between the two cities become heightened when comparing change in residential stock over time. In this respect the cities are almost mirror images. Figures 9.3a and 9.3b show that while policy intervention promotes a sharp increase in average residential values in Tel Aviv in relation to the stable values achieved when no policy is exercised, such an increase is caused in Jerusalem by the absence of intervention. Under the no-policy scenario in Jerusalem demand decreases, while housing supply rebounds and service supply only rises slightly. This increase cannot be attributed just to the growing number of commercial venues. As building values are closely related to amount

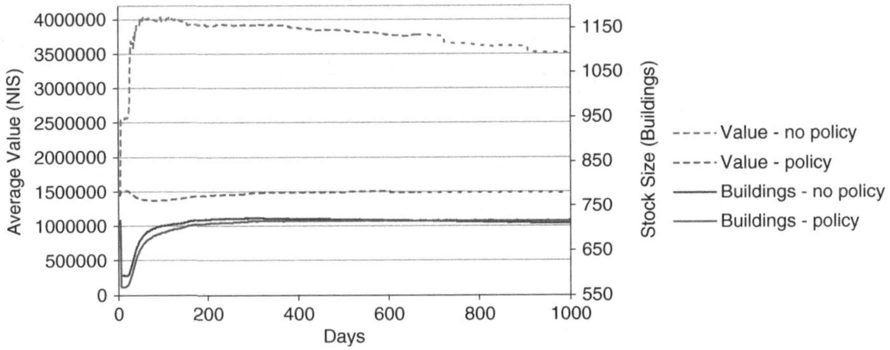


*Figure 9.3a* Changes to residential stock over time, for Jerusalem

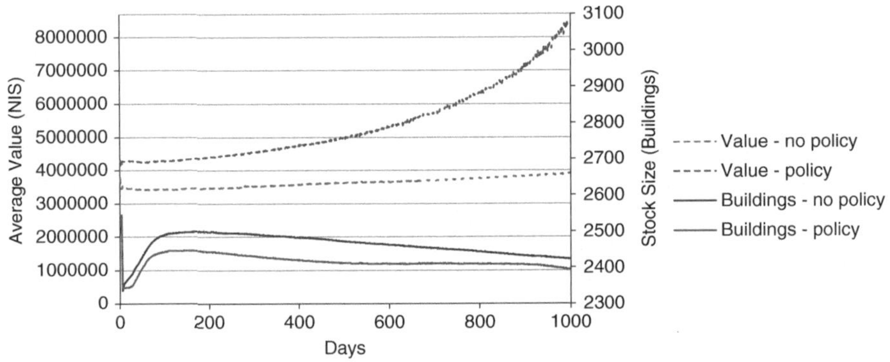


*Figure 9.3b* Changes to residential stock over time, for Tel Aviv

of floor space, the change can be explained as previously commercial buildings with much floor space becoming residential, thereby increasing average values. In the Tel Aviv policy scenario, on the other hand, such an explanation is not applicable, since policy prevents such flexibility of land use. The increase can only be attributed to a short-fall in supply of housing. This cancels out any reduction in property values through a decrease in demand and service supply. These two explanations, grounded in opposing scenarios (Jerusalem policy, Tel Aviv no policy) suggest that in the aftermath of an earthquake the Jerusalem CBD could potentially change its nature whereas in Tel Aviv the CBD is likely to maintain its current function.

Given the potential for the development of new clusters on non-residential activity in the wake of a disaster, we test for a change in urban morphology reflected in a shift in CBD location. Table 9.3 shows that such a change rarely happens. The strength of the Tel Aviv's CBD as an emerging global center is

*Table 9.3* Changes to CBD

| *Region* | *Scenario* | *State* | *N* | *Average non-residential floor space around CBD* | *Total non-residential floor space* | *CBD movement (Meters)* |
|---|---|---|---|---|---|---|
| Jerusalem | No policy | Initial State | 35 | 4,575.97 | 924,279.59 | 0 |
| | | Average Final State – all | 35 | 3,217.63 | 722,116.30 | 20.38 |
| | | Average Final State – CBD movement | 4 | 3,421.54 | 726,528.41 | 178.33 |
| | | Average Final State – no CBD movement | 31 | 3,191.32 | 721,547.00 | 0 |
| | Policy | Average Final State – all | 35 | 3,940.65 | 804,589.79 | 5.1 |
| | | Average Final State – CBD movement | 1 | 3,301.20 | 779,024.24 | 178.33 |
| | | Average Final State – no CBD movement | 34 | 3,959.45 | 805,341.72 | 0 |
| Tel Aviv | No policy | Initial State | 35 | 50,298.71 | 2,977,984.40 | 0 |
| | | Average Final State – all | 35 | 49,856.00 | 2,922,949.19 | 0 |
| | | Average Final State – CBD movement | 0 | | | |
| | Policy | Average Final State – all | 35 | 49,860.01 | 2,826,227.80 | 0 |
| | | Average Final State – CBD movement | 0 | | | |

reflected in the relative stability of the CBD location and in the volume of non-residential floor-space, which decrease only marginally. In Jerusalem, where the CBD serves a more limited market, change in location is registered a number of times, mostly under the no-policy scenario. This suggests that the dispersal of agent movement in the aftermath of a catastrophe is rarely strong enough to generate a change in urban morphology, even in second-tier CBDs such as that in Jerusalem.

The results of the no-policy scenario in Tel Aviv and the policy-on scenario in Jerusalem show evidence of bouncing back and could be construed as interventions promoting resilience. As noted, the Jerusalem no-policy scenario presents a conflicting picture, while the Tel Aviv policy scenario results in a weaker outcome in terms of activity and population. However, the results so far do not present any evidence regarding the stability of these outcomes. In fact, a period of less than three years post-earthquake is probably not long enough for the city to recover entirely. The outcomes presented so far may therefore have captured the city in

*Table 9.4* Rebounding and stabilization of scenarios

| *Scenario* | | *No policy* | | | | *Policy* | | | |
|---|---|---|---|---|---|---|---|---|---|
| *Region* | *Measure* | *Achievement of pre-shock values* | | *Stabilization* | | *Achievement of pre-shock values* | | *Stabilization* | |
| | | *frequency* | *average duration* | *frequency* | *average duration* | *frequency* | *average duration* | *frequency* | *average duration* |
| Jerusalem | Population | 0/35 | | 35/35 | 858 | 0/35 | | 35/35 | 860 |
| | Total Residential Value | 0/35 | | 20/35 | 835 | 33/35 | 246 | 8/35 | 916 |
| | Average Residential Value | 0/35 | | 22/35 | 843 | 33/35 | 54 | 9/35 | 918 |
| | Total Non-Residential Value | 1/35 | 12 | 35/35 | 324 | 31/35 | 122 | 2/35 | 937 |
| | Average Non-Residential Value | 22/35 | 145 | 32/35 | 819 | 24/35 | 40 | 1/35 | 940 |
| Tel Aviv | Population | 16/35 | 54 | 33/35 | 806 | 1/35 | 206 | 26/35 | 879 |
| | Total Residential Value | 32/35 | 168 | 2/35 | 930 | 35/35 | 95 | 0/35 | |
| | Average Residential Value | 27/35 | 65 | 1/35 | 904 | 11/35 | 43 | 0/35 | |
| | Total Non-Residential Value | 0/35 | | 34/35 | 515 | 0/35 | | 35/35 | 281 |
| | Average Non-Residential Value | 24/35 | 459 | 0/35 | | 0/35 | | 0/35 | |

temporary disequilibrium. To address this issue, we test for convergence of different indicators over time at the end of each simulation. An index is said to be stable if the values of its moving average over the last 50 iterations (or more) show insignificant changes. The results of this analysis are presented in Table 9.4. The two bouncing-back scenarios, which indeed reach pre-shock values, do not stabilize around these values, and therefore do not represent resilience, as defined here. The only scenario to show stabilization is the Jerusalem no-policy scenario, in which the city experiences large-scale transformation. This new stable state, while not reflecting the traditional bounce-back concept of resilience, suggests a reorganization of the system embodied in bouncing-forward to a new equilibrium state.

## 9.6 Conclusions

This chapter has presented an agent-based model of urban resilience. Resilience is conceived here as the ability of a system to regain pre-shock equilibrium in the wake of an unanticipated event. The results of the simulation for Jerusalem and Tel Aviv suggest that policy directed at rebounding to pre-shock state does not do well and may even inhibit stability. Elsewhere, we have suggested that cities contain a self-organizing mechanism that facilitates recovery when equilibrium is disrupted and that needs to be considered when policy is formulated (Grinberger & Felsenstein, 2014). The results here reinforce this conclusion, showing how policy interventions lead to unexpected results. The mechanism leading to the formation of such variations can be described in terms of centripetal and centrifugal morphological forces (Fujita & Krugman, 2004). The urban system is the result of previous rounds of investments that generate agglomeration advantages (centripetal forces) in the formation of consumption centers. An unanticipated shock galvanizes centrifugal forces into action by making some places less accessible and dispersing movement and consumption. Progressive rounds of investment are characterized by the tension between centripetal and centrifugal forces. This process may result in rejuvenation of existing urban structures (bouncing back) if agglomerations are strong enough or in the formation of new morphological equilibria ('bouncing forward'). This can result in the emergence of newly formed competitive centers or in the dispersal of activity as in the case of the Jerusalem no-policy scenario.

Interpreting the results of the scenarios this way, we can suggest that the Jerusalem CBD, characterized by a large public sector presence and substantial government intervention, does not manage to develop the critical mass for agglomeration economies to develop. Consequently, when liberal intervention is used to jump-start development in the aftermath of a disaster, this is insufficient to counter the influence of centrifugal forces pushing for dispersal. Due to similar weakness, intervention directed at restoring the previous state fails to display any real recovery, as the city does not reach a stable state. Tel Aviv, on the other hand, displays an almost exact mirror image. While none of the scenarios reaches equilibrium, the policy scenario converges towards bounce-back by inhibiting the work of centrifugal forces, while the no-policy scenario results in a low functioning unstable equilibrium. The role of Tel Aviv as the economic enter of the country, shaped by centripetal market forces, correlates with these results. Due to the magnitude of its agglomeration, a large-scale shock is insufficient to push it off its pre-shock development trajectory.

The fact that the same basic process has led to almost mirror images between the two cities and that none of the results may resemble the 'desired' outcome implies that a procedural check-list approach to urban recovery is insufficient. Our findings show that the rejuvenation goal of having a city 'bounce back' is hard to attain. None of our four simulated scenarios stabilize on pre-shock conditions. This is not surprising since, as suggested earlier, this view of resilience neglects the possibility of a set of unstable equilibria. In fact, over the test period of less than three years, the only stable state witnessed (the Jerusalem no-policy scenario) reflects an 'extended' understanding of resilience that stresses the ability to reorganize when a large enough shock appears (Folke et al., 2002). These results would seem to cast a shadow over the feasibility of the stability view of resilience. The take-away message for urban recovery praxis would seem to be that resilience is not just about the absorption and containment of a change. It is equally about the ability to direct change and exploit the opportunities it presents.

Given the magnitude of threats, both natural and anthropogenic, facing contemporary urban spaces, agent-based simulation of disaster recovery is likely to continue to attract attention. To enhance this form of modeling, more could be done to improve the somewhat mechanistic behavior that underpins agent activity. In addition, agent-based simulation needs to be informed by the wealth of information becoming available via the incipient big data/crowdsourcing revolution. The delivery of agent-based simulation outputs via internet-based mapping also needs to be considered. However, these caveats notwithstanding, the challenges of urban rejuvenation are likely to form an exciting arena for applied spatial modeling in the future.

## Acknowledgement

This research is based on work done in the DESURBS (Designing Safer Urban Spaces) research project funded by the European Commission Seventh Framework Programme 2007–2013 under Grant Agreement No. 261652.

## References

Alberti, M. & Marzluff, J. M. (2004) Ecological resilience in urban ecosystems: linking urban patterns to human and ecological functions, *Urban Ecosystems*, 7(3), 241–265.

Alberti, M., Marzluff, J. M., Shulenberger, E., Bradley, G., Rayn, C., & Zumbrunnen, C. (2003) Integrating humans into ecology: opportunities and challenges for studying urban ecosystems, *BioScience*, 53(12), 1169–1179.

Alfasi, N. & Fenster, T. (2005) A tale of two cities: Jerusalem and Tel Aviv in an age of globalization, *Cities*, 22(5), 351–363.

Allan, P., Bryant, M., Wirsching, C., Garcia, D., & Teresa Rodriguez, M. (2013) The influence of urban morphology on the resilience of cities following an earthquake, *Journal of Urban Design*, 18(2), 242–262.

Brown, D. G., Riolo, R., Robinson, D. T., North, M., & Rand, W. (2005) Spatial process and data models: toward integration of agent-based models and GIS, *Journal of Geographical Systems*, 7(1), 25–47.

Campanella, T. J. (2008) Urban resilience and the recovery of New Orleans, *Journal of the American Planning Association*, 72(2), 141–146.

Chang, S. E. (2010) Urban disaster recovery: a measurement framework and its application to the 1995 Kobe earthquake, *Disasters*, 34(2), 303–327.

Chang, S. E. & Rose, A. Z. (2012) Towards a theory of economic recovery from a disaster, *International Journal of Mass Emergencies and Disasters*, 32(2), 171–181.

Cruz, S. S., Costa, J. P. T., de Sousa, S. Á., & Pinho, P. (2013) Urban resilience and spatial dynamics. In A. Eraydin & T. Tasan-Kok (Eds.), *Resilience Thinking in Urban Planning* (pp. 53–69). Springer, Netherlands.

Davoudi, S. (2012) Resilience: a bridging concept or a dead end? *Planning Theory & Practice*, 13(2), 299–307.

Deppisch, S. & Schaerffer, M. (2011) Given the complexity of large cities, can urban resilience be attained at all? In B. Müller (Ed.), *German Annual of Spatial Research and Policy 2010* (pp. 25–33). Springer, Berlin-Heidelberg.

Fleischauer, M. (2008) The role of spatial planning in strengthening urban resilience. In H. Pasman & E. Kirillov (Eds.), *Resilience of Cities to Terrorist and Other Threats* (pp. 273–298). Springer, Netherlands.

Folke, C. (2006) Resilience: the emergence of a perspective for social-ecological systems analyses, *Global Environmental Change*, 16(3), 253–267.

Folke, C., Carpenter, S., Elmqvist, T., Gunderson, L., Holling, C. S., & Walker, B. (2002) Resilience and sustainable development: building adaptive capacity in a world of transformations, *AMBIO: A Journal of the Human Environment*, 31(5), 437–440.

Fujita, M. & Krugman, P. (2004) The new economic geography: past, present and the future, *Papers in Regional Science*, 83(1), 139–164.

Godschalk, D. R. (2003) Urban hazard mitigation: creating resilient cities, *Natural Hazards Review*, 4(3), 136–143.

Grinberger, A. Y. & Felsenstein, D. (2014) Bouncing back or bouncing forward? Simulating urban resilience, *Proceedings of the Institution of Civil Engineers: Urban Design and Planning*, 157(3), 115–124.

Holling, C. S. (1973) Resilience and stability of ecological systems, *Annual Review of Ecology and Systematics*, 4, 1–23.

Kartez, J. D. (1984) Crisis response planning: toward contingent analysis, *Journal of the American Planning Association*, 50(1), 9–21.

Kartez, J. D. & Lindel, M. K. (1987) Planning for uncertainty: the case of local disaster planning, *Journal of the American Planning Association*, 53(4), 487–498.

Lichter, M. & Felsenstein, D. (2012) Assessing the cost of sea-level rise and extreme flooding at the local level: a GIS-based approach, *Ocean & Coastal Management*, 59, 47–62.

Macal, C. M. & North, M. J. (2005) Tutorial on agent-based modeling and simulation. In *Proceedings of the 37th Conference on Winter Simulation, WSC '05* (pp. 2–15). Winter Simulation Conference, Orlando, FL.

Martin, R. (2012) Regional economic resilience, hysteresis and recessionary shocks, *Journal of Economic Geography*, 12(1), 1–32.

Müller, B. (2011) Urban and regional resilience – a new catchword or a consistent concept for research and practice? In B. Müller (Ed.), *German Annual of Spatial Research and Policy 2010* (pp. 1–13). Springer, Berlin-Heidelberg.

North, M. J., Collier, N. T., Ozik, J., Tatara, E. R., Macal, C. M., Bragen, M., & Sydelko, P. (2013) Complex adaptive systems modeling with Repast Simphony, *Complex Adaptive System Modeling*, 1(1), 1–26.

Olshanky, R. B., Hopkins, L. D., & Johnson, L. (2012) Disaster and recovery: processes compressed in time, *Natural Hazards Review*, 13(3), 173–178.

Prasad, N., Ranghieri, F., Shah, F., Trohanis, Z., Kessler, E., & Sinha, R. (2009) *Climate Resilient Cities: A Primer on Reducing Vulnerabilities to Disasters*. World Bank, Washington, DC.

Salamon, A., Katz, O., & Crouvi, O. (2010) Zones of required investigation for earthquake-related hazards in Jerusalem, *Natural Hazards*, 53(2), 375–406.

Simon, H. (1952) A behavioral model of rational choice, *Quarterly Journal of Economics*, 69(1), 99–118.

UNISDR (United Nations International Strategy for Disaster Reduction) (2012) *How to Make Cities More Resilient – A Handbook for Local Government Leaders*. UNISDR, Geneva, Switzerland.

# Part III

# Population dynamics and population ageing

# 10 The United Kingdom's multi-ethnic future

## How fast is it arriving?

*Philip Rees, Pia Wohland and Paul Norman*

### 10.1 The context for ethnic population projections

Knowledge of the future population is very useful for economic and social planning at international, national and subnational scales. The Population Division of the United Nations (UN) produces projections of the populations of UN member states every two years (United Nations 2014). These populations are used in the forecasting of key indicators such as per capita national income, in which Gross Domestic Product estimates for countries are divided by the forecast population. The United Nations produces GDP forecasts as well, but they are only for the short term. National Governments, on the other hand, have a requirement to forecast GDP and population over a term longer, enough to compute the assets and liabilities of state pension schemes. The forecast of the UK's Gross Domestic Product by the Office of Budget Responsibility (OBR 2012, 2013a) relies heavily on the National Population Projections (NPP) of ONS (the Office for National Statistics), where choice of mortality scenario affects the numbers surviving to pensionable age and the choice of migration variant influences the growth path of national debt (OBR 2013b). For a country like the United Kingdom with its high net immigration the United Nations' Projections are of little use because of their assumption that rates will trend to zero.

International migration flows into the UK are a product of labour demand, the later addition of family members to prior labour migrant, the demand for higher education in the origin regions, the desire for safe havens by refugees and asylum seekers and the search for basic employment by people in poverty. One of the factors creating labour demand in European countries, including the UK, is their entry since the early 1970s into the Second Demographic Transition (van de Kaa 1987), characterised by below replacement fertility and increasing life expectancy. The net balance of immigration over emigration into higher income European countries has two effects: the population growth increases and the age structure becomes more youthful, relative to the position without these international migration flows (Kupiszewski and Kupiszewska 2010). Coleman (2006) has termed this combination of low fertility, increasing longevity and inward migration the Third Demographic Transition.

This transition has further ramifications for the United Kingdom beyond a contribution to labour supply and a postponement of population decline. The composition of national populations in terms of country of birth becomes more diverse with the continuation of inward migration. But the process of compositional change does not stop with the immigration of the foreign born. If the new immigrants settle and raise families, a new generation of native-born children of foreign-born parents will be created. Later this new generation will marry or partner and have children who will be the grandchildren of the foreign-born immigrants, though many family histories will be more complex than this. In most European countries there is no official tracking of the country of birth ancestry of these new generations. Further generations are assigned to the native-born part of the population. However, communities with such foreign birth origins continue to retain cultural and spatial characteristics associated with their origins, while at the same time assimilating into or integrating with the native-born population. In the UK it was recognised in the 1970s that there was a need to monitor groups of distinct ethnic origin because of discrimination in employment and housing markets. An ethnic classification was introduced in the 1980s into the UK Labour Force Survey, using a self-reporting question. After an attempt to introduce an ethnic question into the 1981 Census failed because of some ethnic minority anxiety, a question was used in the 1991 Census, with positive ethnic minority support, and has been used, in modified form, in the 2001 and 2011 Censuses. In the UK we can therefore examine the progress of the ethnic transition from 1991 onwards and use this knowledge to forecast the country's ethnic composition (Jivraj and Simpson 2015).

Why is projecting ethnic group populations important? It is necessary for monitoring ethnic group numbers in order to measure the degree of disadvantage that groups face and discrimination on grounds of ethnicity after controlling for other factors that determine achievement or well-being. Of course, other dimensions of difference, such as disability or sexuality or social class, also produce disadvantage and discrimination, but the projection of the population by these dimensions is less developed. Although most focus is on the monitoring of minority ethnic groups, majority group disadvantages can also be present, as in the poorer performance of White working-class or lower-class boys in school examinations, compared with immigrant pupils in similar socio-economic circumstances.

It could be argued that projections of ethnic group populations are unnecessary because the latest census provides a wealth of information to support investigations. However, censuses are only administered every decade in the UK and, if we add the time needed for processing, production and quality checking of outputs, on occasion it is necessary to use data that are from one to thirteen years out of date. Projections fill the gap, though these will inevitably drift away from the actual path of population development. Projections also supply information for the near future that many government policies require (e.g. central government to local agency funding allocations).

For what purposes is ethnic population information needed? The health needs of ethnic groups may differ in ways related to genetics (e.g. sickle cell anaemia) or life style (e.g. food consumption in relation to cardio-vascular disease), though socio-economic deprivation plays the most important role in determining health status. Businesses also benefit from knowledge of the ethnic composition of their potential consumers where tastes differ in food or clothing choices. Politicians need to know about the future ethnic compositions of constituency electorates because voting preferences vary by ethnic group. In 2014 the 'think tank' Policy Exchange commissioned a reworking for Parliamentary constituencies of previously published ethnic projections for election years between 2015 and 2035 (Rees and Clark 2014). Another important use for ethnic population outputs is to provide context variables for individual-level studies. There is a need to measure impact of spatial community characteristics on behaviour or conditions in addition to individual-level variables. Information on future ethnic compositions of local populations is needed for planning and consultation purposes, as is testified by their production for a number of years by the Greater London Authority (GLA 2014). Ethnic group projections also have a role to play in informing public debate about the way national and subnational populations are changing. Applied geographers and policy makers benefit by taking into account the future ethnic composition of populations of interest. This is particularly important when dealing with local populations, the ethnic composition of which may vary considerably from very homogeneous to very diverse. The important work of Bill Frey of the Brookings Institution about the United States shows how valuable an analysis of the changing demography and geography of America's racial groups is for understanding current and future political debates about welfare policy (Frey 2015).

If ethnic population projections are to fulfil these many roles, they must of the highest quality. The aim of this chapter is to pause between a completed round of such projections based on the 2001 Census (Wohland et al. 2010, 2014; Rees et al. 2011, 2012a, 2012b) and a new round of projections based on the 2011 Census (Rees et al. 2015) to evaluate how well the 2001-based projections reproduced the 2011 Census results and to interpret the differences.

Section 10.2 provides some brief background on 2001–2011 changes in the ethnic composition of the population. In section 10.3, we summarise the way we implemented the projections and discuss key results. In section 10.4, we compare our projections with the results of the 2011 Census and suggest reasons for the differences observed. Section 10.5 discusses how fast the ethnic transition in England and Wales's population is happening.

## 10.2 The changing ethnicity of the population

The population of the England and Wales[1] has been changing in ethnic composition since 1945 as a result of the immigration of a wide range of origin groups. Many of these groups have settled, raised families, inter-married with other British people. As a result, the *White* share of the England and Wales population has declined from 93% in the 1991 Census to 91% in the 2001 Census and 85% in 2011 Census (Jivraj 2012, from Census statistics). The *White British* share of the 2011 Census population was 80% of the 85. So the (not White) ethnic minority share of the England and Wales population increased from 7% in 1991 to 9% in 2001 and 15% in 2011. These statistics indicate some speeding up of the transition during the 2001–2011 period compared with 1991–2001.

By ethnic group we mean people who identify themselves as belonging to a sub-population with national origins outside the United Kingdom, both those born abroad and their descendants. Ethnicity is based on self-identification when answering a survey or census question. Ethnic status has legal recognition in the Equality Act 2010 (EHRC 2013), which replaces earlier Acts with a unified framework for monitoring and acting on evidence of disadvantage or discrimination on grounds of gender, ethnicity, disability or sexual orientation. It is important to compare the numbers of ethnic group members in work compared with the numbers available for work, for example.

The census has been the main source of information on local and national ethnic populations but society can benefit from more frequent knowledge of changing ethnic composition. Updates can be provided through two methods. The first method is to roll forward ethnic group populations from the latest census to successive mid-years, using estimates of the components of population change, which has been done for mid-year 2002 to 2009 for local authorities by the Office for National Statistics (ONS 2011a). The second method is to carry out a representative social survey each year, such as ONS's Annual Population Survey (APS) (ONS 2013a). ONS is currently carrying out an assessment of the reliability of their ethnic population estimates (ONS 2011b) and has suspended the production of local estimates, pending an evaluation against the results of the 2011 Census (ONS 2011c).

We also need a view about how the ethnic composition of the population of England and Wales is likely to change in the future. As ONS are doing, we have embarked on an evaluation of our projections through comparison with the results of the 2011 Census. The main research question we try to answer is 'How well did we do in projecting ethnic group populations for local authorities in England?'

## 10.3 Methods, data and results of the ethnic projections

The projection model used is a bi-regional cohort component model (Rees et al. 2012a). This means we project the population disaggregated by age and sex, accounting for migration within the United Kingdom as outflows from each area and as inflows from the rest of the country. The model is applied separately to each ethnic group using suitable estimates of rates and flows. However, the groups are connected when babies are born to mothers and fathers from different groups and are placed by their parents in mixed ethnic categories. We use single years of age to 100+ and the 16 ethnic groups in the 2001 Census. The coverage of the projections is the United Kingdom. For spatial units, we use 352 local authorities[2] in England together with Wales, Scotland and Northern Ireland (355 zones in total).

Selecting from a wide set of projections, we focus here on two scenarios (Table 10.1) which use assumptions aligned to those in the National Population

*Table 10.1* ETHPOP scenario projections: assumptions

| *Scenario Projection* | *Assumptions* |
|---|---|
| TREND | Fertility, mortality and international migration assumptions follow those of ONS's 2008-based National Population Projections factored to reflect local authority differences. Internal migration assumptions are based on the 2001 Census updated using NHS Patient Register data to 2008 |
| UPTAPER | Uses the same assumptions as the TREND projection by changes the model for projecting emigration from assumptions about emigration flows to assumptions about emigration rates which are multiplied by local authority populations |

Sources: Wohland et al. (2010) and Rees et al. (2012a)

Projections (2008-based) but which differ in one model feature. The first scenario (TREND) projects the future flows of emigrants along with an equivalent immigration series, which together match the ONS net international migration assumption of +180,000 net international migrants per annum. The second scenario (UPTAPER) projects emigration as a product of emigration rates and the changing populations at risk in the 355 zones. Since 2008 the net immigration level has been both higher (2009–2011) and lower (2012) than 180,000 (ONS 2013b). The argument for the second approach is that we should make the emigration flows functions of the population at risk resident in the UK zones, which will change over the projection. The argument against such an approach is that emigrants face many barriers to migration, some determined by destination country policy, just as immigration to the UK is influenced by UK government policy.

Plate 6 shows the results of these two projections for the UK starting from a base population at mid-year 2001. The graphs show the projected changes for the 16 ethnic groups relative to the population at mid-year 2001. The White British and White Irish populations fail to grow in the UPTAPER projection and increase only a little in the TREND projection, falling back towards mid-century. The Black Caribbean population grows quite slowly, experiencing high emigration, low fertility and a loss of children to the mixed White and Black Caribbean group. The other minority ethnic populations have much higher future trajectories, experiencing growth of 2 to 3.2 times in the UPTAPER projections between 2001 and 2051 and 2.5 to 6 times in the TREND projections. These projections show that the diversity of the UK population will increase substantially over the first half of this century.

Figure 10.1 shows diversity patterns for local authorities in England, measured using the Simpson Index of Diversity (DI), mapped using a population cartogram,[3] in which the extent on the map of a zone is proportional to a population at risk (2001 population). The DI is computed as one minus the sum over ethnic groups of the squared proportions of each ethnic group. The minimum DI is zero (no diversity) when only one group is present in an area. The maximum diversity depends on the number of groups considered. With 16 groups each having an equal share of the population, the DI would be 0.9375. The ethnic

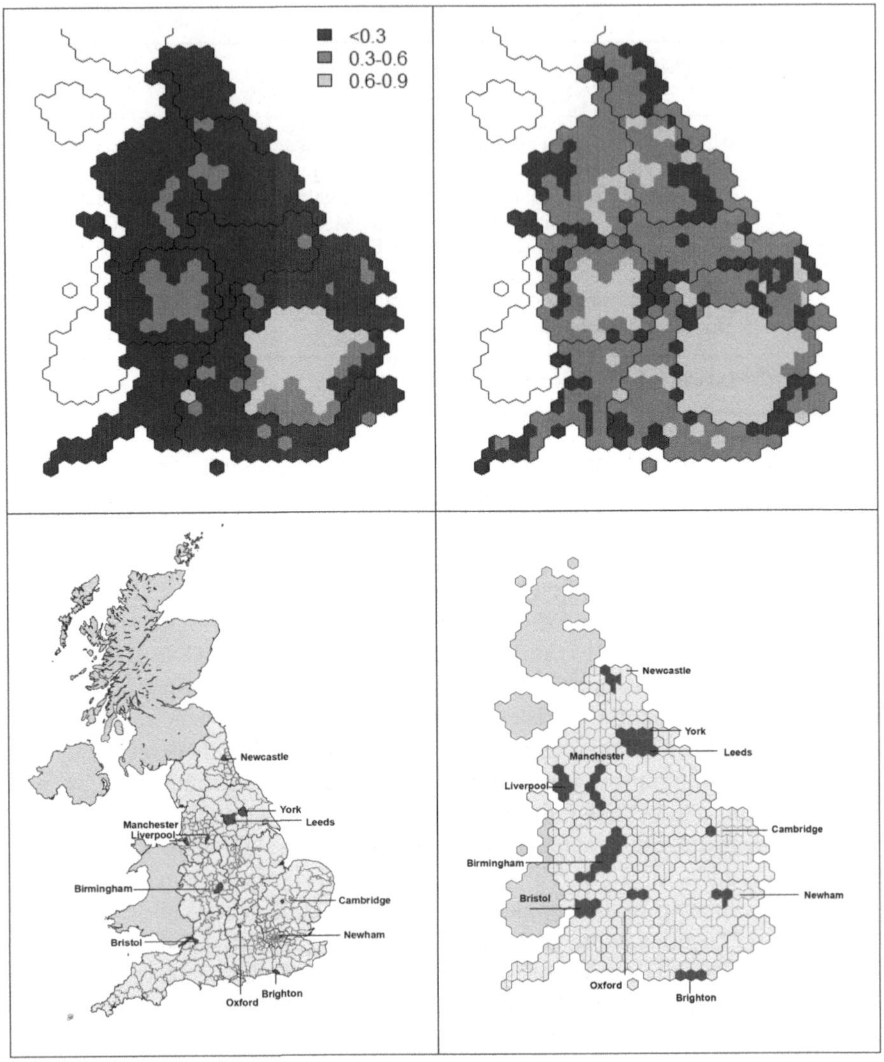


*Figure 10.1* Ethnic diversity for local authorities in England: TREND projection, 2001 and 2051

diversity of the local authority populations increases substantially over the 50 years. By 2051, under the TREND projection, almost all areas are in the top two diversity quartiles, which were found only in London and some big cities in 2001. In the UPTAPER projection the increase in diversity is more subdued.

A note of caution is necessary before we accept this picture of the future ethnic composition of the UK population. The projections are liable to error from three sources: (1) the estimates of the ethnic fertility, mortality, internal and international migration indicators used as projection inputs may be wrong; (2) the assumptions about the future behaviour of the inputs may be wrong; and (3) the model may be wrongly specified. So, it is sensible to validate the projections against the recent 2011 Census. We report on this evaluation in the next section.

## 10.4 Comparison of Census 2011 and ETHPOP projected populations

To compare the 2011 Census results and the ETHPOP projections, we harmonise the ethnic group definitions and the boundaries of the local authorities (LAs). Table 10.2 shows the correspondence of the 16 ethnic groups in the 2001 Census and the 18 ethnic groups in the 2011 Census in England and Wales. We merge the two new groups in 2011 into the matching larger group in 2001. To harmonise the zones used in the two censuses we aggregated the 2001 Census local authorities into unitary county authorities created in April 2009 and used in the 2011 Census. In 2001 England and Wales was made up of 353 zones; in 2011 there were 327 zones. The final adjustment made was to combine the 2011 TREND and UPTAPER projected populations because these figures bracketed the 2011 Census population for England and Wales as a whole and because the arguments for and against each approach are unresolved. We did not correct for the three-month timing difference between the Census (27 March) and the projections (30 June/1 July).

*Table 10.2* The correspondence of ethnic groups in the 2001 and 2011 Censuses

| *2001 Census Ethnic Group (16 Groups)* | *2011 Census Ethnic Group (18 Groups)* |
|---|---|
| White: British | White: English/Welsh/Scottish/Northern Irish/British |
| White: Irish | White: Irish |
| White: Other white | White: Gypsy or Irish Traveller<br>White: Other White |
| Mixed: White and Black Caribbean | Mixed/multiple ethnic group: White and Black Caribbean |
| Mixed: White and Black African | Mixed/multiple ethnic group: White and Black African |
| Mixed: White and Asian | Mixed/multiple ethnic group: White and Asian |
| Mixed: Other Mixed | Mixed/multiple ethnic group: Other Mixed |
| Asian or Asian British: Indian | Asian/Asian British: Indian |
| Asian or Asian British: Pakistani | Asian/Asian British: Pakistani |
| Asian or Asian British: Bangladeshi | Asian/Asian British: Bangladeshi |
| Chinese or Other Ethnic Group: Chinese | Asian/Asian British: Chinese |
| Asian or Asian British: Other Asian | Asian/Asian British: Other Asian |
| Black or Black British: Black African | Black/African/Caribbean/Black British: African |
| Black or Black British: Black Caribbean | Black/African/Caribbean/Black British: Caribbean |
| Black or Black British: Other Black | Black/African/Caribbean/Black British: Other Black |
| Chinese or Other Ethnic Group: Other Ethnic | Other ethnic group: Arab<br>Other ethnic group: Any other ethnic group |

In Table 10.3 we compare results for England and Wales from the 2011 Census with the average of our two ETHPOP projections for five broad ethnic groupings. Our projection of the All Groups population is very close (an error of less than three in 10,000). Our projection of the White groups is close (3% error) but the projected population is greater than the census population. This means that we under-projected the growth of the ethnic minority groups. Our projection of the Asian groups was 18% under the 2011 Census figure. For the Black groups the under-projection was also 18%. The under-projection of the Mixed groups was 17%. The projection of the 'Other' groups was the worst at 23%. Overall we have under-projected the growth of ethnic minority populations and hence the pace at which the population of England and Wales has diversified. Instead of the White/Ethnic Minority population split being 88.5%/11.5% in 2011 (our projections), the Census measured the composition as 86.0% White and 14.0% ethnic minority.

Table 10.4 shows comparisons for the 16 harmonised groups. Within the White groups, the White British group is over-projected by 3% and the White Irish group by 25%. We may have under-estimated the emigration of White British people and over-estimated their survival into old age. For the White Irish group we may have under-estimated the extent that their children born in the 2000s were assigned White British ethnicity and some of the group may have changed their identities between censuses. The White Other group we under-projected by 7%, probably because we under-estimated the level of immigration from other European countries and over-estimated emigration.

Within the Mixed groups, under-projections of the White and Black Caribbean and White and Black African groups were twice as large as the under-projection of White and Asian and Other Mixed groups. These groups grow because children are born to parents of two different ethnicities. The only information available on this process came from a 2001 Census commissioned table. Mixed

*Table 10.3* The England and Wales populations of five ethnic groupings: Census 2011 and ETHPOP 2011

| *Ethnic grouping (2011 definitions)* | *Census Population CD 2011* (thousands) | *Average of TREND and UPTAPER Projections, MY2011* (thousands) | *Difference = Average Projection minus Census* (thousands) | *100 × (Difference/ Census)* (%) |
|---|---|---|---|---|
| All Groups | 56,076 | 56,057 | −18 | −0.03 |
| White | 48,209 | 49,609 | 1,400 | 2.90 |
| Black | 1,865 | 1,521 | −344 | −18.44 |
| Asian | 4,214 | 3,469 | −744 | −17.66 |
| Mixed | 1,224 | 1,017 | −207 | −16.90 |
| Other | 564 | 440 | −124 | −21.92 |

Sources: Census 2011 – ONS (2013c), Crown Copyright.
Projections – ETHPOP (2013), funded by ESRC.
Notes: CD = Census date (27 March 2011), MY = Mid-YEAR (30 June/1 July).

*Table 10.4* The England and Wales populations of 16 harmonised ethnic groups: Census 2011 and ETHPOP 2011

| *16 Harmonised Ethnic Groups* | *Census CD2011 Populations* (thousands) | *Average of TREND and UPTAPER Projections MY2011* (thousands) | *Difference = Average minus Census* (thousands) | *100 × (Difference/ Census)* (%) | *Difference Multiplier = (Census Change/ ETHPOP Change) 2001-2011* |
|---|---|---|---|---|---|
| All Groups | 56,076 | 56,057 | −18 | 0.0 | 1.00 |
| White British | 45,135 | 46,572 | 1,437 | 3.2 | 0.97 |
| White Irish | 531 | 663 | 132 | 24.8 | 0.80 |
| White Other | 2,544 | 2,374 | −169 | −6.7 | 1.07 |
| White and Black Caribbean | 427 | 331 | −96 | −22.5 | 1.29 |
| White and Black African | 166 | 132 | −34 | −20.2 | 1.25 |
| White and Asian | 342 | 299 | −43 | −12.6 | 1.14 |
| Other Mixed | 290 | 256 | −34 | −11.9 | 1.13 |
| Indian | 1,413 | 1,394 | −19 | −1.3 | 1.01 |
| Pakistani | 1,125 | 990 | −134 | −11.9 | 1.14 |
| Bangladeshi | 447 | 369 | −78 | −17.4 | 1.21 |
| Chinese | 393 | 361 | −33 | −8.3 | 1.09 |
| Other Asian | 836 | 355 | −481 | −57.6 | 2.36 |
| Black Caribbean | 595 | 641 | 46 | 7.8 | 0.93 |
| Black African | 990 | 754 | −236 | −23.8 | 1.31 |
| Other Black | 280 | 126 | −154 | −55.0 | 2.22 |
| Other Ethnic Group | 564 | 440 | −124 | −21.9 | 1.28 |

Source: Census 2011 – ONS (2013c), Crown Copyright.
Projections – ETHPOP (2013), funded by ESRC.
Notes: CD = Census date (27 March 2011), MY = Mid-YEAR (30 June/1 July).

partnerships may have increased from the 2001 level because more opportunities for mixing became available.

Within the Asian groups, the Indian group were only slightly under-projected (by 1%), but the Pakistani and Bangladeshi groups were under-projected by 12% and 17% respectively. It is likely we under-estimated the strength of continuing immigration associated with marriage for these groups and over-estimated the falls in their fertility. The Chinese group was under-projected by 8%; there was a growing influx of students from China from 2001, which we may have under-estimated. The Other Asian group, with varied origins in the smaller countries in Asia, was the most under-projected of all the groups at 58%. We likely under-estimated the inflows from crisis countries such as Iraq, Afghanistan, Iran and the student intake from emerging countries in South-East and East Asia (Malaysia, Thailand, Vietnam, South Korea).

The Black Caribbean group was over-projected by 8%. We may have under-estimated the return emigration stream to the West Indies. This is an important process for the older members who arrived in England and Wales in the 1950s and who entered the retirement ages in the 2000s. We under-projected the Black African group by 24%, probably under-estimating immigration. Immigration from sub-Saharan Africa is composed of increasing student numbers (e.g. from Nigeria and Ghana), flows of refugees and asylum seekers (e.g. from Somalia, Eritrea and Zimbabwe). The Other Black group is under-projected by 55%. People in other Black groups may have changed their identity between censuses and new groups may have started immigration in the 2000s from Latin America and the South West Pacific, for example. Finally, the residual group, the Other Ethnic Group was under-projected by 22%, reflecting the increasing diversity of origins of immigration to England and Wales.

Separate comparisons can be made for age groups (not reported here in detail). The patterns of over-projection (White groups) and under-projection (BAME groups) are repeated for children (ages 0–15), working age adults (ages 16–64) and the old (ages 65+). The relative differences from census populations are a little higher for the childhood ages than for all ages, suggesting some under-estimation of ethnic minority fertility. For the working ages, the relative differences are a little lower than for all ages but with the same pattern across groupings. For the older ages, we have over-projected the White groups by more than the total population and over-projected the Black groups as well. For the Asian grouping the older ages are closely projected but for the Mixed and Other Ethnic groups there is substantial under-projection. These differences suggest we have been too optimistic in our mortality decline assumptions and again failed to capture increasing immigration of Other Ethnic groups.

So far we have compared Census and ETHPOP results for England and Wales. Data are available for repeating this analysis across all 326 local authorities in England. Here we just look at the pattern for All Groups, the White British, Indian and Other Asian groups in a series of maps (Figure 10.2), based on population cartograms, using the method of Gastner and Newman (2004). The indicator used in the cartograms is the ratio of Census 2011 Population to ETHPOP 2011 Populations. Darker shades indicate that the Census population is higher than the ETHPOP projection, i.e. we have under-projected. Lighter shades indicate the Census population is lower than the ETHPOP projection, i.e. we have over-projected.

In the top-left map for All Groups, we see that the majority of the population lives in local authorities which are reasonably well projected (mid-grey). Most local authorities which have been under-projected are found in Greater London. However, Greater London also contains Boroughs in which the populations have been over-projected, including Westminster, Barking, Dagenham and Barking and Kingston upon Thames. It is likely that these differences stem from errors in the projection of an internal migration component between 2001 and 2011 or in the estimation of the immigration and emigration components between 2001 and 2005, before revisions improving reliability were introduced.

The top-right map presents the projection errors for the White British group. The most extreme values are seen in Great London, with over-projection in most London Boroughs, except for outer boroughs in the south east and south centre

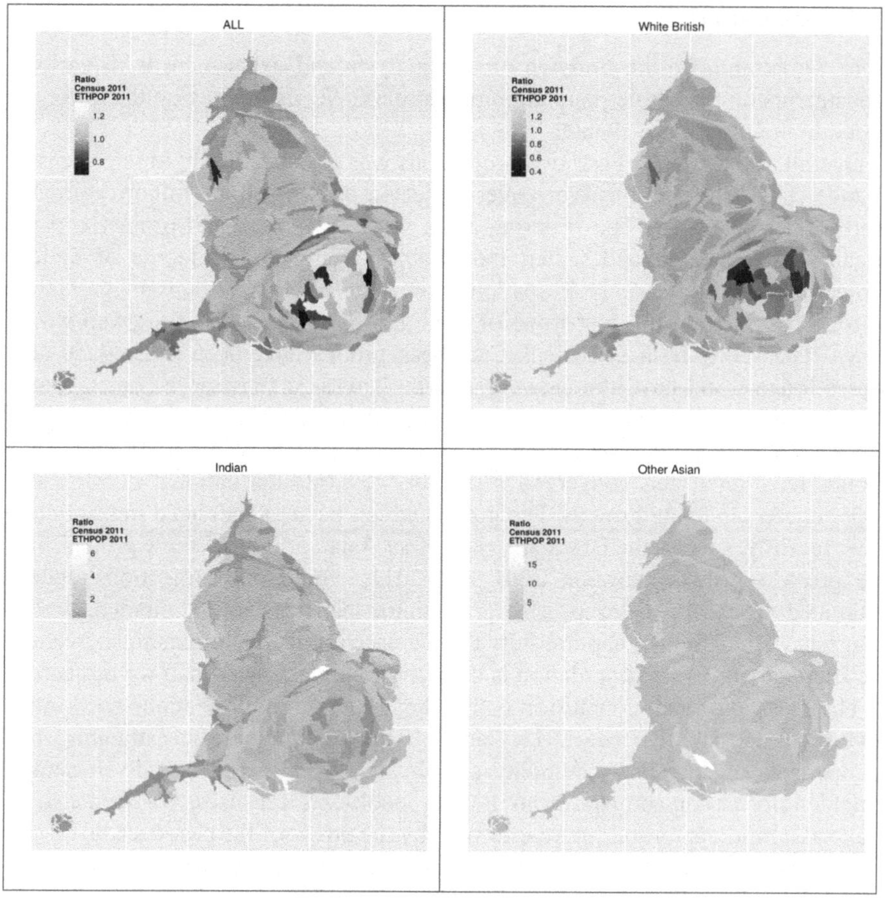


*Figure 10.2* Maps of local authorities showing over-projection (darker shades) and under-projection (lighter shades) for selected ethnic groups

of Greater London. Clearly, our projections have under-estimated the increase in ethnic diversity in London, in which the share of the White British population has shrunk. The Indian group (bottom-left map) was, by and large, well projected. There are some local authorities around the conurbation cores in which the group was under-projected (lighter shades). In some London Boroughs there was some over-projection, along with some remoter rural local authorities in Cornwall, Devon, Norfolk, Cumbria and the North East. It is likely that the internal migration pattern has shifted over the decade in ways we were unable to capture. Our projections perform worst for the Other Asian group (bottom-right map). The local authorities where the under-projection is greatest are in southern England outside Greater London. However, there are many local authorities in the Midlands and northern England as well where the Other Asian population was under-projected.

## 10.5 Discussion

Here we summarise the findings of our analysis and experiment with various adjustments of our ethnic population projections. We also reflect on the question posed in the title of the chapter.

Overall, the average of our two projections was aligned closely to inter-census change. The White groups were over-projected and the ethnic minority groups were under-projected. The ordering from least to most under-projected was: Asian, Black, Mixed and Other, though the differences in degree of under-projection were not large. Examination of the detailed ethnic groups suggests that as time passes from the first wave of immigration the growth of a group slows down. The White Irish and Black Caribbean groups have been in England and Wales longest and have the slowest growth. This is a function of much lower current immigration and ageing of the settled populations of the group. The biggest wave of South Asian immigration was in the 1960s and 1970s and these groups have aged and converged in fertility towards the national norm. They experienced moderate growth. The groups which have experienced immigration most recently, such as the Black African, Other Asian or Other Ethnic groups, had the greatest growth between 2001 and 2011, which our projections under-estimated. The Mixed groups also grew more than we had anticipated, which suggests we had did not capture fully the increase in mixed partnerships. Overall, the England and Wales population is diversifying much faster than we projected.

The most important conclusion is that the ethnic group projections need to be revised in the light of the 2011 Census. This will involve re-estimating the components of change by ethnicity and locality using a full decade of demographic information and data from the two 'book-end' censuses. We need also to make assumptions for the future informed by the errors of the past decade, developing fully the transition theory suggested in the previous paragraph. This work is underway in an ESRC-funded project (Rees et al. 2015a).

Is there any way to fix our ETHPOP projections until revisions can be effected? Table 10.5 presents some alternatives applied to the 2051 average of the TREND and UPTAPER projections (second column). The third column applied the difference ratio for 2001–2011 given in the last column of Table 10.4 to the 2051 projected populations over five decades. The results are implausible: the White British population is halved to 27 million and the Other Asian population grows to 25 million. More realistic are the adjustments in the fourth column which add or subtract five times the 2001–2011 error. The results are more plausible but it is unlikely that the same errors will occur over the next four decades, as happened in the previous decade. The fifth column of Table 10.5 simply uplifts the 2051 projected populations by the 2011 error, which is probably too conservative an adjustment. We have carried out further experiments with uprating the ETHPOP projections using knowledge of the 2001–2011 differences for the Policy Exchange (Rees and Clark 2014) and as part of a PhD thesis (Clark 2015).

So, the answer to the question posed in the title of this chapter is that 'our multi-ethnic future' is arriving a good deal faster than we thought. There is an urgent

*Table 10.5* Alternative corrections to the ETHPOP 2051 average projection, England and Wales

| *16 Harmonised Ethnic Groups* | *ETHPOP populations Average 2051* (thousands) | *ETHPOP populations multiplied by the multiplier 5 times and adjusted to All groups total* (thousands) | *ETHPOP populations uplifted by 5 × difference 2001-2011* (thousands) | *ETHPOP populations uplifted by difference 2001-2011* (thousands) |
|---|---|---|---|---|
| All Groups | 74,477 | 74,600 | 74,569 | 74,496 |
| White British | 57,604 | 27,008 | 50,417 | 56,167 |
| White Irish | 1,540 | 279 | 881 | 1,408 |
| White Other | 3,974 | 3,074 | 4,819 | 4,143 |
| White and Black Caribbean | 720 | 1,411 | 1,200 | 816 |
| White and Black African | 308 | 522 | 476 | 341 |
| White and Asian | 668 | 717 | 883 | 711 |
| Other Mixed | 563 | 581 | 735 | 598 |
| Indian | 2,380 | 1,395 | 2,474 | 2,399 |
| Pakistani | 1,916 | 1,984 | 2,587 | 2,050 |
| Bangladeshi | 669 | 953 | 1,057 | 746 |
| Chinese | 619 | 523 | 782 | 651 |
| Other Asian | 632 | 25,113 | 3,036 | 1,112 |
| Black Caribbean | 754 | 284 | 522 | 707 |
| Black African | 1,211 | 2,592 | 2,391 | 1,447 |
| Other Black | 231 | 6,857 | 1,002 | 385 |
| Other Ethnic Group | 691 | 1,306 | 1,309 | 815 |
| Non-White British | 16,873 | 47,591 | 24,152 | 18,329 |

Sources: Census 2011 – ONS (2013c), Crown Copyright.

Notes: The multiplier for 2001–2011 is shown in the sixth column of Table 10.4. The difference is shown in the fourth column of Table 10.4.

need to revise the ethnic group projections, using proper cohort-component methods rather than extrapolations. Plans for revising the model, estimates and assumptions for a set of 2011 Census-based ethnic projections are set out in Rees et al. (2015a and 2015b). In brief, these new projections will adopt a movement rather than a transition perspective, handle internal migration as origin-destination flows, propose new geographically weighted methods for estimating ethnic mortality, update the estimates of ethnic fertility and produce a better specified set of assumptions and projections. One illustration of the need for up-to-date statistics on ethnic population data by age and sex is the ongoing 'Stop and Search' controversy in England and Wales. There are concerns that Black and Asian Minority people are being stopped and searched by police forces far more than White people (Morris 2015). The statistic used for this assessment is the ratio of stops and searches by ethnicity between December 2014 and April 2015 divided by the 2011 Census population. The population data refer to 3.5 years before the stops and no attempt is made at age standardisation. The debate about whether police forces are abusing their powers is being conducted with inadequate data.

## Notes

1 The ETHPOP projections are for the UK but we focus here on England and Wales where ethnic group definitions can be harmonised easily.
2 Two local auhthorities with very small populations are merged with a larger neighbour: City of London with City of Westminster and Isles of Scilly with Penwith.
3 The boundaries were supplied by Bethan Thomas of the University of Sheffield.

## References

Clark, S. (2015) Modelling the impacts of demographic ageing on the demand for health care services, PhD dissertation, School of Geography, University of Leeds. Online at: http://etheses.whiterose.ac.uk/11676/

Coleman, D. (2006) Immigration and ethnic change in low-fertility countries: a third demographic transition, *Population and Development Review* 32, 401–446.

EHRC (2013) *Equality Act.* Equality and Human Rights Commission. Online at: www.equalityhumanrights.com/legal-and-policy/equality-act/.

ETHPOP (2013) ETHPOP database. Online at: www.ethpop.org/.

Frey, W.H. (2015) *Diversity explosion: how new racial demographics are remaking America.* Brookings Institution Press, Washington, DC.

Gastner, M.T. and Newman, M.E.J. (2004) Diffusion-based method for producing density-equalizing maps, *Proceedings of the National Academy of Sciences of the United States of America*, 101(20): 7499–7504.

GLA (2014) *GLA demographic projections*. Greater London Authority. Online at: http://data.london.gov.uk/dataset/gla-demographic-projections.

Jivraj, S. (2012) How has ethnic diversity grown 1991–2001–2011? Briefing in the *Dynamics of Diversity: Evidence from the 2011 Census* series, Prepared by the ESRC Census on Dynamics of Ethnicity (CoDE). Online at: www.ethnicity.ac.uk/census/869_CCSR_Bulletin_How_has_ethnic_diversity_grown_v4NW.pdf.

Jivraj, S. and Simpson, L. (2015) *Ethnic identity and inequalities in Britain: the dynamics of diversity*. Policy Press, Bristol.

Kupiszewski, M. and Kupiszewska, D. (2010) *Reference scenarios, Deliverable 5.* DEMIFER: Demographic and Migratory Flows affecting European Regions and Cities. The ESPON 2013 Programme. Online at: www.espon.eu/export/sites/default/Documents/Projects/AppliedResearch/DEMIFER/FinalReport/DEMIFER_Deliverable_D5_final.pdf.

Morris, N. (2015) Stop and search: nearly every police force is guilty of ‘racism’, *The Independent*, Friday 7 August, pp.1, 10–11.

OBR (2012) *Fiscal sustainability report, Annex B*. Office for Budget Responsibility, London. Online: http://cdn.budgetresponsibility.independent.gov.uk/FSR2012WEB.pdf.

OBR (2013a) *Fiscal sustainability report, Annex B*. Office for Budget Responsibility, London. Online: http://budgetresponsibility.org.uk/wordpress/docs/2013-FSR_OBR_web.pdf.

OBR (2013b) Migration variants and net debt. Slide 38, in Chote, R. (2013) Fiscal sustainability report 2013. Presentation. Office for Budget Responsibility, London. Online at: http://budgetresponsibility.org.uk/wordpress/docs/FSR-presentation2013.pdf.

ONS (2011a) *Population estimates by ethnic group 2002–2009.* Date: 18 May, Coverage: England and Wales, Theme: Population. Online at: www.ons.gov.uk/ons/rel/peeg/population-estimates-by-ethnic-group–experimental-/current-estimates/index.html.

ONS (2011b) *Assessment of reliability of the population estimates by ethnic group.* Online at: www.ons.gov.uk/ons/rel/peeg/population-estimates-by-ethnic-group–experimental-/current-estimates/index.html.

ONS (2011c) *Population estimates by ethnic group: important note on reliability of estimates for subnational areas.* Online at: www.ons.gov.uk/ons/rel/peeg/population-estimates-by-ethnic-group–experimental-/current-estimates/index.html.

ONS (2013a) *Annual Population Survey (APS).* NOMIS Official Labour Market Statistics. Online at: www.nomisweb.co.uk/articles/676.aspx.

ONS (2013b) *Statistical bulletin: migration statistics quarterly report, August 2013.* Online at: www.ons.gov.uk/ons/rel/migration1/migration-statistics-quarterly-report/august-2013/msqr-august-2013.html.

ONS (2013c) 2011 *Census data for England and Wales on Nomis.* Online at www.nomisweb.co.uk/census/2011.

Rees, P., Wohland, P., Norman, P. and Boden, P. (2011) A local analysis of ethnic group population trends and projections for the UK, *Journal of Population Research* 28, 149–184.

Rees, P., Wohland, P., Norman, P. and Boden, P. (2012a) Ethnic population projections for the UK, 2001–2051, *Journal of Population Research*, 29(1): 45–89. DOI: 10.1007/s12546-0111-9076-z.

Rees, P., Wohland, P. and Norman, P. (2012b) The demographic drivers of future ethnic group populations for UK local areas 2001–2051, *Geographical Journal*, 179(1), 44–60. DOI: 10.1111/j.1475-4959.2012.00471.x.

Rees, P. and Clark, S. (2014) *The projection of ethnic group populations aged 18 and over for Westminster Parliamentary Constituencies in Great Britain for election years 2015, 2020, 2025, 2030 and 2035.* A Report to the Policy Exchange, 10 Storey's Gate, London SW1P 3AY. School of Geography, University of Leeds.

Rees, P., Wohland, P., Norman, P. and Lomax, N. (2015a) *Evaluation, revision and extension of ethnic population projections – NewETHPOP.* Funded by ESRC, Grant Ref ES/L013878/1, 1 Jan 2015 to 31 May 2016.

Rees, P., Wohland, P., Norman, P., Lomax, N. and Clark, S. (2015b) Ethnic population projections for the UK and local areas, 2011–2101: a second chance to get them right. Seminar presented at the Australian Demographic and Social Research Institute, Australian National University, Canberra, Tuesday 7 July.

United Nations (2014) *World population prospects: the 2012 revision.* Methodology of the United Nations, Population Estimates and Projections. Publication ESA/P/WP.235, United Nations, New York. Online at: http://esa.un.org/Wpp/Documentation/pdf/WPP2012_Methodology.pdf.

United Nations (2015) *World economic situation and prospects 2015.* United Nations, New York. Online at: www.un.org/en/development/desa/policy/wesp/.

van de Kaa, D. (1987) Europe's second demographic transition, *Population Bulletin* 42(1), 64pp.

Wohland P., Rees P., Norman P., Boden P. and Jasinska M. (2010) *Ethnic population projections for the UK and local areas, 2001–2051.* Working Paper 10/02, School of Geography, University of Leeds, June 2010. Online at: www.geog.leeds.ac.uk/fileadmin/documents/research/csap/10-02.pdf.

Wohland, P., Norman, P. and Rees P. (2014) *ETHPOP database* (projected populations, presentations and publications). Online at: www.ethpop.org/.

# 11 Decomposition of life expectancy at older ages and prospects for ageing populations[1]

*Leslie Mayhew and David Smith*

## 11.1 Introduction

One of great success stories in the United Kingdom (UK) is that people are living longer. Male life expectancy at birth is now almost 80 years, having advanced 14 years since 1950, thanks to improvements in health and wellbeing, fewer accidents, better health care and higher standards of living. This success in turn presents the country with a huge economic opportunity if extra years are spent in prosperity and good health, but economic danger if not.

Indeed, realising the full potential of older citizens will be central to the government's response to changing economic circumstances and the drive to build a strong, fair economy for the twenty-first century. However, the challenges posed by an ageing society come at a cost in terms of pensions, the higher cost of health and social care, and infrastructural change. Such changes affect the wider economy, public expenditure and taxes, which impacts on individuals, especially in younger generations. These effects manifest in obvious ways, such as having to work until you are older, increases in pension age, higher taxes, more doctor visits, and so on.

It is becoming increasingly clear, through recent policy changes, that people will be required to take reasonable steps in planning for their financial needs in old age and to become less dependent on the state. This new era will mean facing up to an increasingly broader range of personal circumstances. For instance there will be people with quite different mixes of income and assets or family circumstances upon which to draw for care needs and in different areas of the country.

These changes find increasing resonance in a range of government policies from pensions to social care, especially in terms of cost curtailment and transfers of responsibility from government to individuals. The old assumption that the state would look after citizens from cradle to grave can no longer be relied upon, although its demise has never been officially confirmed. An ageing population coupled with shifting government policy resonates with applied geography and spatial modeling in different ways.

First, geographers should beware of taking official forecasts of the older population at face value, as this chapter will show. Rapidly rising life expectancy will mean new coping mechanisms are required to help individuals and families understand the true costs of ageing. It will mean, for example, that available resources, such as pension savings, need to last longer. There will be spatial adjustments in habitation patterns, with not only many extra older households, but also whole communities of older people. The increasing uncertainty over how long a population or an individual will live will impact services affecting the whole care economy, types of employment and also services. Geography is well equipped to address the details but only if it is able to forecast ageing populations with reasonable accuracy and the social and spatial ramifications are understood.

The problem is that life expectancy is not a deterministic process or as predictable as the grains of sand in an egg timer. Rather, it is derived from an average of the experiences of deceased individuals in a period of time, or the experiences of cohorts of individuals born in certain years. In practice, it is easier to predict life expectancies for populations than it is for individuals as some randomness is removed. Even so, we know that official forecasts of life expectancy have underestimated the rates of improvement, often spectacularly, with consequent impacts on medium- to long-term government planning (e.g. Booth 2006; Shaw 2007; Mayhew and Smith, 2012).

Improvements in life expectancy at older ages are occurring at a fast pace but have particularly accelerated in recent decades. Figure 11.1 shows the trend in life expectancy for 60-year-old men and women for England and Wales since records began in 1841. It shows a steady growth for both sexes, especially after 1900, though the improvement for women has been faster. The gap between men and women widened from under one year in the 1800s before plateauing at just over 4.5 years in the 1970s. Since then, the gap between the sexes has narrowed considerably and now stands at just under three years. This is expected to narrow further or even converge before 2030 if current trends are extrapolated; moreover, this trend applies to other start ages, not only aged 60 (e.g. see Mayhew and Smith 2014, 2015).

There are several reasons for this turnaround in male fortunes, among the most notable of which is the demise in male smoking habits and shifts in employment

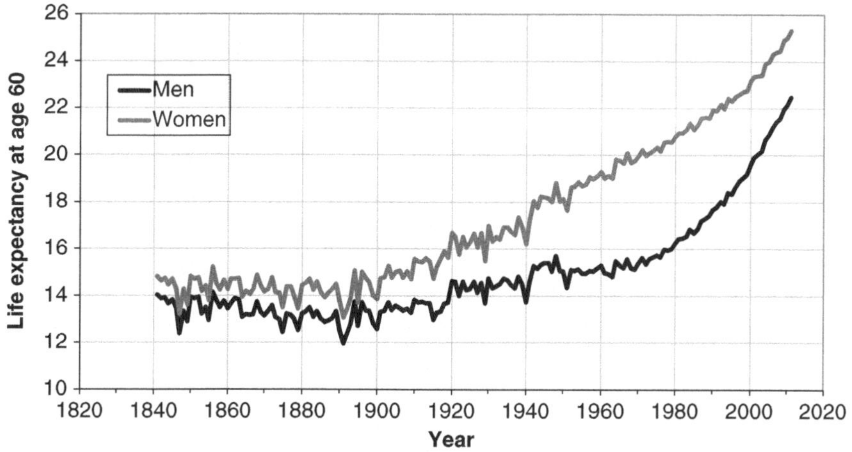


*Figure 11.1* Changes in male and female life expectancy in England and Wales at age 60 from 1841 to 2009

Source: Human Mortality Database (2009)

patterns away from hazardous occupations (e.g. see Preston and Wang, 2006; Murphy and Di Cesare, 2012; or Trovato 2005). Another factor affecting both sexes has been the reduction in year-on-year fluctuations caused mainly by wars, epidemics and cold winters. This has meant that short-term trends in life expectancy have become easier to predict, but it still leaves three important unanswered questions: what will life expectancy be in the medium term (10–20 years, say)? Will it go on rising indefinitely? And what will be the variation in age of death (Mayhew and Smith 2012, 2016)?

The corollary is that if life expectancy does continue to grow, in which age groups will the growth come – people in their 60s, 70s or older, or even centenarians say? In short, the need for better information about life expectancy is demonstrable at both the population and individual level – for framing government policies to planning personal finances and building new communities. In addition, policies must be durable, especially anything to do with pensions, health and social care, or housing. In this chapter, we present a new method for forecasting life expectancy based on decomposition techniques. The key advantages are more certainty over which age groups are affected and more accurate information about possible limits to life. We call this the 'jam-jar model' of life expectancy, for reasons which are explained later.

In what follows, section 11.2 sets out the methodology, while in section 11.3, we provide our main results. In particular, we demonstrate how regularities in ten-year life expectancies can be used to forecast future life expectancy and discuss the assumptions and limitations of the method therein. An important feature of our approach is the exploitation of a 'natural ceiling in the data', progress towards which is statistically calibrated using a logistic function. In the concluding section, we discuss the implications of our findings in the wider context of society and in terms of other methods for forecasting life expectancy.

## 11.2 Methodology

It is standard practice to measure life expectancy at different ages, e.g. from birth, age 30, or some other age. For example, life expectancy for England and Wales (E&W) males at birth in 2009 was 78.04 years (Human Mortality Database 2015), suggesting that at age 60 there should be 18.04 years of life remaining. However, the data show that life expectancy at age 60 was actually 22.04 years because of survival and selection effects. We, therefore, seek a method for calculating life expectancy within discrete age limits that addresses this anomaly. The aim is to allow us to reassemble component life expectancies to produce estimates of whole life expectancy over any desired age range.

Hickman and Estell (1969) proposed a similar idea, which they termed 'partial life expectancies'. In their context, they argued that partial life expectancies are relevant to discussions on the economic costs of illness, since 'partial life expectancies may be related to the ages at which the economic contributions … are usually greater' (1969: 2244). Pollard (1982), posing a different question, showed that the change in expectation of life can be expressed as a weighted function of mortality changes at individual ages plus interaction effects. Arriaga (1984) develops this further, setting out the basic equations for measuring temporary expectancies using a discrete life table approach rather than the continuous methods of Pollard.

In this chapter, we proceed similarly, but our aims are different. Our definition of life expectancy, is similar to Arriaga's temporary life expectancy but our focus is on trends in life expectancy within specified age intervals. In particular, a technique is presented that gives the expectation of a person aged 60, 70, 80, 90 etc. reaching ages 70, 80, 90, 100, etc. Under our approach, a ten-year life expectancy at age 60, for example, means that everyone survives to age 70; a two-year expectancy at 60 means that a person can only expect to live two of a possible ten years, and so on. It should thus be easier to pinpoint in which age intervals future increases in life expectancy are more likely to occur by fitting bounded functions to trends in the data from any start age, whether at birth or some later age, in our case 60. Appendix A sets out the decomposition methodology in detail.

The results, discussed in the next section, strongly indicate that contributions to life expectancy have transferred in a predictable wave-like fashion to older ages as opposed to a process in which each age interval has contributed equally. For example, in 1950 when male life expectancy at 60 in E&W was 15.1 years, the age range 80–90 only contributed 9.1% to this figure, but by 2009, when life expectancy was 22.1 years, they contributed 18.5%. In fact, we can imagine each decade of life as a 'jam-jar' which fills to the brim with life years, with extra life years being added to each decade's jam-jars at different rates, filling the early ones first, until all are full.

## 11.3 Results

Table 11.1 shows the results of applying decomposition to male and female life expectancy at age 60 in England and Wales. The contribution from each decade of life is given in the rows for each given calendar year starting in 1950. Column totals give overall life expectancy at age 60 calculated by adding together the ten-year expectancies above. A final column shows the gain in years over a 60-year time span from 1950 to 2010.

Although in earlier decades improvements in life expectancy for women were greater than for men, the results show an overall seven-year gain for both men and women in the period under review. For men, gains between ages 60 and 70 have been less than one year for the key reason that the likelihood of reaching age 70 is already high and close to the ten-year limit. Most of the gains have been made between ages 70 to 80 and 80 to 90 at 2.6 and 2.7 years respectively, where there has been most scope for improvements. A similar story applies to women, except that more of the gains have occurred in the 80 to 90 age bracket than between 70 to 80.

Looking along the rows of each table, we observe a gradually increasing contribution across the decades, but once we reach ages above 90, the rate of progress slows considerably. For the age bracket 100+ there are no significant gains at all, with the exception of women in 2010. This raises the important question of

*Table 11.1* Male and female ten-year life expectancies from age 60 for England and Wales: 1950 to 2010

(a) Men

| *age range* | *1950* | *1960* | *1970* | *1980* | *1990* | *2000* | *2010* | *gain 1950 to 2010 (years)* |
|---|---|---|---|---|---|---|---|---|
| 60–70 | 8.6 | 8.6 | 8.7 | 8.8 | 9.0 | 9.3 | 9.5 | 0.9 |
| 70–80 | 5.0 | 5.0 | 5.0 | 5.5 | 6.0 | 6.8 | 7.6 | 2.6 |
| 80–90 | 1.4 | 1.5 | 1.5 | 1.8 | 2.3 | 3.1 | 4.1 | 2.7 |
| 90–100 | 0.1 | 0.1 | 0.1 | 0.2 | 0.3 | 0.5 | 0.9 | 0.8 |
| 100+ | 0.0 | 0.0 | 0.0 | 0.0 | 0.0 | 0.0 | 0.0 | 0.0 |
| total | 15.1 | 15.2 | 15.3 | 16.3 | 17.6 | 19.7 | 22.1 | 7.0 |

(b) Women

| *age range* | *1950* | *1960* | *1970* | *1980* | *1990* | *2000* | *2010* | *gain 1950 to 2010 (years)* |
|---|---|---|---|---|---|---|---|---|
| 60–70 | 9.2 | 9.3 | 9.3 | 9.4 | 9.4 | 9.6 | 9.7 | 0.5 |
| 70–80 | 6.4 | 6.8 | 7 | 7.2 | 7.5 | 7.9 | 8.4 | 2.0 |
| 80–90 | 2.3 | 2.8 | 3.2 | 3.6 | 4.1 | 4.6 | 5.4 | 3.1 |
| 90–100 | 0.2 | 0.4 | 0.5 | 0.6 | 0.9 | 1.1 | 1.5 | 1.3 |
| 100+ | 0.0 | 0.0 | 0.0 | 0.0 | 0.0 | 0.0 | 0.1 | 0.1 |
| total | 18.1 | 19.3 | 20.0 | 20.8 | 21.9 | 23.2 | 25.1 | 7.0 |

whether we can ever expect future increases in that age group; in the meantime, it is evident that most future gains will occur in younger age bands which will, using the jam-jar analogy, 'fill up' to their maximum extent more rapidly.

### *11.3.1 Fitting trend lines to ten-year life expectancies*

The picture presented above suggests a systematic shift in survival at higher ages. Having calculated ten-year life expectancy for each decade of life, it is important to know how longevity will behave in the future and when it will reach its natural limit in each decade of life. As is evident from Figure 11.1, simply projecting existing trends does not have the right mathematical properties to perform this role because it is unbounded and, hence, could reach uncharted levels within a few years.

Any fitted trend needs to be a non-decreasing function (if we are modelling Western economies where we expect life expectancy to continue to increase with time). We also need to ensure there is a natural limit to life expectancy, meaning that we do not project infinite life spans. By working with ten-year age segments, we know there is a definite limit within each age interval and so we can concentrate on age intervals still showing scope for growth. If growth trends are well established, it should be possible to test functions that are good fits to the data but also trend towards this natural upper limit.

We therefore proceed as follows. Life expectancy in ten-year age bands is calculated using observed data for each calendar year. The results are presented in Figure 11.2, in which each set of data points represents one of five ten-year age bands starting at age 60 for each year since 1950. The dotted extensions are trend

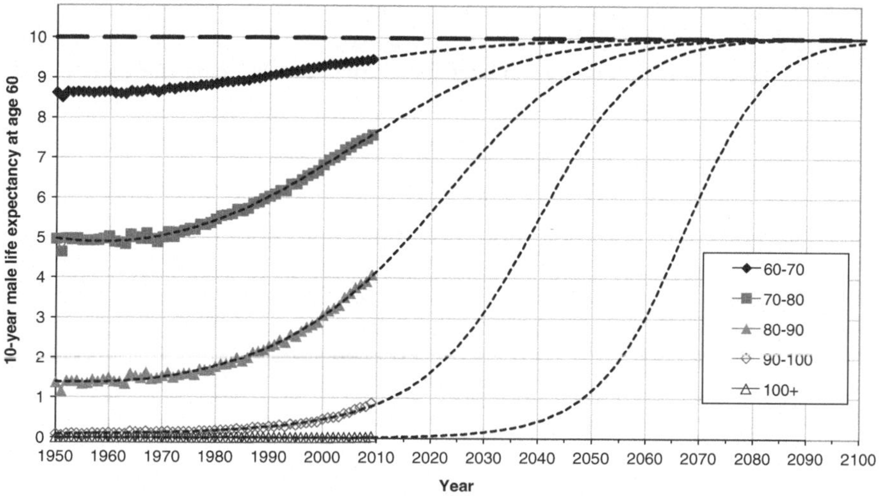


*Figure 11.2* Trends in England and Wales male life expectancy by ten-year segments from age 60. Dotted line shows extrapolated trends based on function described in text

lines based on the procedure, which is outlined and justified below. For each band, we have assumed that the maximum life expectancy of ten will be reached asymptotically over time, although clearly this would imply, for reasons already given, the elimination of all deaths if projected over an infinite period of time.

For the youngest age band (ages 60 to 70), the impact of the limit is minor as the life expectancy is already at or close to ten, and has been for many years. For older age bands, the true limit may, in reality, be less than ten and hence an element of uncertainty will always remain as we project forward; however, this is expected to have only a minor impact on short- to medium-term projections, which are likely to be of most use in practice. At older ages there are signs of more rapid progress, which supports the argument for retaining the current limit rather than reducing it from ten.

### *11.3.2 Basic projection model*

Long-range projections of life expectancy are obviously sensitive to both the projection methodology and assumptions. As already noted, we sought a function for which we could impose a limit that could not be breached, rather than a function that finds its own limit. Mayhew and Smith (2012) discuss the problem of demographic agencies imposing arbitrary limits to life expectancy, which subsequently proved to be too low. By allowing the data to impose the trend, we showed that forecasts could have been better. However, we could also see our predictions starting to become unlikely post-2030 as the trend of life expectancy increased at an increasing rate.

By conceptualising life as a series of ten-year blocks, we are not forcing the data to fit this limit through our judgement, but rather it is a natural limit based on the design of our model. We are then able to chain-link expectations of life together to get an aggregate life expectancy projection built on the component parts. Because of its convenient properties, and the ease with which the parameters can be estimated, the most useful function we tested is a form of the logistic function which can be written as follows:

$$y_i = \frac{Ae^{f_i(t,n)}}{1+e^{f_i(t,n)}} \tag{11.1}$$

Where $f_i(t, n)$ is a polynomial equation of order $n$, which generally takes the value of one or two. That is:

$$f_i(t) = a_i + b_i t \tag{11.2}$$

or

$$f_i(t) = a_i + b_i t + c_i t^2 \tag{11.3}$$

where $a_i$, $b_i$ and $c_i$ are parameters to be estimated, $t$ is calendar year, and $y$ is life expectancy in age interval $i$. $A$ is a constant defined by the user taking all age intervals to be equal and in this case is set to 10 years.

Re-arranging yields:

$$\ln\left[\frac{y_i}{A - y_i}\right] = f_i(t) = a_i + b_i t \tag{11.4}$$

or

$$\ln\left[\frac{y_i}{A - y_i}\right] = f_i(t) = a_i + b_i t + c_i t^2 \tag{11.5}$$

In these equations, $a_i$, $b_i$ and $c_i$ are estimated using multiple linear regression. Note that higher order polynomials ($n > 2$) could also be considered depending on the patterns observed in practice. By fitting this model to data in each age band from 1950 to 2009, we analysed how future life expectancy may be expected to develop, assuming the underlying trend is unchanged. In low mortality countries, such as those considered here, we found that first or second-order polynomial equations generally performed best. In the case of England and Wales, we used second-order polynomials as these gave the best fit to the data.

By fitting the model to data in each age band, we analysed how future life expectancy may be expected to develop assuming the underlying trend is unchanged. Table 11.2 shows the regression parameters for both sexes. All coefficients are statistically significant at the 99% level of probability. As can be seen, values of $R^2$ are mostly above 0.95, indicating very good statistical fits to the data. Returning to the jam-jar analogy, on combining the life expectancies in each 'jam-jar' we arrive at total life expectancy at age 60, which is represented in the chart

*Table 11.2* Fitted parameter values for the projection model applied to data for England and Wales for men and women

(a) Men

| *Age band* | α | β | $\beta^2$ | $R^2$ |
|---|---|---|---|---|
| 60–70 | 1,506.04 | −1.53753 | 0.00039289 | 0.995 |
| 70–80 | 1,790.56 | −1.82854 | 0.00046682 | 0.995 |
| 80–90 | 1,909.54 | −1.95528 | 0.00050005 | 0.989 |
| 90–100 | 1,934.80 | −1.99756 | 0.00051432 | 0.974 |
| 100+ | 1,246.23 | −1.33228 | 0.00035287 | 0.947 |

(b) Women

| *Age band* | α | β | $\beta^2$ | $R^2$ |
|---|---|---|---|---|
| 60–70 | 830.93 | −0.85020 | 0.00021814 | 0.971 |
| 70–80 | 630.84 | −0.65222 | 0.00016874 | 0.980 |
| 80–90 | 327.28 | −0.35180 | 0.000094043 | 0.985 |
| 90–100 | 0.80 | −0.03391 | 0.000016223 | 0.974 |
| 100+ | −540.43 | 0.49157 | −0.00011207 | 0.947 |

in Figure 11.3 and Table 11.3. This shows how each 'jam-jar' is in the process of filling up, starting in 1950, and with extra years (i.e. more 'jam') being added between 1950 and 2010. It is noteworthy that the projected added years between 2010 and 2030 will be greater than the added years from 1950 to 2010, emphasising the rapidity of the process. The unshaded tips of the columns show the available years for improvement in each decade of life post-60. Once ten years is reached, there are no available years left and so the 'jam-jar' is full, but the chart also shows there are still many available years left in the 80 to 90 and 90 to 100+ age range.

The extrapolated trends in Table 11.3 and Figure 11.4 are obtained by adding together the forecast in each age bracket. They show that male (and female) life expectancy could continue to grow rapidly in coming decades, particularly in certain age brackets. Clearly, as with any forecasting technique, the process of projection becomes more speculative the further ahead one looks, but we believe

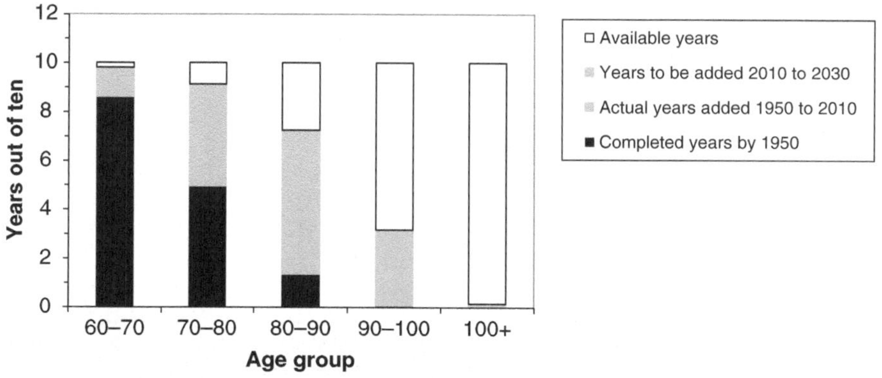


*Figure 11.3* Jam-jar model of life expectancy at age 60 (England and Wales, men)

*Table 11.3* Completed and available years by age interval and calendar period (men)

| *Age interval* | *1950 (completed years)* | *added years between 1950 and 2010* | *added years between 2010 and 2030* | *available years* |
|---|---|---|---|---|
| 60–70 | 8.6 | 0.8 | 0.3 | 0.2 |
| 70–80 | 5.0 | 2.6 | 1.5 | 0.9 |
| 80–90 | 1.4 | 2.7 | 3.2 | 2.7 |
| 90–100 | 0.1 | 0.8 | 2.3 | 6.8 |
| 100+ | 0.0 | 0.0 | 0.1 | 9.8 |

it is realistic for these trends to maintain their course until at least 2040 – not withstanding any unexpected shocks, such as pandemics or wars.

Our extrapolations so far are clearly based on a normative assumption that a person can only have a maximum life expectancy of 110 years (i.e. 60 + 50 = 110), but any presumed age horizon and whether people will survive beyond it remains an open question. Some argue, for example, that the limit could be higher (e.g. Wilmoth 2000; Vaupel 2010). This uncertainty is plainly truer at higher ages, where ten-year life expectancy has advanced least, but it may take decades to establish if this view is correct or not, which is why there is an interest in studying super centenarians (those aged 110 or older).

For super centenarians, the Human Mortality Database truncates data due to the tiny number of survivors, so no extrapolation is possible using our model at present. It is therefore important that the upper limit is not regarded as an absolute, but one which can be altered in the light of any new evidence and updated accordingly. The implication from Figure 11.4, with the data at our disposal, is

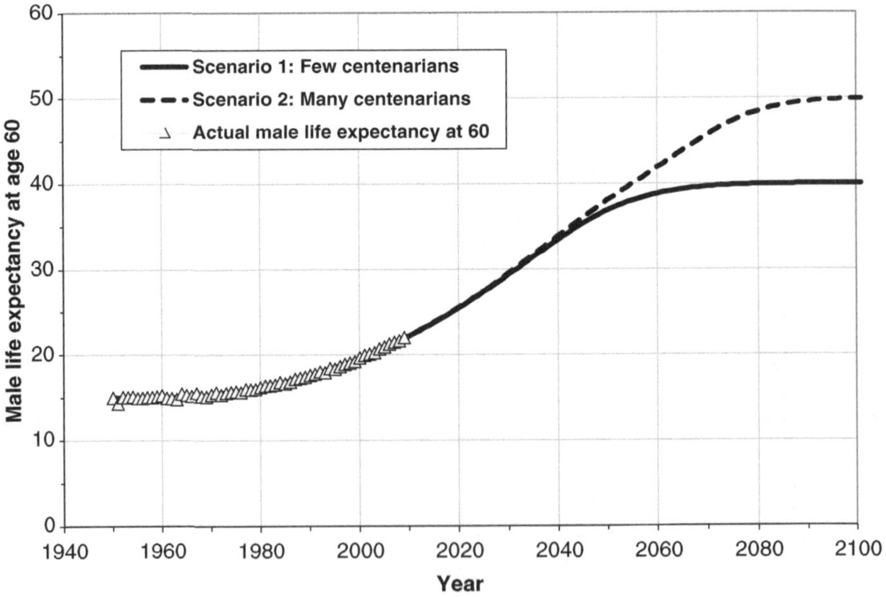


*Figure 11.4* Projected male life expectancy at age 60 for England and Wales

that life expectancy at age 60 would start to level out after 2050, although this is clearly speculative as there are two possible scenarios. The first is that there will not be any appreciable growth in the number of centenarians and the second is that there will be! This results in two limits and not one.

Note also that if the question is 'what is life expectancy in the age interval 60 to 100?' rather than 'what is the upper limit?', then including occurrences of survival above 100 becomes redundant. Because of the paucity of data from age 100 and above, this is potentially a more reliable way of comparing future life expectancy in different countries (see below). Arguably, it is the forecasts of life expectancy in intermediate decades that are of more practical value for demographers and governments.

## 11.4 Discussion and conclusions

If the process of ageing described is correct, then it will also be accompanied by a convergence in age of death. This will give individuals, government policy makers, pension providers, and insurers more certainty with which to plan. This in turn could arguably be beneficial to individuals and to society as a whole, but there are downsides as well as upsides to this suggestion. For example, the extra years from age 80 to 90 are important because it depends on whether they are spent in good health. However, the jury is still out on whether extra years of life expectancy are actually being spent in good health or in disability.

In fact, the decade of life from 80 to 90 is fast becoming the new pivotal years in terms of care provision, which can be characterised in terms of four types of older household. These depend on whether it is a couple household or single dweller household and on whether there are care needs or not. From our results, life expectancy in this age segment has grown by nearly three years for both genders and that the genders are starting to converge. Depending on the age difference, we know that couple households will generally be more resilient than single dwellers, especially if the single dweller is disabled and hence more likely to require state support and intervention.

It follows that trends in the prevalence of each household type could become a key driver for determining future care needs. The fact that male life expectancy is catching up with female life expectancy, and that both are living longer, suggests that older couple households will grow in prevalence and remain in their homes for longer. In turn, this will impact on housing supply and, hence, house prices unless house building keeps abreast of demand. This could point to a new policy opportunity for Government, namely to provide financial inducements to downsize at younger ages, for example, when the children reach adulthood.

For geographers and planners there are, hence, a number of important challenges as well as implications. Estimates of life expectancy will depend on factors such as gender, where people live, and also lifestyles – especially whether a person smokes – and also risk factors, such as levels of obesity. We know that these can vary by socio-economic group, which often gives rise to spatially distinctive patterns in towns and surrounding areas (indeed, postcodes are commonly used as risk factors by insurers). The migration patterns of older people are often triggered by life-cycle factors such as retirement, the death of a partner, or moving into retirement village or care home. Geographers and planners need to take notice of these trends as they impact on local housing markets and employment structures.

Will decomposition be helpful for forecasting populations? Currently, our decomposition methodology produces a single forecast of life expectancy for each age band. This essentially deterministic approach could be supplemented using regression confidence intervals, which would show ranges of uncertainty. This is likely to be most useful at higher age intervals for obvious reasons. Stochastic formulations are another possible direction for development. Because the method is new, it is too early to compare its forecasting potential with the greatly expanding literature on demographic forecasting based either on life expectancy or mortality (e.g. for reviews see Booth 2006; or Rafferty et al. 2013). Conceptually, however, there are clear differences of approach.

Our model is based on trends in what were termed 'partial life expectancy', from which mortality rates may be derived as required. It requires a long-time series and unabridged life tables for best results and so far, has only been tested on low mortality countries. In the widely used Lee-Carter model, it is the other way around, with trends in mortality rates forming the starting point (Lee and Carter 1992). Whereas Lee-Carter is stochastic, our model is able to predict turning points using the logistic function and exploits natural age limits found in each age segment. This is an important difference since, as Bongaarts (2005) notes, the invariance in the rate of decline of mortality assumed by Lee-Carter produces implausible results after just a few decades.

Using the logistic function, Bongaarts is able to project the force of mortality more accurately than the Lee-Carter method. Nevertheless, drawing general conclusions of the relative merits of each approach, and comparing it with our own, is somewhat complicated and also, arguably, premature because of the many Lee-Carter variants available (e.g. see Renshaw and Haberman 2003, 2006; or the 'rotation' approach in Li et al. 2013). In contrast to Lee-Carter, deterministic methods are widely used for forecasting life expectancy as opposed to mortality, with the United Nations, which produces world estimates by country, being a notable example of this practice (United Nations, 2009). More recently, Rafferty et al. (2013) have proposed a stochastic UN variant for which superior accuracy is claimed.

Both the United Nations (2009) and Rafferty et al. (2013) base their forecasts on the double logistic function. This assumes life expectancy will decline over a period, before continuing asymptotically at a linear rate. Our method also uses a logistic variant to forecast life expectancy but within discrete ten-year age bands which are upper-bounded. In other words, we do not consider the possibility of ever-increasing life spans, nor do we rule it out. By working with, in this case, ten-year age segments, we know there is a definite limit within each age interval and so we can concentrate on age intervals that are still showing growth.

Put another way, decomposition can address the question 'what is life expectancy in an arbitrary age interval, say, between 60 and 100?' rather than 'what is the upper limit to life expectancy?' The maximum attainable is 40 years in this particular example, and so the question of life expectancy above 100 becomes redundant. Although the method is flexible in this respect, and works equally well for men and women, available data from which to establish trends is clearly more problematic, and users can choose whether to include the highest age brackets in their reporting, for which there may be a lack of obvious trend.

In conclusion, most future growth in life expectancy in retirement will come between ages 70 and 100. Life expectancy beyond 100 years of age is increasing very slowly and so will not contribute as much as was thought. Age at death will tend to increasingly cluster in the early 90s as the age of death of men and women converge (Wilmoth and Horiuchi, 1999; Canudas-Romo, 2008; Mayhew and Smith 2014). Decomposition gives us more precision and flexibility in identifying and unpicking these future trends, which is beneficial both for governments and individuals alike.

This brings us back to the more fundamental question of whether we are starting to plan for older age better than we were. Later retirement requires a capacity to work for longer and it may also mean downsizing one's home at an earlier stage, i.e. before a person has reached the nursing home stage, requiring a forced home sale in order to pay for care. This could be a boost to the person's income in retirement or allow them to pay for luxury items, such as a new car or a world cruise. If this kind of planning gains traction, it could lead to a significant transformation in the housing market as well as growth in the provision of professional financial advice, including tax and inheritance planning.

In parallel, greater responsibility will fall on the individual to make choices, to pay for services, and to seek help and advice. We already know that average pension savings are small and must last for longer so that using the value in the home by releasing equity becomes more important, especially if ill health increases living costs (Mayhew and Smith 2016). Better planning tools, including decomposition approaches to life expectancy, as described here, will help to anticipate and mitigate these effects, in which Geographers have an important role to play.

## References


Arriaga, E. (1984) Measuring and explaining the change in life expectancies, *Demography*, 21(1), 83–96.

Bongaarts, J. (2005) Long range trends in adult mortality models and projection methods: fitting the logistic model to the force of mortality, *Demography*, 42(1), 23–49.

Booth, H. (2006) Demographic forecasting: 1980 to 2005 in review, *International Journal of Forecasting*, 22(3), 547–581.

Canudas-Romo, V. (2008) The modal age at death and the shifting mortality hypothesis, *Demographic Research*, 19, 1179–1204. Available at: www.demographic-research.org/Volumes/Vol19/30/.

Hickman, J.C. and R.J. Estell (1969) On the use of partial life expectancies in setting health goals, *American Journal of Public Health*, 59(12), 2243–2250.

Human Mortality Database (2009) *Human Mortality Database*. University of California, Berkeley (USA), and Max Planck Institute for Demographic Research (MPIDR). Available at: www.mortality.org.

Lee, R. and L. Carter (1992) Modelling and forecasting US mortality, *Journal of the American Statistical Association*, 87, 659–671.

Li, N., R. Lee and P. Gerland (2013) Extending the Lee-Carter method to model the rotation of age patterns of mortality decline for long-term projections, *Demography*, 50, 2037–2051.

Mayhew, L. and D. Smith (2012) A new method of projecting populations based on trends in life expectancy and survival, *Population Studies*, 67(2), 157–170. DOI: 10.1080/00324728.2012.740500.

Mayhew, L. and D. Smith (2014) Gender convergence in human survival and the postponement of death, *North American Actuarial Journal*, 18(1), 194–216.

Mayhew, L. and D. Smith (2015) On the decomposition of life expectancy and limits to life, *Population Studies*, 69(1), 73–89. DOI: 10.1080/00324728.2014.972433.

Mayhew, L. and D. Smith (2016) An investigation into inequalities in adult lifespan. International Longevity Centre – UK, London. http://www.ilcuk.org.uk/index.php/publications/publication_details/an_investigation_into_inequalities_in_adult_lifespan

Mayhew, L., D. Smith and D. O'Leary (2016) Paying for Care Costs in Later Life Using the Value in People's Homes. The Geneva Papers (2016) 0, 1–23. doi:10.1057/gpp.2015.34.

Murphy, M. and M. Di Cesare (2012) Use of an age-period-cohort model to reveal the impact of cigarette smoking on trends in twentieth-century adult cohort mortality in England and Wales, *Population Studies*, 66(3), 259–277.

Pollard, J. (1982) The expectation of life and its relationship to mortality, *Journal of the Institute of Actuaries*, 109, 225–240.

Preston, S. H. and H. D. Wang (2006) Sex mortality differences in the United States: the role of cohort smoking patterns, *Demography*, 43(4), 631–646.

Rafferty, A., J. Chunn, P. Gerland and H. Ševčíková (2013) Bayesian probabilistic projections of life expectancy, *Demography*, 50, 777–801.

Renshaw A. and S. Haberman (2003) Lee-Carter mortality forecasting: a parallel generalized linear modelling approach for England and Wales mortality projections, *Journal of the Royal Statistical Society: Series C*, 52(1), 119–137.

Renshaw A. and S. Haberman (2006) A cohort-based extension to the Lee-Carter model for mortality reduction factors, *Insurance: Mathematics and Economics*, 38(3), 556–570.

Shaw, C. (2007) Fifty years of United Kingdom national population projections: how accurate have they been? *Population Trends*, 128, 8–22.

Trovato, F. (2005) Narrowing sex differential in life expectancy in Canada and Austria: comparative analysis, *Vienna Yearbook of Population Research*, Austrian Academy of Sciences, Vienna, 17–52.

United Nations, Department of Economic and Social Affairs, Population Division (2009) *World Population Prospects: The 2008 Revision, Highlights*, Working Paper No. ESA/P/WP.210.

Vaupel, J. (2010) Biodemography of human ageing, *Nature*, 464(7288), 536–542.

Wilmoth, J.R. (2000) Demography of longevity: past, present, and future trends, *Experimental Gerontology*, 35, 1111–1129.

Wilmoth, J.R. and S. Horiuchi (1999) Rectangularization revisited: variability of age at death within human populations, *Demography*, 36(4), 475–495.

## Appendix A: Changes in life expectancy within discrete age intervals[2]

### Segmenting life expectancy by age

We can derive the future expectation of life for a life currently aged $x_1$ by calculating the area under the population curve and dividing by the starting population (i.e. so that we turn the population curve into an individual survival curve for a standard member of the population).

Figure 11.5 shows a survival curve divided into segments based on age. Each value of $l_x$ is joined by a straight line on the assumption that population decreases linearly between two ages. The standard equation for expectation of future life at age $x_1$ is

$$e_{x_1} = \frac{1}{l_{x_1}} \sum_{y=x_1}^{x_5} l_y - \frac{1}{2} = \frac{1}{l_{x_1}} \sum_{y=x_2}^{x_5} l_y + \frac{1}{2}$$

In effect, what we are calculating is a series of rectangles and triangles. Assuming that we are increasing age by 1 each time, i.e. $x_{n+1} - x_n = 1$, then the area of each rectangle is simply the height or number of people alive at age $x_{n+1}$, and the area of each triangle is $\frac{1}{2}(l_x - l_{x+1})$. Therefore

$$\begin{aligned} e_{x_1} &= \frac{1}{l_{x_1}} \left\{ \begin{matrix} \left(l_{x_2} + \frac{1}{2}(l_{x_1} - l_{x_2})\right) + \left(l_{x_3} + \frac{1}{2}(l_{x_2} - l_{x_3})\right) + \\ \left(l_{x_4} + \frac{1}{2}(l_{x_3} - l_{x_4})\right) + \left(l_{x_5} + \frac{1}{2}(l_{x_4} - l_{x_5})\right) \end{matrix} \right\} \\ &= \frac{1}{l_{x_1}} \left\{ \sum_{y=x_2}^{x_5} l_y + \frac{1}{2}(l_{x_1} - l_{x_5}) \right\} \\ &= \frac{1}{l_{x_1}} \sum_{y=x_2}^{x_5} l_y + \frac{1}{2} \end{aligned}$$

(because $l_{x_5} = 0$ )

### Contribution to expected life

We can now break this future lifetime into two parts – from ages $x_1$ to $x_3$ and from $x_3$ to $x_5$ – and see how much of the total expected future lifetime each section gives.

Define:

$e_{x_n(x_i : x_j)}$ as the future expected life of someone currently aged $x_n$ between the ages of $x_i$ and $x_j$

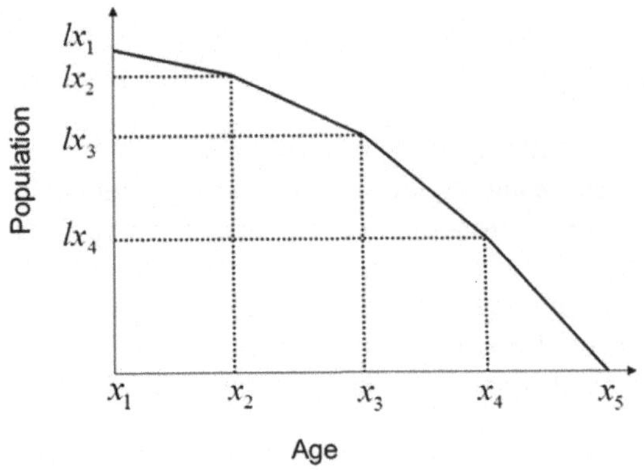


*Figure 11.5* Survival curve S(x) divided into segments

Hence

$$
\begin{aligned}
e_{x_1(x_1:x_3)} &= \frac{1}{l_{x_1}}\left\{\left(l_{x_2} + \frac{1}{2}\left(l_{x_1} - l_{x_2}\right)\right) + \left(l_{x_3} + \frac{1}{2}\left(l_{x_2} - l_{x_3}\right)\right)\right\} \\
&= \frac{1}{l_{x_1}}\left\{\sum_{y=x_2}^{x_3} l_y + \frac{1}{2}\left(l_{x_1} - l_{x_3}\right)\right\}
\end{aligned}
$$

And

$$
\begin{aligned}
e_{x_1(x_3:x_5)} &= \frac{1}{l_{x_1}}\left\{\left(l_{x_4} + \frac{1}{2}\left(l_{x_3} - l_{x_4}\right)\right) + \left(l_{x_5} + \frac{1}{2}\left(l_{x_4} - l_{x_5}\right)\right)\right\} \\
&= \frac{1}{l_{x_1}}\left\{\sum_{y=x_4}^{x_5} l_y + \frac{1}{2}\left(l_{x_3} - l_{x_5}\right)\right\}
\end{aligned}
$$

And we can see that

$$e_{x_1} = e_{x_1(x_1:x_3)} + e_{x_1(x_3:x_5)}$$

We can extend this idea to as many ages as we want. For example, if we have a population that we are studying from age 60, where the population dies out at age 110, then:

$$e_{60} = \frac{1}{l_{60}} \sum_{y=61}^{110} l_y + \frac{1}{2}$$

And we can break this population down into contributions to expected life from each 10 years of age, i.e. ages 60–70, 70–80, …, 100–110.

$$e_{60(60:70)} = \frac{1}{l_{60}} \left\{ \sum_{y=61}^{70} l_y + \frac{1}{2}(l_{60} - l_{70}) \right\}$$

$$e_{60(70:80)} = \frac{1}{l_{60}} \left\{ \sum_{y=71}^{80} l_y + \frac{1}{2}(l_{70} - l_{80}) \right\}$$

$$e_{60(80:90)} = \frac{1}{l_{60}} \left\{ \sum_{y=81}^{90} l_y + \frac{1}{2}(l_{80} - l_{90}) \right\}$$

$$e_{60(90:100)} = \frac{1}{l_{60}} \left\{ \sum_{y=91}^{100} l_y + \frac{1}{2}(l_{90} - l_{100}) \right\}$$

$$e_{60(100:110)} = \frac{1}{l_{60}} \left\{ \sum_{y=101}^{110} l_y + \frac{1}{2}(l_{100} - l_{110}) \right\}$$

And

$$e_{60} = e_{60(60:70)} + e_{60(70:80)} + .... + e_{60(100:110)}$$

## A different way at looking at contributions

When we calculate a term such as $e_{60(70:80)}$ we are calculating a value based on a life currently aged 60 surviving to age 70 and then contributing these expected years. We can therefore express this in a different way.

Instead of looking at $e_{60(70:80)}$, we can look at $e_{70(70:80)}$, i.e. the amount of expected life that a person who is now aged 70 can expect to live over the next 10 years.

$$e_{70(70:80)} = \frac{1}{l_{70}} \left\{ \sum_{y=71}^{80} l_y + \frac{1}{2}(l_{70} - l_{80}) \right\}$$

We can see that:

$$e_{60(70:80)} = \frac{1}{l_{60}} \left\{ \sum_{y=71}^{80} l_y + \frac{1}{2}(l_{70} - l_{80}) \right\} = \frac{l_{70}}{l_{60}} \frac{1}{l_{70}} \left\{ \sum_{y=71}^{80} l_y + \frac{1}{2}(l_{70} - l_{80}) \right\} = \frac{l_{70}}{l_{60}} e_{70(70:80)}$$

This, of course is intuitive as the term on the left is the expected life between the ages of 70 and 80 of someone currently aged 60, while the term on the right is the expected life between the ages of 70 and 80 of someone currently aged 70 multiplied by the probability that someone currently aged 60 reaches the age 70, i.e. a conditional probability.

Hence:

$$e_{60} = e_{60(60:70)} + \frac{l_{70}}{l_{60}} e_{70(70:80)} + .... + \frac{l_{100}}{l_{60}} e_{100(100:110)}$$

## Notes

1 Sections of this chapter rely on material which is reproduced with the permission of the publishers based on Mayhew and Smith (2015) ‘On the decomposition of life expectancy and limits to life’. *Population Studies*, 69(1), March, 73–89.

2 Reproduced with permission from Mayhew and Smith (2015) ‘On the decomposition of life expectancy and limits to life’, *Population Studies*, 69(1), March, 73–89.

# 12 Using agent-based modelling to understand crime phenomena

*Nick Addis*

## 12.1 Introduction

Comprising a myriad of different disciplines, crime is a complex, multi-faceted construct that incorporates a multitude of different components, including the offender(s), victim(s), environment in which a crime(s) takes place, and any other features/individuals associated with an offence. However, the heterogeneous nature of each of these components results in a particularly complex crime system. Indeed, Epstein and Axtell (1996) note that a number of the processes inherent within *any* social system are complex and cannot be compartmentalised into individual processes. This raises inevitable questions about how we can begin to understand crime. One key approach to develop our understanding of crime is the analysis or modelling of crime.

## 12.2 Modelling crime systems

While crime has been acknowledged as a particularly complex system, there exists a large body of literature that suggests crime clusters in time and space (Johnson and Bowers, 2004; Ratcliffe and Rengert, 2008). Therefore, to try to understand the crime system, it is important to analyse or model crime phenomena and their underlying dynamics. Indeed, there exists a range of approaches used to model crime. For example, the use of 'thematic' mapping methods remains popular in enabling crime phenomena to be illustrated across a geographical space (Chainey and Ratcliffe, 2005). Such techniques may also be used over time to highlight the presence of any temporal patterns in the data. However, while these approaches provide a clear visualisation of data, they remain limited in terms of the extent of information that can be portrayed.

Advances in computing and Geographic Information Systems (GIS) mean that more scientific approaches can now be utilised to understand patterns of crime and their distribution more effectively. These include statistical and mathematical techniques, such as 'Kernel Density Estimation', which can be used to explore crime density and highlight 'hotspots' of crime. However, these techniques are predominantly descriptive in their approach. Approaches such as 'regression' offer a clear advantage over those that can only describe or illustrate patterns of crime, in that they seek to explore the impact of explanatory variables (Chainey and Ratcliffe, 2005). However, Chainey and Ratcliffe (2005) note how standard linear regression models are problematic when focusing on geographical spatial data because variables are assumed to exert equal influence across an area despite the fact that this is unlikely to be the case. Furthermore, while methods such as geographically weighted regression account for the variance in independent variables across spatial areas, such methods still seek to account for crime at the aggregate rather than individual-level. Therefore, the richness of data at the individual level is lost using this approach.

Moving beyond these approaches, further computational techniques, such as microsimulation, enable the exploration of behaviour at the individual level. In particular, the microsimulation approach involves the generation of large individual-level data sets through matching known individual data to aggregate area data sets, enabling researchers to undertake 'test-bed' experimentation across different scenarios. However, Malleson *et al.* (2010) argue how the above methods are predominantly unable to model features of crime at the individual level, i.e. victims and offenders, which is particularly key when looking to explore the validity of criminological theory. Furthermore, Birks *et al.* (2012) note that even where modelling approaches are able to explore offending at the 'micro-level' perspective, it remains particularly difficult to link individual-level behaviour with larger patterns of crime. This helps illustrate one of the focal areas with which computational modelling approaches, such as 'Agent-Based Modelling', an advanced simulation computational technique, command authority over existing modelling approaches through their focus on individual-level behaviours and how these link with broader behavioural patterns.

## 12.3 Agent-Based Modelling

The Agent-Based Modelling approach evolved through earlier computational approaches to modelling behaviour and changes in state, based on the nature of the environment. Such computational models were initially introduced in the 1940s by Von Neumann and his work on Cellular Automata (von Neumann, 1966). Agent-Based Models (ABMs), comprise of decision-making 'agents', who interact both with other agents and their environment, based on a series of behavioural rules and learning through their actions (Macy and Willer, 2002; Janssen and Ostrom, 2006;). This technique enables users to build computational models that directly represent the individual agents present in complex social systems. The key features of the ABM approach are outlined in Table 12.1.

Schelling developed one of the first key examples of an Agent-Based Model in his work on segregation (Schelling, 1971) using pennies and dimes on a chessboard. His finding of segregated penny and dime 'neighbourhoods' demonstrated how following simple rules at the local level led to the emergence of more 'macro' global patterns. Epstein and Axtell (1996) note that at the time of this work, the capability of ABMs was limited by the computational resources available, and that it is only with recent advances in computing power that ABMs have become more advanced. For a more detailed overview of the development of the Agent-Based Modelling approach, the reader is directed to Bonabeau (2002).

*Table 12.1* Features of the ABM approach

| *Feature* | *Description* |
|---|---|
| *Heterogeneous Agents* | The ABM approach enables model agents to be heterogeneous (Epstein, 1999), behaving based on their individual sets of rules. |
| *Model Environment* | Users are able to replicate the physical and social aspects of the environment under consideration (Gilbert, 2008), including the traits of different agents, which in the context of crime may be particularly relevant for testing the impact of crime prevention policies, or the impact of physical features on risk of victimisation. |
| *Agent Interactions driven by Rules* | The integration of rules within an ABM is one of the key aspects that distinguish this approach from other modelling techniques. Rules can be applied to individual agents or groups of agents, and will determine the nature of their interactions with other agents and the environment around them. |
| *Ontological Correspondence* | This concept relates to the direct correspondence between agents within a model and individuals within the real world (Gilbert, 2008). Therefore, ABMs that focus on crime problems in the real world are likely to increase the ontological correspondence between 'real' and 'simulated' worlds. |
| *Bounded Rationality* | In developing any ABM it is important for agents to be as similar to individuals in the real world as possible. Therefore, it is important for agents to demonstrate a *realistic* level of cognitive reasoning and rationality and not an *unrealistic* sense of 'infinite' rationality (see DeHaan and Vos, 2003). |
| *Learning* | Agents are able to learn and subsequently modify their behaviour over time (Axelrod, 1997); learning may be across individuals or across a population (Gilbert, 2008). |

Source: Gilbert (2008)

## 12.4 Agent-Based Modelling of crime

During recent years there has been a surge of growth in Agent-Based Modelling research across the social sciences. Indeed, Gilbert (2008) highlights a range of fields in which ABMs have been effectively applied, including consumer behaviour, opinion dynamics, and urban models. This approach has also become particularly popular within the field of crime, with a number of ABMs developed in recent years across crime types, including burglary (Birks *et al.*, 2012; Malleson *et al.*, 2013), street robbery (Liu *et al.*, 2005; Groff, 2008), and policing strategies of drug markets (Dray *et al.*, 2008a, 2008b). Early crime ABMs appear to have been more abstract in nature, with greater focus on behavioural dynamics; although over time such ABMs have become more sophisticated, utilising realistic geographical environments. There appear to be two distinct strands as to the purpose of Agent-Based Models of crime:

1) To support the task of theorising – to explore, verify, and 'test' criminological theory.
2) To support policy-makers – through exploring the impact of criminal justice policy, acting as a 'test-bed' to examine the impact of crime reduction initiatives.

However, prior to discussing any ABM, it is important to consider the spatial environment upon which the model is based.

### *A note on spatial environment*

The spatial environment is a fundamental component of any crime ABM. With regards to model environments, there are two main schools of thought:

- Abstract environment
- Realistic environment

Models utilising an abstract environment enable the user to focus more on the underpinning behavioural dynamics of the system. Conversely, models that utilise more realistic environments allow users to explore the behavioural dynamics within the context of a realistic geographical space. Indeed, Malleson (2012) notes the importance of ensuring an ABM model environment is an accurate representation of the system it aims to depict, and there exist a wide range of programming platform tools available which make the process of integrating a GIS much more achievable than previously. However, Elffers and van Baal (2008) warn against the use of a GIS in an ABM that may result in an overly complex model which detracts from our understanding of model dynamics. Furthermore, Axelrod (1997) notes the importance of considering the ultimate aim of the model which will help determine whether an abstract or realistic model environment is required.

The next section of this chapter will illustrate two distinct examples of ABMs that have been effectively applied to real-world crime phenomena and informed through collaborations with criminal justice agencies. These case studies will demonstrate how ABMs that utilise either abstract or realistic spatial environments can be used effectively to help aid our understanding of crime. They have been chosen because they are two important examples of ABMs that demonstrate how collaborations with criminal justice agencies may help enhance the power and effectiveness of ABMs developed, while in turn illustrating how these models may be utilised by such agencies. The first of these case studies will illustrate how an ABM was used to explore the impact of the 'heroin drought' of Australia in late 2000/early 2001.

## 12.5 The heroin drought in Australia, 2000–2001

Between late 2000 and early 2001, there was a global shortage in opiate production (United Nations Office on Drugs and Crime, 2007), resulting in a substantial heroin drought in Australia (Dietze *et al.*, 2004). This was felt particularly across local Australian drug markets, resulting in both opiate price rises, and diminishing heroin purity across local markets (Day *et al.*, 2003). However, the number of injecting drug users in Melbourne remained relatively stable as a consequence of injecting drug users turning to drugs such as amphetamines, benzodiazepines,

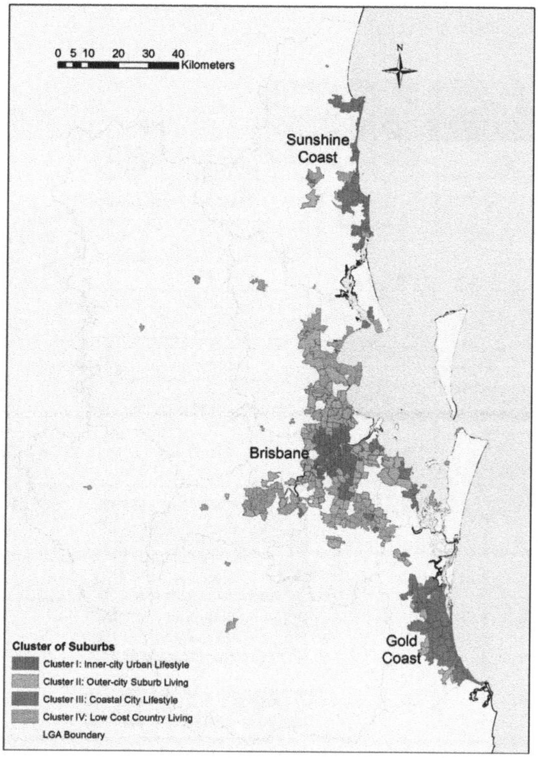


*Plate 1* Typology of private rental housing market in South East Queensland

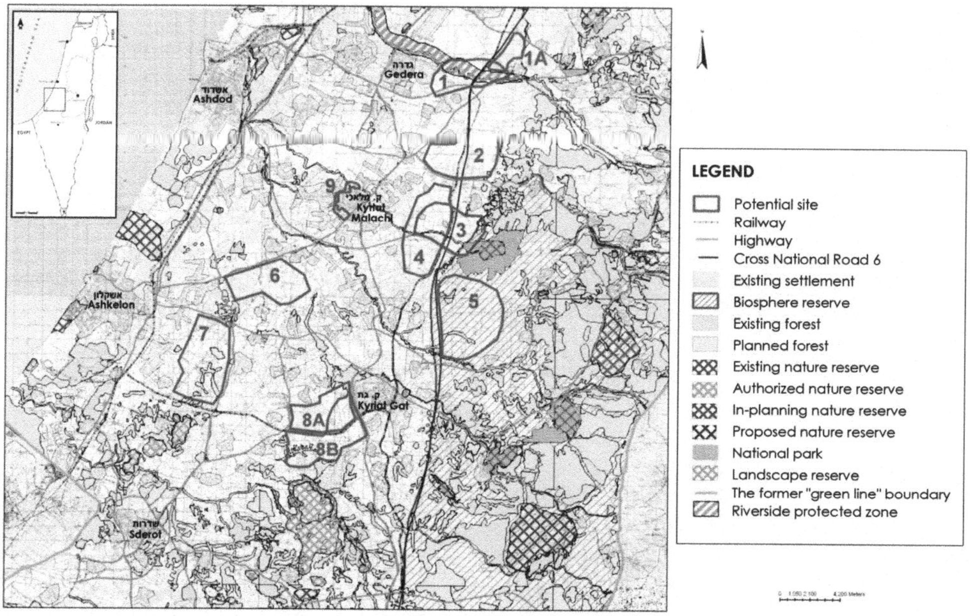


*Plate 2* Protected green areas in the southern part of the search zone

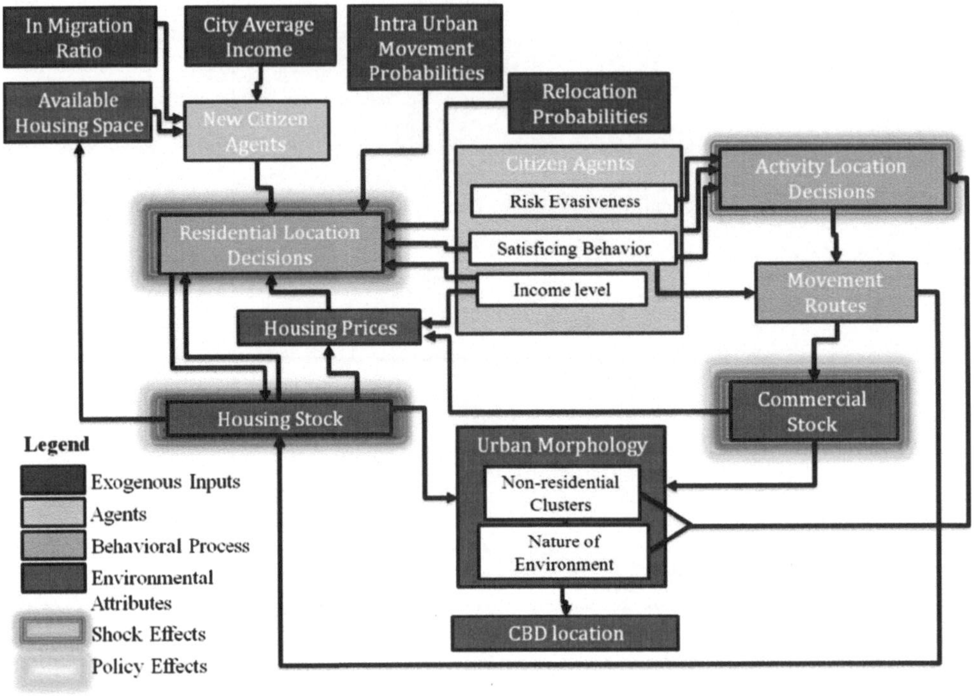


*Plate 3* A conceptual representation of ABM of an earthquake in a city

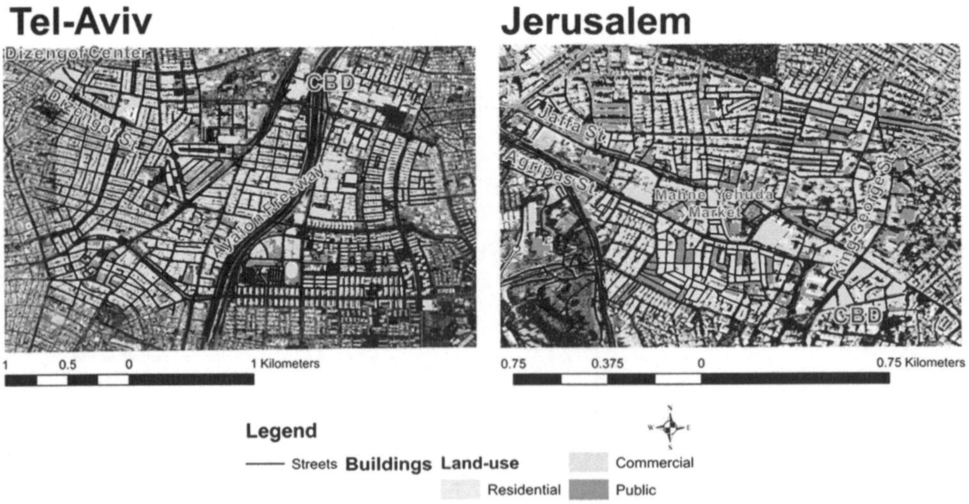


*Plate 4* Study areas

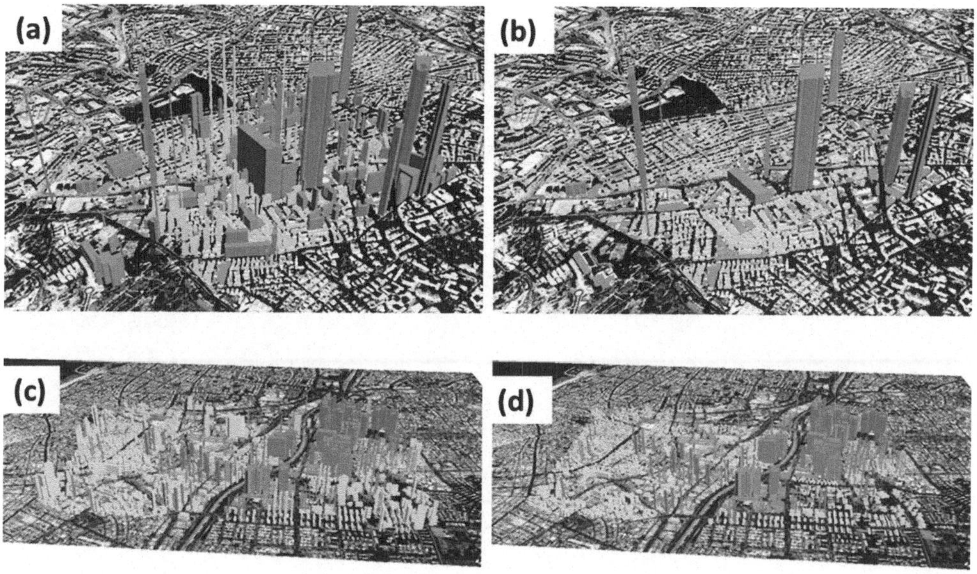


*Plate 5* (a) Frequency of land-use change for Jerusalem – no policy. Color represents initial land use (green – residential, purple – non-residential), height represents frequency of different land-use at the end of a simulation (b) Frequency of land-use change for Jerusalem – policy. Color represents initial land use (green – residential, purple – non-residential), height represents frequency of different land-use at the end of a simulation (c) Frequency of land-use change for Tel Aviv – no policy. Color represents initial land use (green – residential, purple – non-residential), height represents frequency of different land-use at the end of a simulation (d) Frequency of land-use change for Tel Aviv – policy. Color represents initial land use (green – residential, purple – non-residential), height represents frequency of different land-use at the end of a simulation

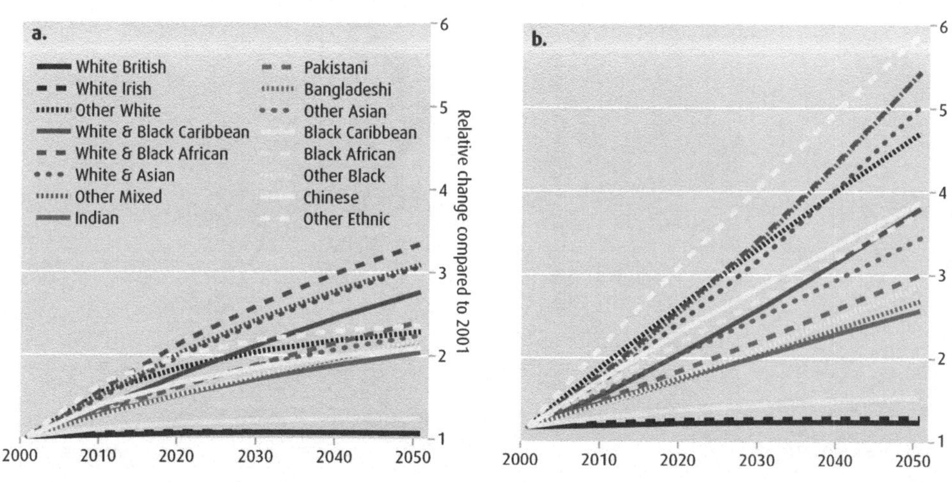


*Plate 6* (a) Projected populations for ethnic groups under the emigration rates scenario (UPTAPER) (b) Projected populations for ethnic groups under the emigration flows scenario (TREND)

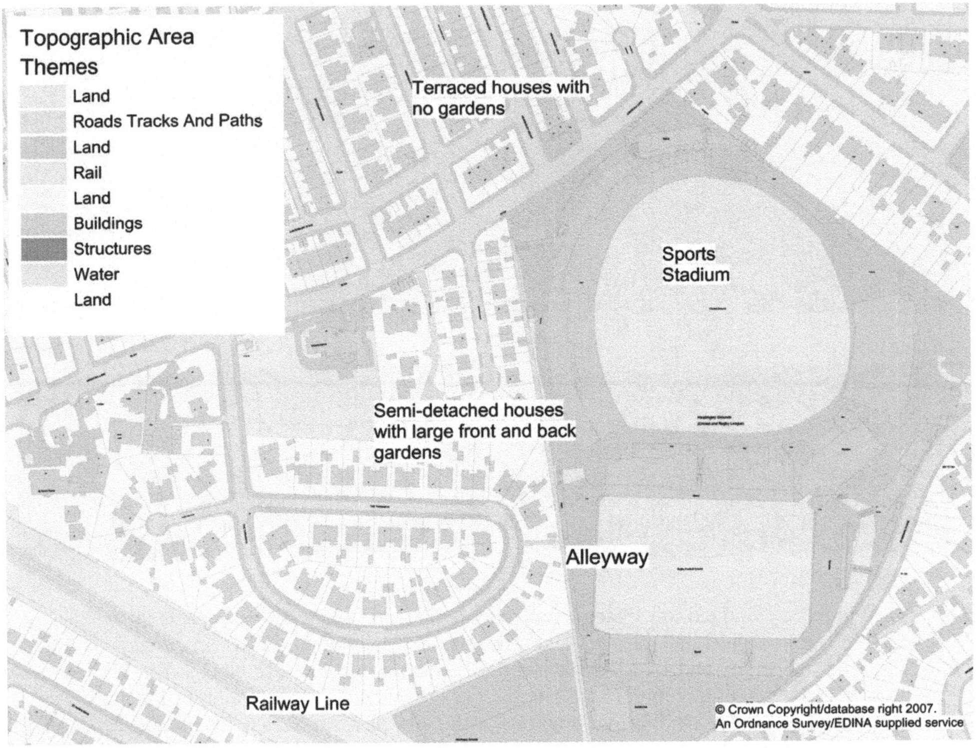


*Plate 7* MasterMap data used in the model environment

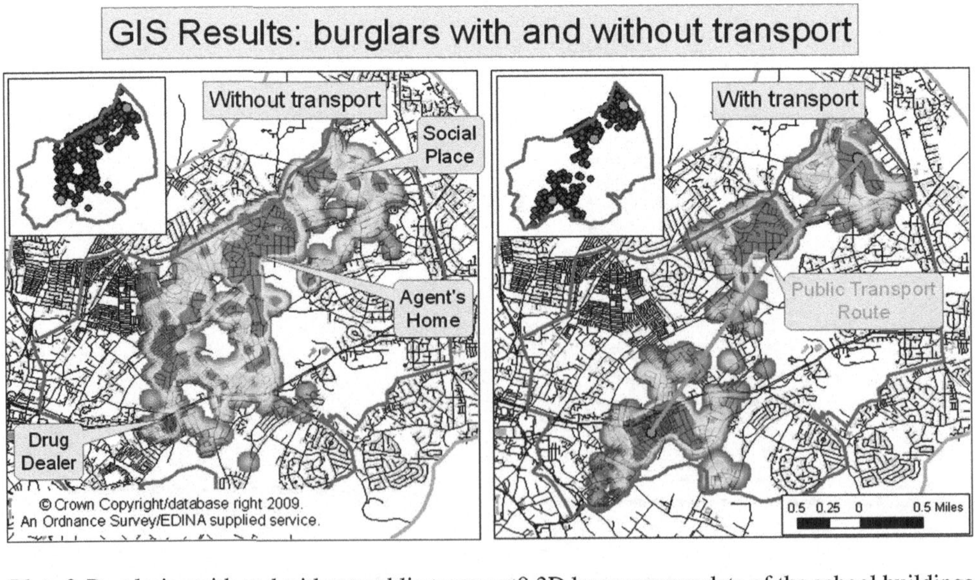


*Plate 8* Burglaries with and without public transport9 3D laser scanner data of the school buildings

*Plate 9* 3D laser scanner data of the school buildings. The upper figure shows the external view – this is not a photograph, but a "point cloud" coloured using photographs taken by the scanner. The blue area (a curtain) and accompanying "shadow" on the left near the tree is in fact a missing data region, where the internal structure of the room (scanned from the inside) shows through . The lower figure shows the internal scan from the opposite direction, illustrating how we can section the data to reveal desks, computers and other classroom objects

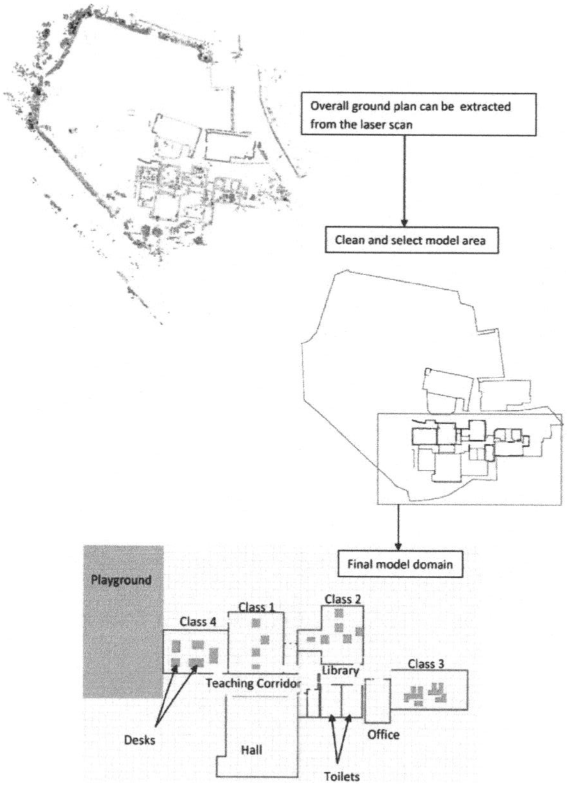


*Plate 10* Workflow showing the creation of the model domain from the initial full 30 scanner data of the whole school. The red rectangle outlines the main set of school buildings, which are shown in the bottom figure labeled with their functions. Only the subset of buildings routinely occupied by children has been extracted . The head's office and boiler room, for example, are not used in the final model domain

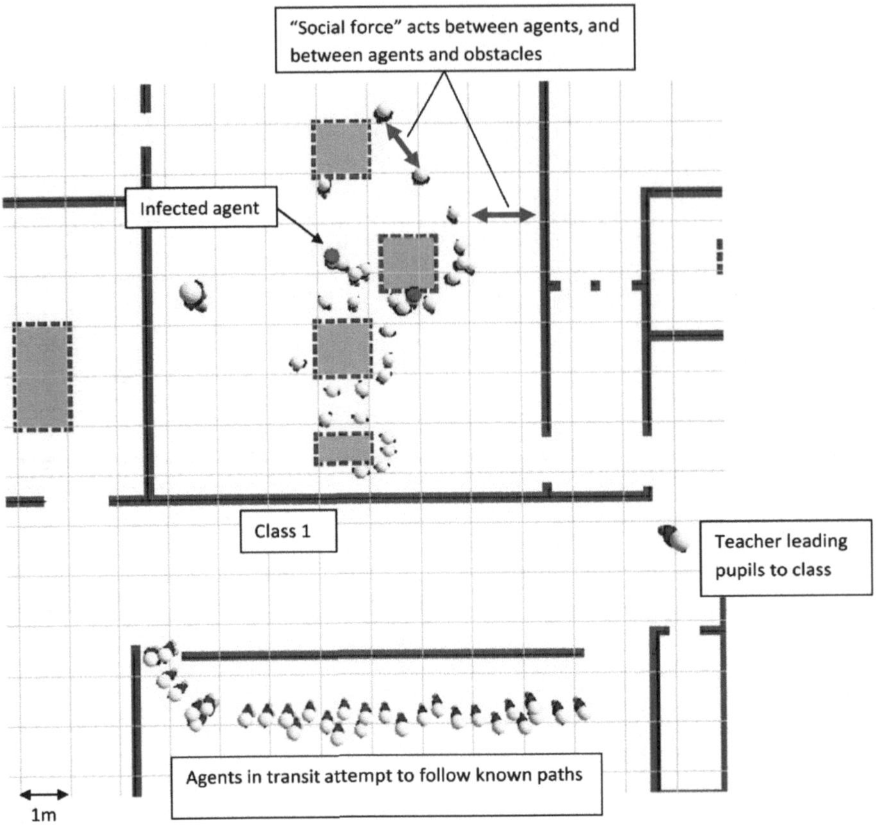


*Plate 11* Snapshot from a visualisation of the running model. This is from early in the day, when children are on their way from playground to class. In the lower part of the figure class 3 are on their way through the hall, as school rules do not permit them to go through the adjacent corridor. Class 1 are already at their desks awaiting assembly. Two children in Class 1 have at this stage acquired the infection

*See www.geog.cam.ac.uk/research/projects/agentbaseddisease for a movie of the model in action*

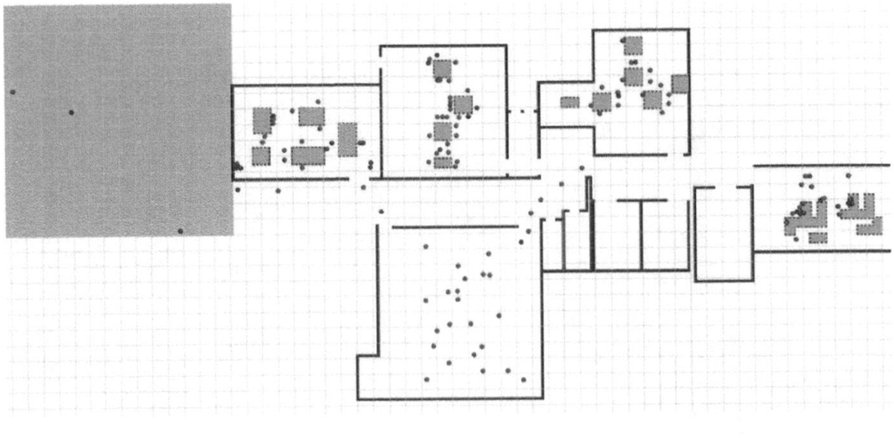

*Plate 12* Locations at which infections take place (red dots) and recoveries (blue dots) inside the school grounds for the run of Figure 15.7. All other recoveries (the majority) take place at home. Note that only one infection takes place in the playground – all others happen in classrooms or the hall, with a few in transit between locations

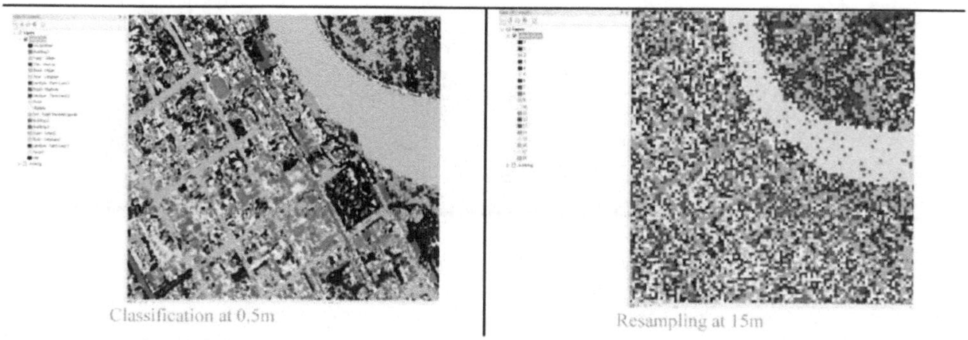


*Plate 13* Resampling process – depicting the classified imagery at 0.5m on the left and how it looks when resampled at 15m on the right

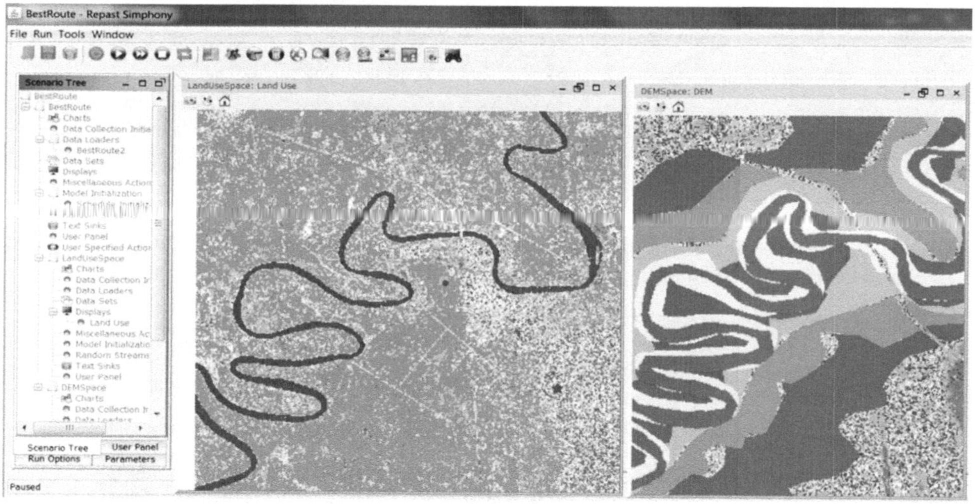


*Plate 14* Model interface showing land use (left) and DEM layers (right)

*Plate 15* Scenario 1 – illustration showing the flooding (the dark area) and route taken by the agent (faded red line) in Repast Simphony Java (left). The coordinates of the route in Repast Simphony Java were converted and overlayed on GeoEye1 0.5m imagery using ArcGIS (right). The agent commenced at the upper left, travelled via the faded red line (on the left image) and red line (on the right) to the 'safe area' near the bottom right

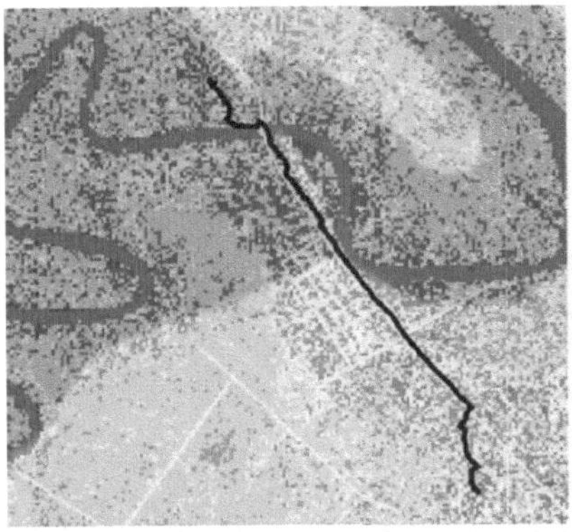

*Plate 16* Illustration showing the flooding and route taken by the agent in Repast Simphony Java (left). The coordinates of the route in Repast Simphony Java were converted and overlayed on GeoEye1 0.5m imagery using ArcGIS (right)

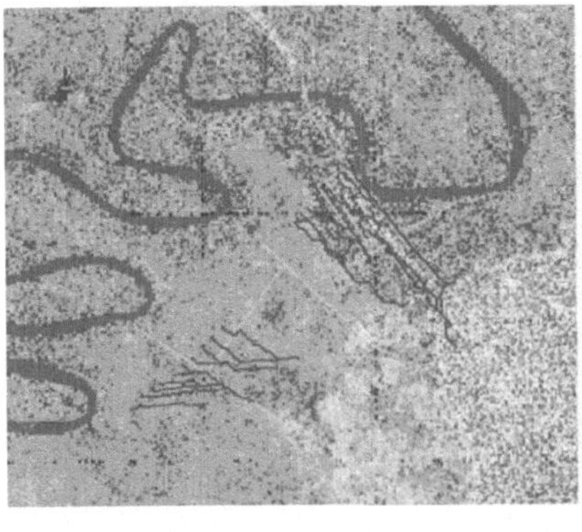

*Plate 17* Illustration showing the flooding and routes taken by the agents in Repast Simphony Java (left). The coordinates of the routes in Repast Simphony Java were converted and overlayed on GeoEye1 0.5m imagery using ArcGIS (right)

and prescribed opioids (Dietze *et al.*, 2004). Dray *et al.* (2008a) recognised how these factors created a situation whereby the impact of different law enforcement strategies on street-level drug markets could be explored through the use of simulation techniques (Dray *et al.*, 2008a; Liu and Eck, 2008). This provided the rationale behind 'SimDrugDrought', an ABM developed by Dray *et al.* (2008a) to explore the impact of three different law enforcement strategies on a range of measures relating to street-level drug markets. These measures include the number of crimes committed, levels of wealth, and the number of overdoses of drug users. In particular, Dray *et al.* (2008a) explored how such markets adapt to heroin supply disruption at the broader, macro level. The SimDrugDrought model is an extension of the earlier 'SimDrug' (Perez *et al.*, 2006) and 'SimDrugPolicing' (Dray *et al.*, 2008b) models, which were designed to understand components of street-level drug markets more effectively.

### *12.5.1 The SimDrugDrought model*

Using three years of data from Melbourne, Australia, Dray *et al.* (2008a) developed the SimDrugDrought model to explore the relationship between individual components of the drug system, including drug users, drug dealers and wholesalers, and the police. The model used an abstract spatial environment, based upon a 'street block' system, separated into five distinct demographic areas. Thus, the model focused on exploring the decision-making processes and the behavioural dynamics of agents over the use of a realistic spatial environment (Dray *et al.*, 2008a).

The model was developed using the 'Common-pool resources and multi-agent systems' (CORMAS) programming platform (Bousquet *et al.*, 1998), which uses the 'Smalltalk' programming language. Model dynamics, interactions and decision-making processes of agents in the model were developed based on a range of official statistics provided by the Australian authorities.

Each grid cell in the model environment had the capacity to facilitate illegal activity (termed 'conductivity'), for example, drug dealing. On initialisation of each simulation, 150 conducive cells are randomly allocated, generating heterogeneity across the model environment. Each grid cell also comprises two additional variables: 'wealth', which denotes the value of goods available at a property (this diminishes following each instance of crime on the cell), and 'risk', which reflects the total number of crimes, overdoses, and drug users at each time step ($t$) for each cell (see Equation 12.1).

$$risk_{cell,t} = 10\left(\sum crime_{cell,t} + \sum overdose_{cell,t}\right) + \sum user_{cell,t} \tag{12.1}$$

The SimDrugDrought model includes different 'actors' (agents), drug users, drug dealers and wholesalers, in addition to patrol officers and outreach workers. The behavioural drivers and motivations of each agent were derived from rule-based heuristics, collectively developed by an inter-disciplinary panel of experts. The population of both drug users and dealers remained stable throughout the model, as based on previous literature (Dietze *et al.*, 2004); if a drug user dies from an

*Table 12.2* Features of the policing interventions applied

| *Parameter* | *Random Patrol* | *Hot Spot Policing* | *Problem-Orientated Policing* |
|---|---|---|---|
| Constable's destination | Randomly selected | Locations with highest risk values during previous three time steps | Locations with highest risk values |
| Outreach worker's destination | Locations with highest overdose rates | Locations with highest overdose rates | Destination of constable |

Source: Adapted from Dray *et al.* (2008a, p. 274)

overdose or successfully completes intervention, they are taken out of the model and replaced with another agent on the next time step. Similarly, on arrest, a dealer is replaced with another dealer agent at the next time step.

As noted earlier, the model explored the impact of three law enforcement/policing strategies on a range of measures relating to street-level drug markets, as based on a meta-analytic review of law enforcement strategies on street-level drug markets (Mazerolle *et al.*, 2006). Table 12.2 outlines the features of the policing strategies implemented.

Simulations were run on a daily time step to capture the behavioural trends of injecting drug users. Each simulation was run over a 36-month period between January 1999 and December 2001, using input data of heroin price, purity, and availability, as reported for Melbourne during this period (Dietze *et al.*, 2004). The model was run 100 times across each of the policing scenarios.

### *12.5.2 Model results*

The model was assessed on the following outcome measures: (1) Number of arrested dealers; (2) Number of users in treatment; (3) Number of crimes committed; (4) Number of crimes aborted; (5) Prevalence of heroin use; (6) Overdose occurrence; (7) Spatial evolution of conducive clusters.

To explore the impact of policing interventions on these measures during the heroin drought, model results were compared with those emerging under 'stable' market conditions using previously published findings from the 'SimDrugPolicing' model (see Dray *et al.*, 2008b). Dray *et al.* (2008a) found a number of key findings:

1) Policing interventions at the street level were successful in disrupting *both* criminal activity and the dynamics of the drug market. In particular, the 'Problem-oriented Policing' approach resulted in the highest percentage of arrested drug dealers (Figure 12.1) and the lowest percentage of crimes, while demonstrating a substantial impact upon heroin use (Figure 12.2), the percentage of users engaging in treatment, and '…the spatial extension of conducive clusters' (Dray *et al.*, 2008a, p. 283). In particular, the problem-oriented approach demonstrated significant differences over the 'Random patrol'

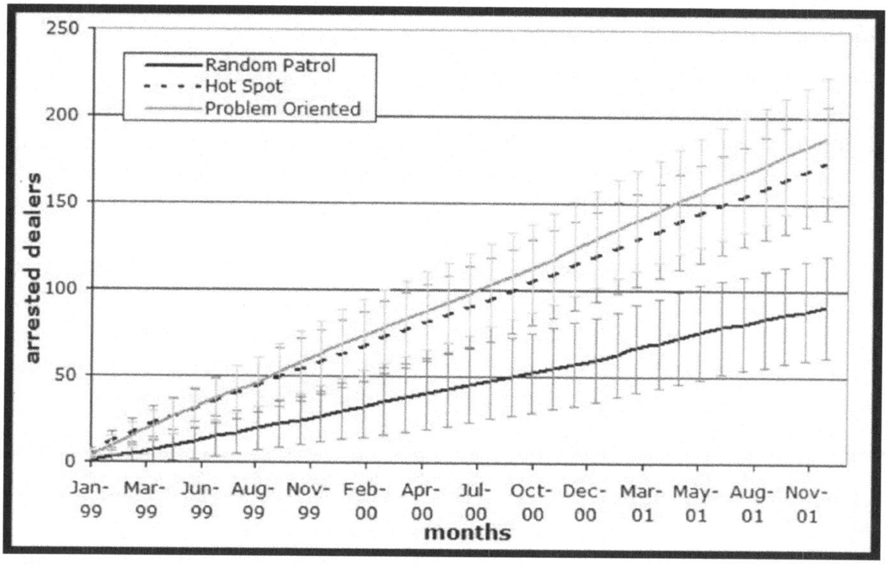


*Figure 12.1* Cumulative number of arrested drug dealers (together with associated standard deviations) across three policing strategies

Source: Dray *et al.* (2008a, p. 276)

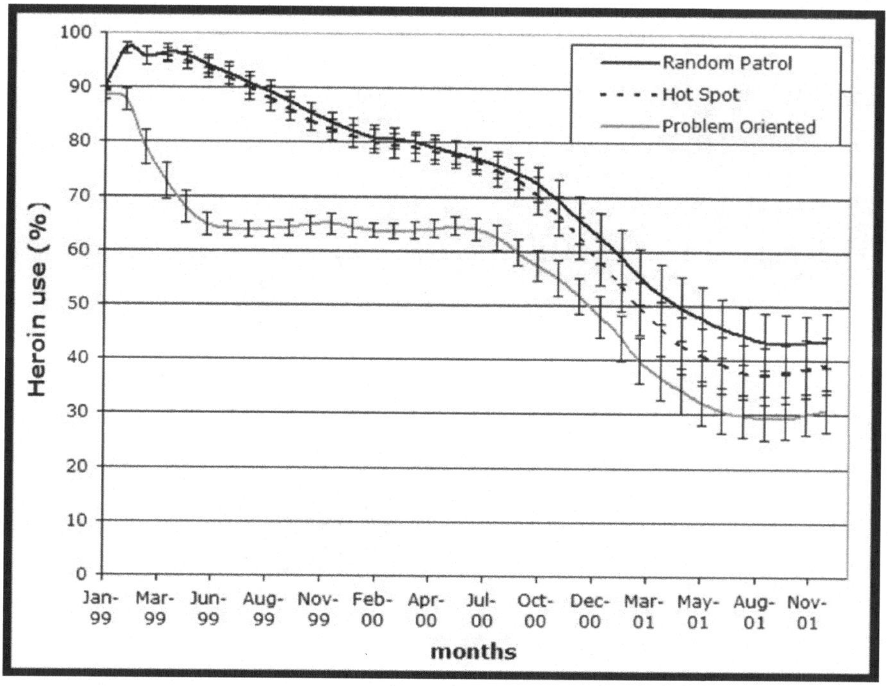


*Figure 12.2* Prevalence of heroin use among users (by %) and associated standard deviations across three policing strategies

Source: Dray *et al.* (2008a, p. 280)

approach (not including overdose occurrence) during the period in question. Furthermore, the 'hot spot' scenario was found to demonstrate a similar capacity to disrupt criminal activity as the 'problem-oriented' approach; although, this approach did not limit individual or social harm to the same extent.

2) Disruptions in the supply chain resulting in displacement from one illegal drug to another are unlikely to impact on street-level responses to localised interventions, with relative responses across each of the policing scenarios consistent with previous results from more 'stable' market conditions (Dray *et al.*, 2008a).
3) Street-level interventions had a greater chance of causing disruption to local markets and drug use than more macro-level initiatives that resulted only in market displacement, with the problem-oriented approach more likely to impact on street-level drug markets.

The results here suggest that street-level drug markets are best explained by features of the community rather than those of the drug market itself. For example, Figure 12.3 illustrates the cells conducive to illegal drug activity on conclusion of the simulation, demonstrating two emerging 'hot spots' that may be viewed as problematic areas (Dray *et al.*, 2008a).

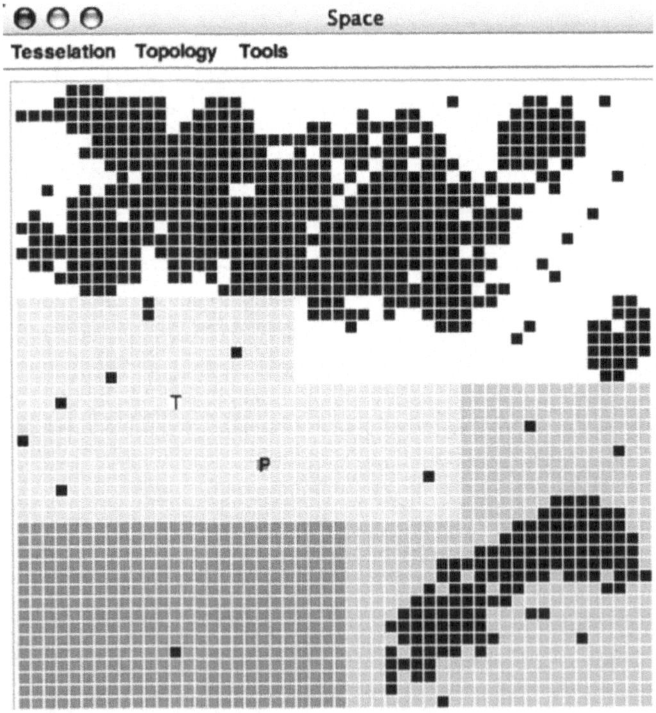


*Figure 12.3* Emerging 'hot-spots' on conclusion of the simulation

Source: Dray *et al.* (2008a, p. 282)

Dray *et al.* (2008a) highlight the value of computer simulation techniques in enabling researchers to explore the complex relationship between criminal activity, intervention from criminal justice agencies, and community-based responses. Notably, the work undertaken by Dray *et al.* (2008a) illustrated how problem-oriented policing was the most effective policing approach of those tested, not only in disrupting criminal activity, but in limiting individual or social harm related to street-level drug markets. Furthermore, Dray *et al.* (2008a) demonstrated how such markets may be best explained through community features rather than features of the drug markets themselves.

As noted previously, the SimDrugDrought model was based on earlier versions of the model, namely 'SimDrug' and 'SimDrugPolicing'. Dray *et al.* (2008a) undertook sensitivity testing with model parameters to ensure the robustness of the model. Dray *et al.* (2008a) also began to validate their results through baseline comparisons with model results emerging under stable market conditions from the SimDrugPolicing model (Dray *et al.*, 2008b). However, the SimDrugDrought model itself does not appear to have been subject to rigorous evaluation or validation testing. Because the model utilised artificial policing scenarios, there was no observed data that could be accurately matched to the conditions created across each of the policing scenarios in order to effectively validate the SimDrugDrought model.

Nevertheless, the data on which the SimDrugDrought model is based was taken from the earlier 'SimDrug' model, which sought to understand the dynamics of illegal drug markets in Melbourne, encompassing the period of the opiate drought in Australia (Perez *et al.*, 2006). Perez *et al.* (2006) were able to validate the results of their SimDrug model using observed data from 1998–2002. Perez *et al.* (2006) found some consistency between simulated and observed data, namely for the proportion of overdoses, levels of drug dealer wealth, and the proportion of drug users who are also drug dealers. Nevertheless, Perez *et al.* (2006) also acknowledged where simulated findings were inconsistent with observed data, for example, total overdose figures or the proportion of users in treatment. Furthermore, they acknowledged where simulated results required further validation against observed data, for example, with regards to total offences committed (Perez *et al.*, 2006). Therefore, while there has been some initial validation of the figures on which the model is based, further validation will strengthen the potential applicability of the model, particularly where such validation can be made more explicit for the current SimDrugDrought model.

Dray *et al.* (2008b), authors of the SimDrugPolicing model, highlight how the models designed here may be further developed through the integration of a more realistic GIS environment, in addition to developing a 'community wide' policing approach beyond individual constables. Furthermore, Dray *et al.* (2008b) acknowledge the dynamic and individual nature of drug users and note that all drug users within a population are unlikely to be uniformly regular users at the time of model initialisation. They therefore acknowledge that these models should be amended to reflect this.

## 12.6 Understanding the burglary problem in Leeds, UK

### *12.6.1 Background*

In recent years, the city of Leeds has witnessed a substantial burglary problem. Between April 2007 and March 2010, the burglary rate in Leeds increased by 13%, despite a 4% fall nationally. Indeed, a number of factors have been identified as underpinning high levels of burglary in Leeds: a high student population, poor quality private housing, a thriving stolen goods market, and a high transient population, which may result in lower levels of community cohesion (Safer Leeds Partnership, 2011). Malleson (2010) developed an ABM of burglary that sought to better understand the burglary problem in Leeds, which, at the time of this work, had one of the highest burglary rates in the UK (Safer Leeds Partnership, 2011). The ABM focused on the East and South East areas of the city, which had recently observed a regeneration project termed 'EASEL' ('East and South East Leeds') shortly prior to this work (further details of this are provided later in the chapter). The extent of the EASEL scheme boundaries are outlined in Figure 12.4.

The model was built in the Java programming language, using the Repast Simphony programming platform (North *et al.*, 2005). This software enabled the integration of a realistic GIS environment into the model. Aspects of the model were used to represent and verify criminological theory, namely Routine Activity, Crime Pattern and Rational Choice theories. This was explored through examining whether the model results generated are consistent with those expected, based on our current theoretical understanding of criminal behaviour.

Malleson (2010) sought to recreate the activities of offenders and victims in the model as closely as possible to develop an ABM that generates realistic crime patterns. Agent (offender) data was therefore based upon data provided by the 'Safer Leeds Partnership' (hereafter referred to as 'Safer Leeds'), a local government collaboration between the police and Leeds City Council to reduce crime and the fear of crime (Safer Leeds Partnership, 2011). Safer Leeds provided data on every individual involved with an offence between April 2000 and March 2004. This was comprised of each individual's address (aggregated to postcode level), together with the crime location. The home locations of offenders in the model were randomly assigned to a property in the output area of the postcode for each individual in the data set. The locations of drug dealers in the model were assigned based on data provided by Safer Leeds of the locations of drug-dealing-related crime. The provision of this crime data enabled the validation of crime patterns generated by the model with actual crime patterns.

One of the key features of the model is the integrated cognitive architecture of agents, with the motivations and behaviours of offender agents being driven by their needs and motives at any given time. Specifically, the model utilised the 'PECS' (physical conditions, emotions, cognitive capabilities, and social status) cognitive architecture, which posits that human behaviour can be modelled based

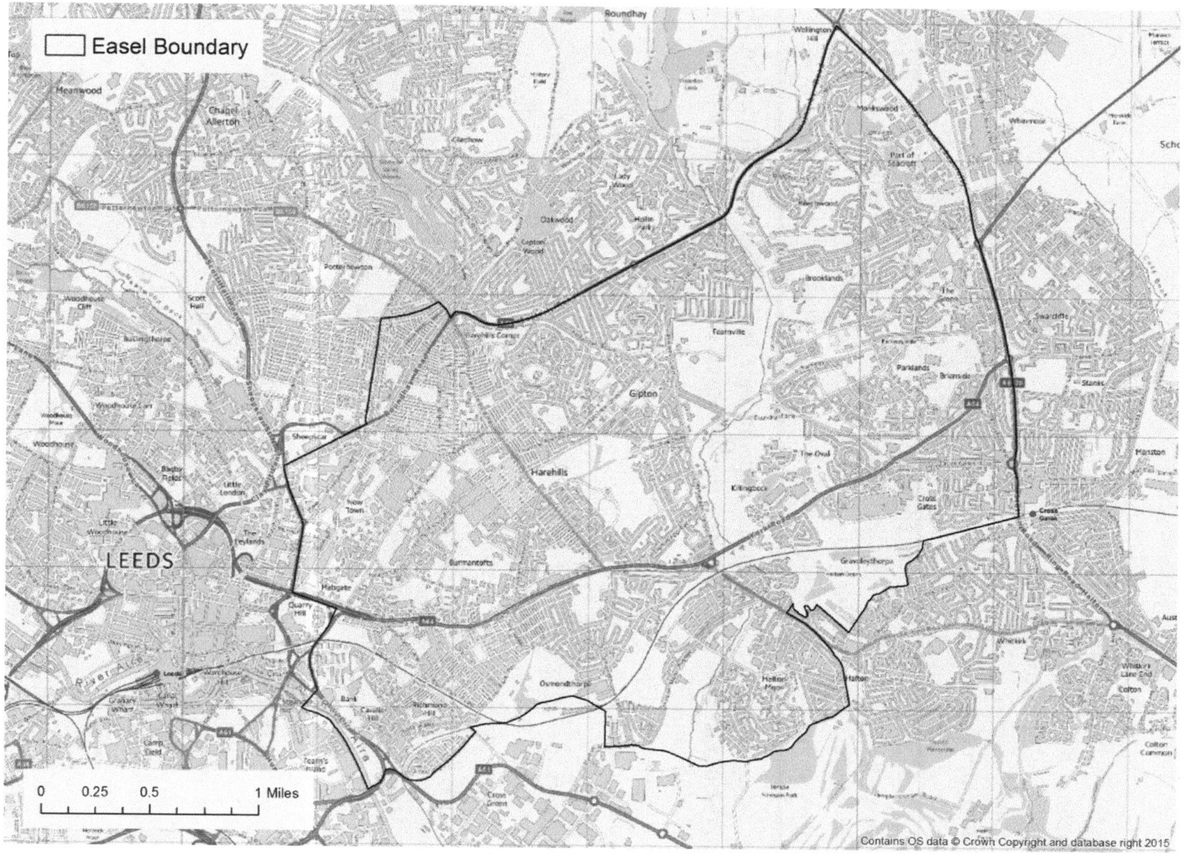


*Figure 12.4* The EASEL regeneration scheme boundary

Source: Malleson *et al.* (2012, p. 61)

on the above four elements (Schmidt, 2000). Agents held a number of motives, the strongest of which at any given time drove a specific behaviour. An overview of this process is illustrated in Figure 12.5.

Malleson *et al.* (2013) disaggregated the offending process into three distinct stages:

- Deciding where to begin searching
- Searching for a target
- Choosing a suitable target

Agents developed a 'cognitive map' of their environment, making a mental note of houses and neighbourhoods passed as part of their daily activities. This enabled agents to develop an 'awareness space' of the areas around their home and places visited regularly (in line with crime pattern theory). Agents can travel from their home to a given location in their awareness space before searching that area for a suitable target. Offender agents will target individual houses in these areas based on an attractiveness value, incorporating features such as affluence, visibility, community cohesion, and distance travelled to the offence. In addition, the model was built to reflect the increased risk of burglary for a property and neighbouring properties for a subsequent period following a burglary offence (Johnson and Bowers, 2004). Should a suitable property arise, the burglar agent would target that property to burgle. An overview of the agent decision-making process is illustrated in Figure 12.6.

In addition to a sophisticated cognitive architecture, Malleson (2010) integrated a detailed model environment based on the city of Leeds. In developing the model environment, Malleson (2010) identified the distinction between burglary risk at the household and community levels. The model environment therefore comprised two layers: individual house features and community-level features. House boundaries were established using Ordnance Survey MasterMap data, which contained other detailed environmental features, such as rivers or buildings (see Plate 7).

Two examples of household-level risk factors used in the model were 'accessibility' and 'visibility', which were calculated through analysing the boundary data of individual households. The inclusion of these factors allowed for the verification of criminological theory, namely Rational Choice and Routine Activity theories.

The model also comprised a transport network, which governed how and where agents were able to navigate the model. A variety of different road types were represented, using the Ordnance Survey Integrated Transport Network (ITN) MasterMap layer. This network enabled agents to navigate the environment at different speeds, based on the transport utilised and available to them. Malleson (2012) also analysed the roads within the transport network to estimate traffic volume for each road and explore the impact of this on the potential burglary risk associated with individual houses.

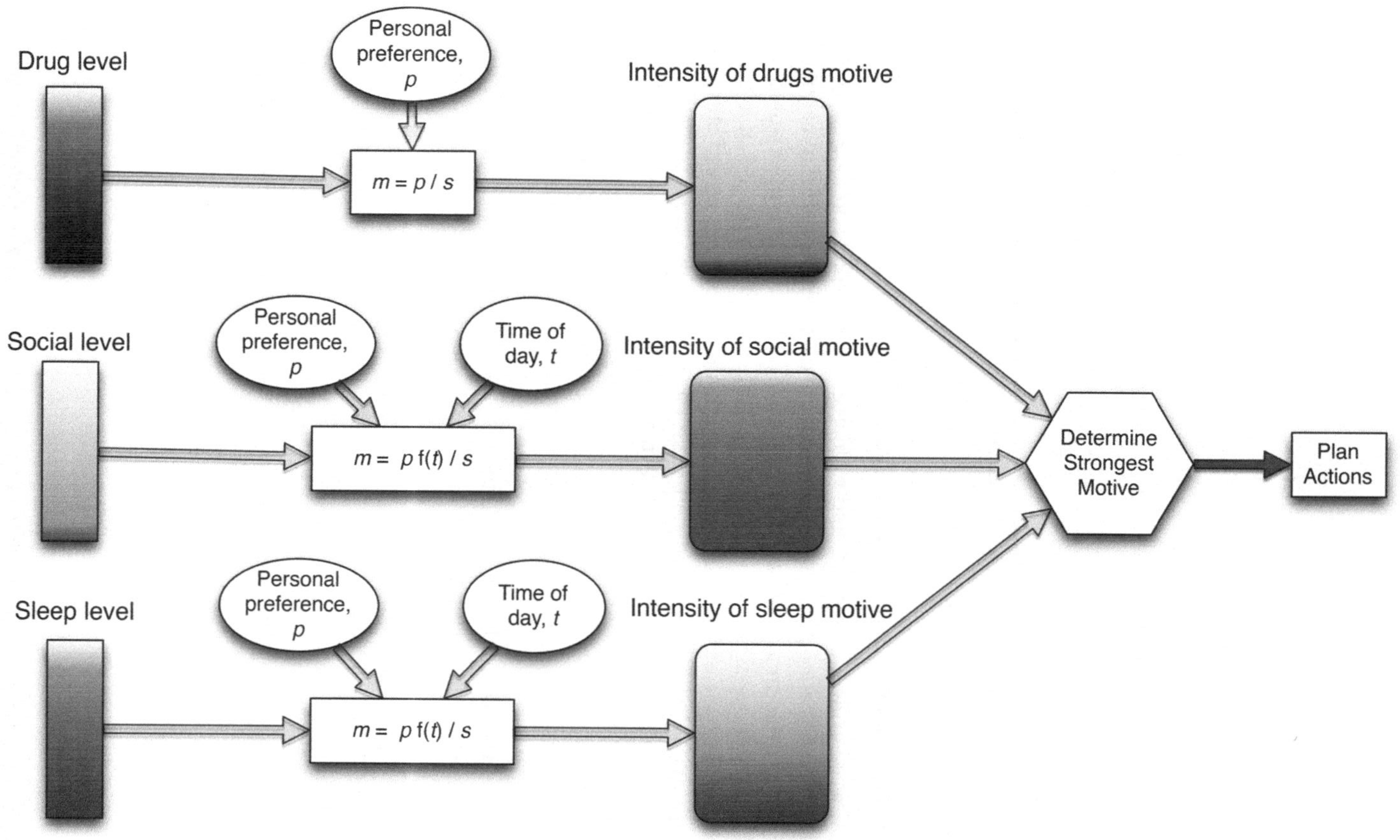


*Figure 12.5* Overview of how state variables, *s*, personal preferences, *p*, and external factors, *t*, are used to determine an agent's strongest motive and subsequent behaviour

Source: Malleson *et al.* (2012, p. 56)

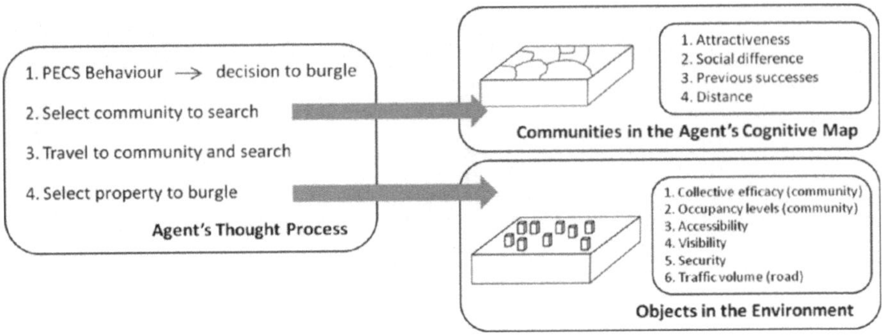


*Figure 12.6* The decision-making process of agents in the model

Source: Adapted from Malleson (2010, p. 124)

### *12.6.2 Model evaluation*

The model was verified, calibrated, and validated to evaluate its predictive ability. The initial verification process was undertaken to ensure the model dynamics were correct and the model behaved as expected. The model results exhibited were in line with those expected. For example, Malleson (2010) found that lower values of visibility, security, and traffic volume resulted in greater risk of burglaries, in line with Rational Choice and Routine Activity theories, as well as Newman's 'defensible space' theory (Newman, 1972). Malleson *et al.* (2012) also found that, in line with crime pattern theory, agents' awareness spaces centre around certain 'anchor' points, with burglaries committed in areas with which offenders were most familiar, for example, clustering around agents' homes.

A further important result was in relation to public transport. Malleson (2010) found that when agents had access to public transport, they would use this for travelling between their home, social nodes, and drug dealers. However, Malleson (2010) also found that the areas and properties passed while using public transport were not added to agents' awareness space, which helped to account for why burglaries were more clustered (see Plate 8). This finding was particularly important in illustrating the value and power of the model. The model also found that agents did not burgle in close proximity to the location of their drug dealers. This appeared to be explained in that as agents visited more nodes, they 'expanded' their awareness space, therefore expanding their knowledge of potential opportunities with which to offend (Malleson *et al.*, 2012).

### *12.6.3 Calibration*

The model was calibrated and validated using real offence data provided by West Yorkshire Police. One of the key findings that emerged during this process was the model's failure to highlight burglary hotspots in the Halton Moor area of Leeds, with no apparent indication of why this was not simulated through the model. Following discussions with Safer Leeds, it emerged that burglary in this area was often driven by the desire to *intimidate* rather than pursue financial gain. Malleson *et al.* (2013) note, therefore, how common assumptions upon which the model has been based, i.e. financial incentive, may be flawed, which will likely bear implications for crime reduction initiatives in areas where financial incentive is not the sole driver for offending.

Furthermore, the calibration process found that the model predicted offending patterns relatively well at the aggregate level; although, what remained unclear was whether the 'right offenders' were committing the 'right crimes' at the individual level. The use of origin–destination matrices enabled a more detailed exploration of offending at the individual level; this illustrated the flows of people travelling between areas to offend, for both observed and simulated data. Upon further analysis, it emerged that offender agents in the model were likely to travel further to crime sites than offenders in the real world (Malleson *et al.*, 2012).

### *12.6.4 Validation*

The model was then validated using crime data from a different year. The validation process suggested that the model did not perform as well as during the calibration process, which may infer that the model was 'over-fitted' to the calibration data. However, Malleson (2010) suggests there may be a number of factors that can account for these results, for example, a lack of up-to-date Census data.

### *12.6.5 Forecasting experiments: Leeds, UK*

The model was also assessed to explore its use as a forecasting tool. As noted earlier, the model was developed based on the South-East of Leeds, which had recently observed the EASEL regeneration scheme to reduce deprivation through the generation of increased opportunities for employment, transport links, new housing, and 'green space' (Malleson *et al.*, 2013). It was anticipated that this initiative would help attract residents to create more stable, affluent neighbourhoods. Malleson *et al.* (2013) therefore sought to explore the impact of the EASEL scheme on burglary in this area.

The physical and social features of the model environment were adapted to represent the changes made through the EASEL scheme. Based on these changes, the model was able to estimate burglary at the individual household level. The model found that property security was a greater indicator of burglary than levels of community cohesion. In addition, the model indicated that properties in close proximity to the EASEL scheme may be at an increased risk of burglary. In particular, it was found that properties on the main road of the northern boundary to the EASEL scheme experienced particularly high burglary rates. Upon further exploration, Malleson *et al.* (2013) suggested that the regeneration scheme may have drawn offenders to that area to burgle, and that the position of properties on a main road, together with features of some properties, such as greater accessibility, increased the risk of burglary for certain properties on this road compared with neighbouring properties. This finding demonstrates the potential value of the Agent-Based Modelling approach to policy makers.

### *12.6.6 Forecasting experiments: Vancouver, Canada*

Malleson (2010) also demonstrated the flexibility of the model, applying it to the city of Vancouver, Canada. The impact of a high-speed rail network was examined in relation to rates of burglary; this was particularly pertinent in light of the forthcoming opening of a new rail line ahead of the 2010 Winter Olympics. Malleson (2010) found that the rail network demonstrated little impact on patterns of burglary, which was a stark contrast to common assumptions about the transport network impacting on burglary rates. Nevertheless, this work demonstrated the value of the ABM approach in exploring the dynamics between residential burglary and the transport system.

The model developed by Malleson (2010) and illustrated within subsequent research (Malleson *et al.*, 2012) is a clear advancement of previous crime (and in particular, burglary) ABMs. The model includes a sophisticated cognitive architecture and realistic GIS environment, while using microsimulated demographics with simulated routine activities for victims and offenders to verify the merits of criminological theory. The model also demonstrates the impact of an urban regeneration scheme on patterns of burglary incidence. Furthermore, the model was subject to rigorous evaluation methods, which remains one of the key drawbacks with existing crime ABMs.

Despite these strengths, Malleson *et al.* (2012) found the model was more accurate in predicting crime patterns at an aggregated rather than individual level. Furthermore, the model does not account for the heterogeneity of offenders with regards to their selection criteria, and Malleson (2010) acknowledged that there was limited information available on aspects of the model that may impact on burglary, including socialising practices and details of co-offending. The author of this chapter is currently undertaking research that will help to address these areas. Namely, the author is undertaking research to help understand the heterogeneity of offenders in Leeds. This will include gathering information on offenders' target selection criteria and their behavioural practices prior to, during, and following their offending, which will allow for the behavioural preferences of offenders to be more accurately reflected in a model. This work is of clear value for policy makers and criminal justice agencies, for example to support and act as a 'test-bed' for crime reduction initiatives.

## 12.7 Conclusions

This chapter has detailed the use of the Agent-Based Modelling approach to real-life crime phenomena, charting a new and growing wave of spatial analysis techniques, moving from analytical approaches to more predictive, exploratory work. The case studies presented here illustrate the value of the ABM approach, together with the benefits associated with using both abstract and realistic model environments. This chapter has illustrated how such crime ABMs can be used to support and understand crime prevention/reduction efforts across different crime contexts. Furthermore, the ABM approach demonstrates a number of key advantages over other existing techniques for modelling crime; most notably, enabling the exploration of the behavioural dynamics of individual offenders and how these may lead to broader crime patterns, while acting as a 'test-bed' to explore the impact of crime prevention/reduction initiatives. The case studies presented demonstrate the capacity of the ABM approach to explore the impact of existing and future policies.

Nevertheless, one of the current key drawbacks to the ABM approach relates to the process of evaluating models. Researchers have highlighted the lack of a standardised approach to evaluate models effectively (Grimm *et al.*, 2005; Malleson, 2013), and Castle and Crooks (2006) note the lack of *specific* analytical methods to evaluate ABM results, particularly for models which are exploratory in nature. Based on the work of Castle and Crooks (2006), the process of model evaluation can be split into the processes of verification, calibration, and validation. However, Malleson (2012) highlights how evaluating the predictive accuracy of ABMs remains a particularly difficult task due to highly complex models and a lack of individual-level information from which to evaluate the model. There remains substantial scope for future work in this area. As these methods remain largely in their infancy, their development may help to create more 'standardised' means of model evaluation (Malleson, 2013). Indeed, these processes are vital to the advancement of simulation methodologies, to ensure researchers are confident in the methods used in this evolving area of research (Townsley and Johnson, 2008; Malleson, 2013).

Finally, the introduction of more accessible programming platforms means that the development of ABMs to explore social/ecological systems is no longer restricted to those with a programming background. Indeed, programming platforms such as the 'Netlogo' software package (Wilensky, 1999) enable researchers without the necessary programming experience to build relatively sophisticated models relatively quickly, allowing researchers to place greater focus on the behavioural dynamics of a model over specific programming syntax.

## 12.8 Future work

In demonstrating the value of the ABM approach, the unavoidable question of 'where next?' is raised with regards to the future of ABM and crime. With the upsurge in ABMs developed across crime phenomenon, one obvious response will be to ensure models are developed through being informed directly from the population under study. The purpose of this would not be to model individual people as such, but to more accurately represent the population of the system. In the context of burglary, this could involve establishing the target selection criteria of offenders, for example, the types of area and property features attractive to offenders. This could be informed directly from offenders themselves, and supported through collaborations with criminal justice agencies.

At the broader level of model development, one direction for future research may be the move towards the prediction of crime. This would follow the trend of the 'Big Data' initiative currently spreading across social science research, which enables the exploration of individual behaviours and movement patterns across large data sets. While this can be invaluable across a number of areas, in the context of crime, modelling the behaviour of people at the individual level is fraught with a range of ethical, moral and logistical issues. Further, this would raise the question of 'to what end' would this serve, as well as issues around how resources could be justifiably invested based on individual offenders. Nevertheless, with regards to the practical application of crime ABMs, there appear to be two key uses, both of which have the ultimate aim of supporting criminal justice policy and operational policing efforts:

1) Medium- to longer-term prediction – exploring how crime patterns emerge over time across different contexts/populations and how this may be used to support operational policing efforts.
2) To continue to act as a 'test-bed' for crime reduction initiatives – to enhance our understanding of crime system dynamics and how this can be applied to support the effective implementation of crime reduction policy.

## References

Axelrod, R. (1997) Advancing the art of simulation in the social sciences. In: Conte, R., Hegselmann, R., and Terna, P. (eds), *Simulating Social Phenomena* (vol. 456). Berlin: Springer-Verlag, pp. 21–40.

Birks, D. J., Townsley, M., and Stewart, A. (2012) Generative explanations of crime: using simulation to test criminological theory. *Criminology* [Online], 50(1), 221–254. Available from: http://onlinelibrary.wiley.com/doi/10.1111/j.1745-9125.20112.00258.x/abstract [accessed 25 October 2014].

Bonabeau, E. (2002) Agent-based modeling: methods and techniques for simulating human systems. *Proceedings of the National Academy of Sciences* [Online], 99(3), 7280–7287. Available from: www.pnas.org/content/99/suppl_3/7280.full [accessed 15 November 2014].

Bousquet, F., Bakam, I., Proton, H., and Le Page, C. (1998) CORMAS: common-pool resources and multi- agent systems. *Lecture Notes in Artificial Intelligence* [Online], 1416, 826–837. Available from: http://link.springer.com/chapter/10.1007/3-540-64574-8_469 [accessed 12 December 2014].

Castle, C. J. E. and Crooks, A. T. (2006) *UCL Working Papers Series, Paper 110: Principles and Concepts of Agent-Based Modelling for Developing Geospatial Simulations* [Online]. London: Centre for Advanced Spatial Analysis, University College London. Available from: http://eprints.ucl.ac.uk/archive/00003342/01/3342.pdf [accessed 2 December 2014].

Chainey, S. and Ratcliffe, J. (2005) *GIS and Crime Mapping*. Chichester: John Wiley & Sons.

Day, C., Topp, L., Rouen, D., Darke, S., Hall, W., and Dolan, K. (2003) Decreased heroin availability in Sydney in early 2001. *Addiction* [Online], 98(1), 93–95. Available from: www.ncbi.nlm.nih.gov/pubmed/12492759 [accessed 5 December 2014].

De Haan, W. and Vos, J. (2003) A crying shame: the over-rationalized conception of man in the rational choice perspective. *Theoretical Criminology* [Online], 7(1) 29–54.

Available from: http://tcr.sagepub.com/content/7/1/29.full.pdf+html [accessed 15 October 2012].

Dietze, P., Miller, P., Clemens, S., Matthews, S., Gilmour, S., and Collins, L. (2004) *NDLERF Monograph, No 6: The Course and Consequences of the Heroin Shortage in Victoria* [Online]. Adelaide: Australasian Centre for Policing Research. Available from: http://ndarc.med.unsw.edu.au/sites/default/files/ndarc/resources/TR.206.pdf [accessed 12 December 2014].

Dray, A., Mazerolle, L., Perez, P., and Ritter, A. (2008a) Policing Australia's heroin drought: using an agent-based model to simulate alternative outcomes. *Journal of Experimental Criminology* [Online], 4(3), 267–287. Available from: http://link.springer.com/article/10.1007%2Fs11292-008-9057-1 [accessed 10 October 2014].

Dray, A., Mazerolle, L., Perez, P., and Ritter, A. (2008b) Drug law enforcement in an agent-based model: simulating the disruption to street-level drug markets. In: Liu, L. and Eck, J. (eds), *Artificial Crime Analysis Systems: Using Computer Simulations and Geographic Information Systems*. Hershey, PA: IGI Global, pp. 352–371.

Elffers, H. and van Baal, P. (2008) Realistic spatial backcloth is not that important in agent based simulation: an illustration from simulating perceptual deterrence. In: Liu, L. and Eck, J. (eds), *Artificial Crime Analysis Systems: Using Computer Simulations and Geographic Information Systems*. Hershey, PA: IGI Global, pp. 19–34.

Epstein, J. M. (1999) Agent-based computational models and generative social science. *Complexity* [Online], 4(5), 41–60. Available from: http://citeseerx.ist.psu.edu/viewdoc/download?doi=10.1.1.118.546&rep=rep1&type=pdf [accessed 5 October 2014].

Epstein, J. M. and Axtell, R. L. (1996) *Growing Artificial Societies: Social Science from the Bottom Up*. Cambridge, MA: MIT Press.

Gilbert, N. (2008) Agent-based Models. In: *Series: Quantitative Applications in the Social Sciences* (vol. 153). Thousand Oaks, CA: Sage.

Grimm, V., Revilla, E., Berger, U., Jeltsch, F., Mooij, W. M., Railsback, S. F., Thulke, H.-H., Weiner, J., Wiegand, T., and Deangelis, D. L. (2005) Pattern-oriented modeling of agent-based complex systems: lessons from ecology. *Science*, 310, 987–991. Available from: www.sciencemag.org/content/310/5750/987 [accessed 2 December 2014].

Groff, E. R. (2008) Characterizing the spatio-temporal aspects of routine activities and the geographic distribution of street robbery. In: Liu, L. and Eck, J. (eds), *Artificial Crime Analysis Systems: Using Computer Simulations and Geographic Information Systems*. Hershey, PA: IGI Global, pp. 226–2512.

Janssen, M. A. and Ostrom, E. (2006) Empirically based, agent-based models. *Ecology and Society* [Online], 11(2), 37. Available from: www.ecologyandsociety.org/vol11/iss2/art37/ [accessed 10 October 2014].

Johnson, S. D. and Bowers, K. J. (2004) The burglary as clue to the future: the beginnings of prospective hot-spotting. *European Journal of Criminology* [Online], 1(2), 237–255. Available from: http://euc.sagepub.com/content/1/2/237.full.pdf+html [accessed 28 October 2014].

Liu, L. and Eck, J. (eds) (2008) *Artificial Crime Analysis Systems: Using Computer Simulations and Geographic Information Systems*. Hershey, PA: IGI Global.

Liu, L., Wang, X., Eck, J., and Liang, J. (2005) Simulating crime events and crime patterns in a RA/CA model. In: Wang G. F. (ed.) *Geographic Information Systems and Crime Analysis*. Hershey, PA: IGI Global, pp. 197–213.

Macy, M. W. and Willer, R. (2002) From factors to actors: computational sociology and agent-based modeling. *Annual Review of Sociology* [Online], 28, 143–166. Available from: www.annualreviews.org/doi/pdf/10.1146/annurev.soc.28.110601.141117 [accessed 5 October 2014].

Malleson, N. (2010) Agent-based modelling of burglary. Unpublished PhD thesis, University of Leeds.

Malleson, N. (2012) Using agent-based models to simulate crime. In: Heppenstall, A. J., Crooks, A. T., See, L. M., and Batty, M. (eds.), *Agent-Based Models of Geographical Systems*. Dordrecht: Springer, pp. 411–434.

Malleson, N. (2013) Calibration of simulation models. In: Bruinsma, G. and Weisburd, D. (eds), *Encyclopedia of Criminology and Criminal Justice*. New York: Springer-Verlag, pp. 243–252.

Malleson, N., Heppenstall, A., and See, L. (2010) Crime reduction through simulation: an agent-based model of burglary. *Computers, Environment and Urban Systems* [Online], 34(3), 236–250. Available from: www.sciencedirect.com/science/article/pii/S0198971509000787 [accessed 15 October 2014].

Malleson, N., Heppenstall, A., See, L., and Evans, A. (2013) Using an agent-based crime simulation to predict the effects of urban regeneration on individual household burglary risk. *Environment and Planning B: Planning and Design* [Online], 40(3), 405–426. Available from: www.envplan.com/abstract.cgi?id=b38057 [accessed 2 December 2014].

Malleson, N., See, L., Evans, A., and Heppenstall, A. (2012) Implementing comprehensive offender behaviour in a realistic agent-based model of burglary. *Simulation* [Online], 88(1), 50–71. Available from: http://sim.sagepub.com/content/88/1/50 [accessed 20 October 2014].

Mazerolle, L., Soole, D. W., and Rombouts, S. (2006) Street-level drug law enforcement: a meta-analytical review. *Journal of Experimental Criminology* [Online], 2(4), 409–435. Available from: http://link.springer.com/article/10.1007%2Fs11292-006-9017-6 [accessed 15 December 2014].

Newman, O. (1972) *Crime Prevention through Urban Design Defensible Space*. New York: Macmillan.

North, M., Howe, T., Collier, N., and Vos, R. (2005) The Repast Simphony development environment. In: Macal, C. M., North, M. J., and Sallach, D. (eds.), *Proceedings of the Agent 2005 Conference on Generative Social Processes, Models, and Mechanisms*, 13–15 October 2005, Argonne National Laboratory, Argonne, IL, USA.

Perez, P., Dray, A., Ritter, A., Dietze, P., Moore, T., and Mazerolle, L. (2006) SimDrug: a multi-agent system tackling the complexity of illicit drug markets in Australia. In: Perez, P. and Batten, D. (eds), *Complex Science for a Complex World: Exploring Human Ecosystems with Agents* [Online]. Canberra: ANU EPress, pp. 193–223. Available from: www.oapen.org/download?type=document&docid=458885 [accessed 4 January 2015].

Ratcliffe, J. H. and Rengert, G. F. (2008) Near-repeat patterns in Philadelphia shootings. *Security Journal* [Online], 21, 58–76. Available from: www.palgrave-journals.com/sj/journal/v21/n1/pdf/8350068a.pdf [accessed 16 November 2014].

Safer Leeds Partnership (2011) *Burglary Joint Inspection: Safer Leeds Partnership* [Online]. Available from: http://democracy.leeds.gov.uk/documents/s55189/Appendix%2012.pdf [accessed 23 October 2014].

Schelling, T. C. (1971) Dynamic models of segregation. *Journal of Mathematical Sociology*, 1, 143–186.

Schmidt, B. (2000) *The Modelling of Human Behaviour*. Erlangen: SCS Publications.

Townsley, M. and Johnson, S. D. (2008) The need for systematic replication and tests of validity in simulation. In: Liu, L. and Eck, J. (eds), *Artificial Crime Analysis Systems: Using*

*Computer Simulations and Geographic Information Systems*. Hershey, PA: IGI Global, pp. 1–18.

United Nations Office on Drugs and Crime (2007) *World Drug Report 2007* [Online]. Vienna, Austria: United Nations Publications, Office of Drugs and Crime. Available from: www.unodc.org/pdf/research/wdr07/WDR_2007.pdf [accessed 3 December 2014].

von Neumann, J. (1966) *Theory of Self-Reproduction Automata*. Urbana, IL: University of Illinois Press.

Wilensky, U. (1999) *Netlogo* [Online]. Evanston, IL: Center for Connected Learning and Computer-Based Modeling, Northwestern University. Available from: http://ccl.northwestern.edu/netlogo/ [accessed 20 November 2014].

# Part IV

# Health care planning and analysis

# 13 Modelling the impact of new community hospitals on access to health care

*Holly Shulman, Graham Clarke and Mark Birkin*

## 13.1 Introduction

Since radical reforms to the UK National Health Service (NHS) in the 1990s, policy makers have tended to favour the promotion of economies of scale in health care provision and funding and have sought to concentrate surgical facilities in large, mainly urban hospitals. Generally, this policy has been considered as a success in areas such as stroke care and major trauma care, where significant benefits of concentrating facilities have been claimed. The creation of an 'internal market' in the 1990s also encouraged a number of hospital managers to consider mergers with nearby hospitals in order to avoid duplication of service provision and allow hospitals to become even more specialised (providing further economies of scale). For example, the Leeds General Infirmary and the nearby St James Hospital (both in the city of Leeds, UK) merged in January 2010.

However, it is widely recognised that as service provision becomes more centralised, accessibility for patients becomes a more crucial consideration. Many geographical studies have shown that as distance from a service location increases, utilisation generally decreases. If service provision is very centralised in city-centre hospitals, for example, then increasing concern is raised about access of patients living in outer urban housing estates or rural areas. Thus, there has been recent interest in UK public health policy in providing more localised facilities for minor treatments, and even potentially for day-care minor operations. There are various versions of this new 'local' model: ambulatory or walk-in centres have been opened to provide a first port of call in order to get advice or a preliminary diagnosis. Minor injury units have also been set up with similar responsibilities, especially for treating minor ailments such as cuts, stings, sprains, etc. To date, many of these new ambulatory centres remain in central locations, having been opened in large hospitals simply to ease pressure on the existing accident and emergency units.

However, in May 2014, the (new) UK NHS Chief Executive Simon Stevens went on record as supporting the policy of opening more local hospitals – smaller community hospitals. These are deemed to be especially important for serving older patients with greater mobility problems. As Chris Ham (Chief Executive of the Kings Fund in the UK) comments, 'we know much of the demand for hospital care these days is for routine acute care for growing numbers of older people … what they want is a really good local accessible hospital for them where they need it' (quoted in BBC News, 30 May 2014). However, perhaps these new hospitals would also be of great benefit to poorer communities living in one of the many outer urban council estates (typical in UK cities), whose residents are more likely to suffer from greater health problems.

The aim of this chapter is to explore access to hospital treatment for one major type of condition which could easily be treated in local (day-care) community hospitals or ambulatory centres – cataracts. Cataracts are the clouding of the lens of the eye where surgery is the only effective treatment, typically affecting an older demographic. Carlisle (2014) summarises a recent round-table discussion among eye health experts that recognised that as the population ages, the demand for cataract surgery will increase dramatically. The bad news is that much cataract treatment is currently very centralised, available only in a selected number of hospitals. However, the good news is that surgery today has routinely fallen from a five-day inpatient stay to a day case. The latter could be ideally suited, therefore, to the idea of new community hospitals or ambulatory centres.

In order to evaluate the impact of new community hospitals we need to build two types of model. The first is a morbidity model for cataracts. This would allow us to compare current geographies of hospital usage against potential need (to explore whether there are areas of the city-region where we suspect patients are not getting the treatment they might require). Spatial microsimulation models can be used as a technique to estimate health care morbidity for each individual and hence small-areas within the study region. The microsimulation models enable us to estimate morbidity based on a person's age, gender, ethnicity, social class, etc. Second, we then need to allocate patients (both the present distributions and those estimated through the microsimulation model) to individual hospitals. Spatial interaction models can be used first to reproduce the known flows of patients between an origin (patient residence) and destination (hospitals) and have a long history of applications in a wide variety of social science areas. By exploring the interaction between residents and hospitals, we can examine how important accessibility to hospitals is for understanding variations in health care demand. Spatial interaction models are also excellent for then analysing the impacts of various 'what-if' questions: i.e. what are the impacts on access if new community-style hospitals are established?

The rest of the chapter is set out as follows. In section 13.2 we describe the data and map current hospital rates for cataracts in our study area. Then we explore the individual characteristics of patients with cataracts in section 13.3, using national survey data. In section 13.4 we describe the microsimulation model for estimating persons likely to have cataracts, drawing upon the analysis in section 13.3. In section 13.5 we build the spatial interaction models and conduct various what-if scenarios for examining the impacts of introducing a number of new local hospitals.

## 13.2 Data, conditions and methods

Figure 13.1 shows the study area selected for this research. It contains most of West Yorkshire in the UK, plus the nearby towns of Harrogate and York in North Yorkshire. The main hospitals for cataract surgery are also marked on the map. The study uses a combination of datasets. First, the UK Hospital Episode Statistics (HES) for 2006/2007 are explored for the hospitals in our region. HES is a comprehensive data source that includes detailed records on all inpatient or day cases of admittance at NHS-based hospitals in England. The HES contains information not only on each patient episode, but also the geographic, economic and demographic characteristics of each patient. Table 13.1 shows the main hospitals providing cataract treatment in the study region.

*Figure 13.1* The study area with key hospital locations

*Table 13.1* Main hospitals for cataract surgery in the study region

| *Site Code* | *Hospital Site* |
|---|---|
| RCF00 | Airedale NHS Trust |
| RAE01 | Bradford NHS Trust |
| RWA16 | Castle Hill Hospital |
| RCD01 | Harrogate District Hospital |
| RR801 | Leeds General Infirmary (LGI) |
| RR813 | St. James University Hospital |
| RR807 | Wharfedale Hospital |
| NTYD6 | Yorkshire Eye Hospital |
| RCB00 | York NHS Trust |

Using this data it is possible to plot the numbers of patients receiving cataract surgery from the census tracts which surround the hospitals (the middle super output areas – MSOAs). Figure 13.2 shows that it is clear that there are widespread geographical variations. The Leeds area has the highest rates, with much lower rates in Bradford and Harrogate. An immediate issue is what is the explanation for such widespread variations? Is it simply variations in demand for surgery based on the geodemographic characteristics of the different areas, or is access (or lack of it) producing a major barrier to treatment? In order to answer this question we need, first, to try to estimate 'demand' or 'need' (the number of persons we might expect to have cataracts and require subsequent surgery), then we can compare to actual hospital treatment rates and attempt to understand the differences (especially the spatial outliers).

To estimate the number of persons with cataracts in the study area we need to understand the key drivers of need. Thus in the next section we explore the characteristics of persons with cataracts reported in the Health Survey of England (HSE). This is an annual survey that asks questions relating to health and health activities, including the presence of cataracts. We shall undertake a statistical analysis of this dataset in section 13.3. The 2001 UK Census was also utilised for analysis. The 2001 Census data was downloaded from the Casweb website. For this research, the unit of analysis of census data chosen was middle super output areas (MSOAs), of which there are 318 for the entire study region, each containing approximately 7,200 people.

## 13.3 Exploring the attributes of patients with cataracts

Both chi-squared and logistic regression analyses were employed to view the strength of the relationship between the different risk factors for cataracts seen in the HSE. The higher the chi-squared value, the higher the explanatory power of this variable is for predicting the risk of cataracts. The higher the Wald statistic, the greater also is the significance of the risk factor. The exp(b) represents the odds ratio. If this value is above one, then this factor leads to increased risk. If the odds ratio is below one, than there is less risk.

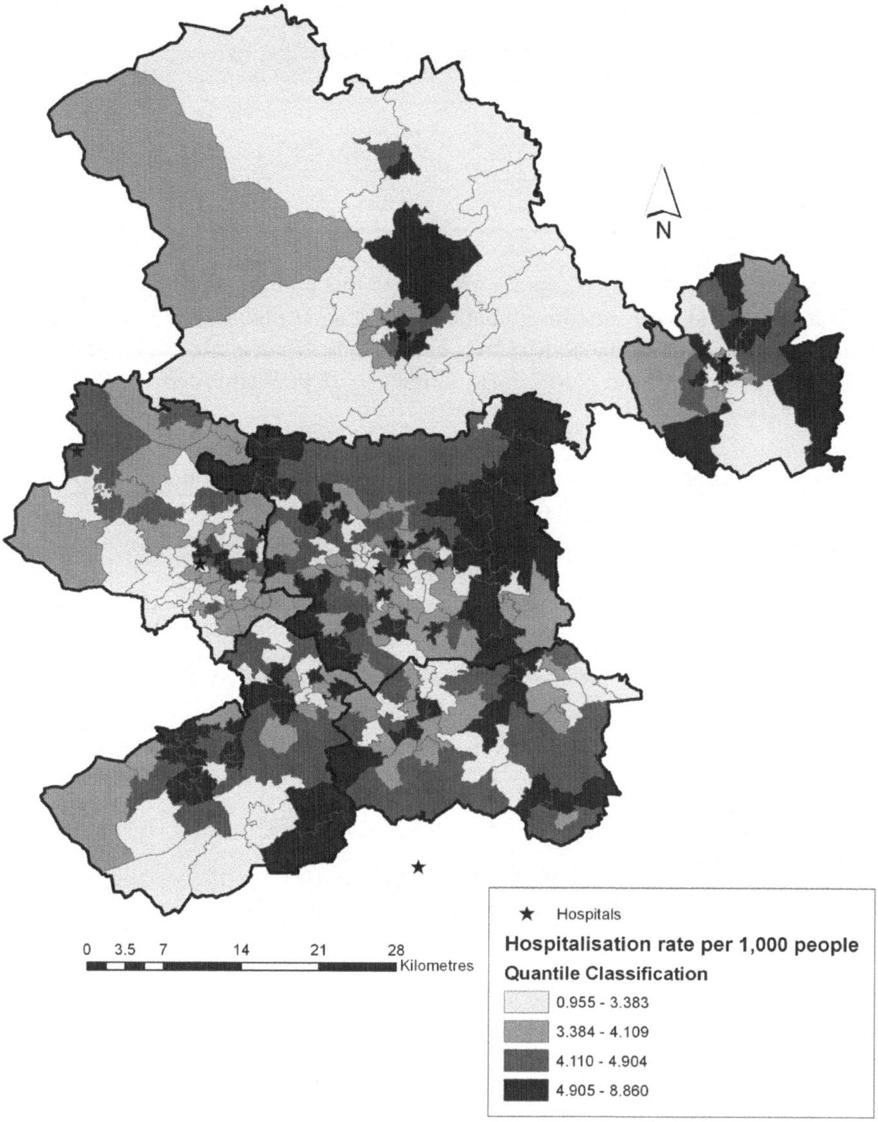


*Figure 13.2* Hospitalisation rates for cataract surgery in the study region

As Table 13.2 and 13.3 demonstrate, age can be identified as a significant predictor of cataracts at the 95% confidence level. The older the person, the more likely it is that they are susceptible to cataracts (as is also demonstrated by the high Wald statistic at the 95% confidence level in the logistic regression analysis). The exp(b) also increases as age increases, which demonstrates again that

*Table 13.2* Results of the chi-squared analysis for HSE 2005 with cataracts as the dependent variable compared to all independent variables at the 95% confidence level

| *Predictors* | *Likelihood Ratio* | *Chi-squared* | *Degrees of freedom* | *Significance* |
|---|---|---|---|---|
| Age | 1,258.39 | 1,557.05 | 5 | 0.000 |
| Gender | 67.27 | 67.64 | 1 | 0.000 |
| Ethnic Group | 44.05 | 30.75 | 4 | 0.000 |
| Self-Assessed General Health | 21.81 | 23.31 | 2 | 0.000 |

older persons are more at risk in comparison to those in younger age categories. The literature also supports this statistical analysis (Hodge et al., 1994; Seddon et al., 1995; Leske et al., 1999; McCarty et al., 1999; Klein et al., 2003; Resnikoff et al., 2004; Mayo Clinic, 2008; NHS Choices, 2009; Waudby et al., 2011). In fact, the Mayo Clinic (2008) state:

> Everyone is at risk of developing cataracts simply because age is the greatest risk factor. By age 65 about half of all Americans have developed some degree of lens clouding, although it may not impair vision. After age 75,

*Table 13.3* Results of the logistic regression for HSE with cataracts as the dependent variable compared to all independent variables at the 95% confidence level

| | *Predictor* | *B* | *S.E.* | *Wald* | *Df* | *Sign.* | *Exp(b)* | *95% CI for Exp(b)* | |
|---|---|---|---|---|---|---|---|---|---|
| | | | | | | | | *Lower* | *Upper* |
| *Ethnicity* | White | | | 6.94 | 3 | 0.074 | | | |
| | Asian or Asian British | −0.59 | 0.53 | 1.26 | 1 | 0.262 | 0.55 | 0.19 | 1.56 |
| | Black or Black British | −1.38 | 0.88 | 2.44 | 1 | 0.118 | 0.25 | 0.04 | 1.42 |
| | Chinese and Other | 0.41 | 0.70 | 0.34 | 1 | 0.56 | 1.52 | 0.38 | 5.95 |
| *Gender* | Men | −0.17 | 0.15 | 1.23 | 1 | 0.27 | 0.85 | 0.63 | 1.14 |
| *General Health* | Good General Health | | | 45.48 | 2 | 0.000 | | | |
| | Fairly Good General Health | −1.10 | 0.21 | 28.53 | 1 | 0.000 | 0.33 | 0.22 | 0.49 |
| | Not Good General Health | −0.09 | 0.19 | 0.24 | 1 | 0.621 | 0.91 | 0.61 | 1.134 |
| *Age* | 16–34 | | | 138.95 | 5 | 0.000 | | | |
| | 35–49 | −3.66 | 0.48 | 57.74 | 1 | 0.000 | 0.03 | 0.01 | 0.07 |
| | 50–59 | −2.85 | 0.34 | 72.32 | 1 | 0.000 | 0.06 | 0.03 | 0.112 |
| | 60–64 | −2.25 | 0.31 | 54.04 | 1 | 0.000 | 0.11 | 0.06 | 0.19 |
| | 65–84 | −2.03 | 0.34 | 29.84 | 1 | 0.000 | 0.13 | 0.06 | 0.27 |
| | 85+ | −1.49 | 0.19 | 64.69 | 1 | 0.000 | 0.23 | 0.16 | 0.32 |
| | Constant | −0.85 | 0.56 | 2.27 | 1 | 0.132 | 0.43 | | |

> as many as 70% of Americans have cataracts that are significant enough to impair their vision.
>
> (Mayo Clinic, 2008)

The second key variable of importance is gender. Although being a woman is not significant in the logistic regression, the chi-squared analysis does highlight the importance of gender in the risk factors for cataracts. According to the literature, women are typically affected more by cataracts than men (Seddon et al., 1995; McCarty et al., 1999; Resnikoff et al., 2004; The Eye Digest, 2009; Waudby et al., 2011). Resnikoff et al. (2004, p. 847), for example, note: 'The number of women with visual impairment, as estimated from the available studies, is higher than men even after adjustment for age. Female/male prevalence ratios indicate that women are more likely to have a visual impairment than men in every region of the world: the ratios range from 1.5 to 2.2.'

As seen in Tables 13.2 and 13.3, ethnicity is also significant at the 95% confidence level for the chi-squared and the logistic regression analyses. The fact that for certain ethnic groupings, the exp(b) was above one is important. For cataracts, the Chinese and Other ethnic category has an exp(b) above one, which indicates that these ethnicities are more at risk for cataracts than others. In the literature, certain ethnic groups have been found to be more at risk of cataracts than other ethnic groups; for example, more so for those of black heritage than Caucasians (Leske et al., 1999; McCarty et al., 1999).

Finally, self-assessed general health is significant at the 95% confidence level for cataracts in the chi-squared analysis. However, in the logistic regression, 'not good' general health is not a significant risk factor, but having 'good' general health or 'fairly good' general health are both significant at the 95% confidence level. This seems odd at first glance, as poor self-reported health tends to be associated with conditions such as diabetes, obesity, smoking and high steroid and alcohol use. Perhaps 'poor' health is simply not as important for cataracts, given more important variables such as age, gender and ethnicity.

The statistical analyses demonstrates the strength of association of each of the potential key variables of those diagnosed with cataracts. We have identified the four most important factors and each of these will be used as constraints or matching variables in the microsimulation analysis below.

## 13.4 Estimating demand for cataract surgery

A growing number of studies have used microsimulation techniques to combine variables to produce new 'missing' data in the field of health care (Clarke and Spowage, 1984; Tomintz et al., 2008; Edwards and Clarke, 2009; Morrissey et al., 2010, 2012; Smith et al., 2011). In this case, we wish to combine the four variables identified as important risk factors for cataracts in section 13.3 to estimate the distribution of persons most likely to need subsequent surgery. Thus an older, female, Chinese person who is in poor health is far more likely to have cataracts than a young, white male in good health. Microsimulation is a technique to link

*Table 13.4* Variables used and their source at OA level

| *Variable* | *Where does it come from?* | *Cross-tabulated with what?* | *Present in Survey, Census or both?* |
|---|---|---|---|
| Age | CAS016 Sex and age by general health and limiting long-term illness (2001 Census Data) | Gender | Both |
| Gender | CAS016 Sex and age by general health and limiting long-term illness (2001 Census Data) | Age | Both |
| Ethnicity | CAST03 Theme table on ethnic group(2001 Census Data) | N/A | Both |
| Self-Assessed General Health | CAS016 Sex and age by general health and limiting long-term illness (2001 Census Data) | N/A | Both |

survey data with small-area census data. In this case, we use simulated annealing, a global optimisation method that can be used to find a solution for combinatorial optimisation problems (for more detail, see Hynes et al., 2008, 2009a, 2009b; Cullinan et al., 2011; Harland et al., 2012; Hermes and Poulsen, 2012a, 2012b). Table 13.4 shows the census variables used to match with the survey data.

The simulated annealing microsimulation model code employed and revised for this model was initially created by Harland et al. (2012). The first step in the simulated annealing methodology is to randomly select a sample record from the health survey (i.e. person with cataracts) to be matched with the Census population for each census output area (middle super output areas: MSOAs), using the constraint variables identified above. Once this step is complete, goodness-of-fit tests can be calculated to see how close the matching procedure has worked. Next, an individual within each area is replaced with another individual to see if a better match can be performed. For each iteration, the goodness-of-fit tests are recalculated to see if the fit has improved (i.e. the error between the two datasets decreases or lessens). If so, that individual is kept and another random individual is replaced and tested again. If the fit is found to be worse, then that individual is not replaced and the algorithm moves on to the next iteration. This process is repeated until the annealing threshold, set at the beginning of the simulation, reaches zero. Those individuals estimated to have cataracts were then aggregated to output areas.

Figure 13.3a shows the estimated proportion of the population that had cataracts based on the spatial microsimulation model, whereas Figure 13.3b again plots the actual hospitalisation rates for 2006/2007 for cataracts. The city of Leeds not only experienced high hospitalisation rates for 2006/2007, but also had a high proportion of persons estimated to have cataracts. North Leeds (as seen in Figure 13.3a) had a higher estimated proportion of the population to have cataracts, whereas it has only a moderately high hospitalisation rate. Bradford and Kirklees show the greatest difference on the whole between actual hospitalisations and those estimated to have cataracts, as the MSOAs in these areas have very low actual hospitalisation rates. As far as the estimation procedure is

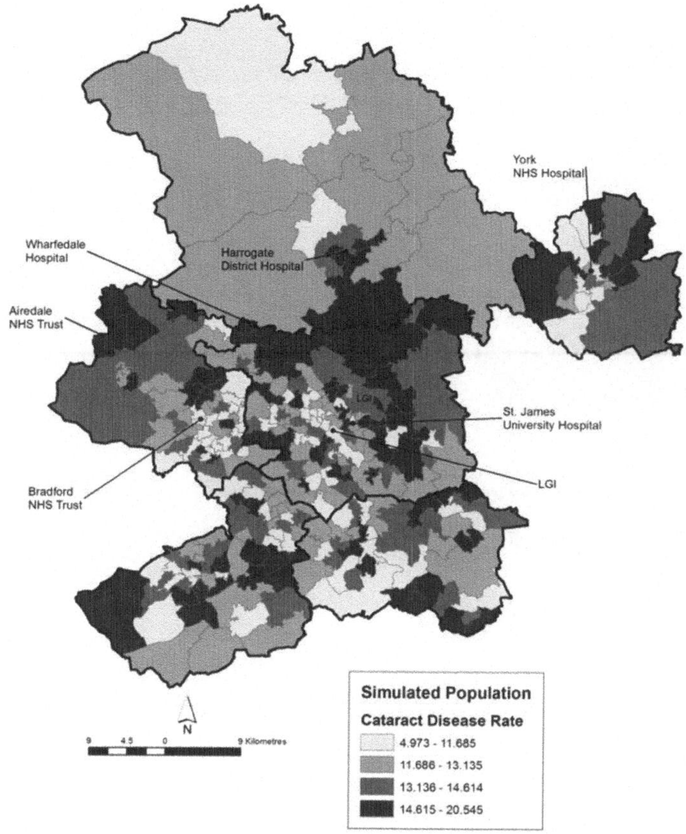


*Figure 13.3a* Simulated rate of cataracts within the population per 1,000 people by MSOA

concerned, Bradford has a very mixed ethnic population and might be expected to have a high proportion of individuals with cataracts due to the high non-white population present in this area.

In Harrogate there were MSOAs where a greater proportion of the population was expected to have cataracts than were treated, given that Harrogate has a very large proportion of elderly residents. Thus, it could be argued that these are areas where need for treatment was/is perhaps not being met adequately. York shows the least difference between actual hospitalisations and estimated populations with cataracts as the values are closer to one.

Thus, there are many areas of the study region where it seems likely that there might be significantly more persons with cataracts than are currently being treated in hospitals. Is it simply because GPs are not referring individuals over long distances? Because cataracts affect such a significantly older population, these people might not be as likely to be referred onwards for treatment as younger patients, as GPs might take into account the accessibility of these

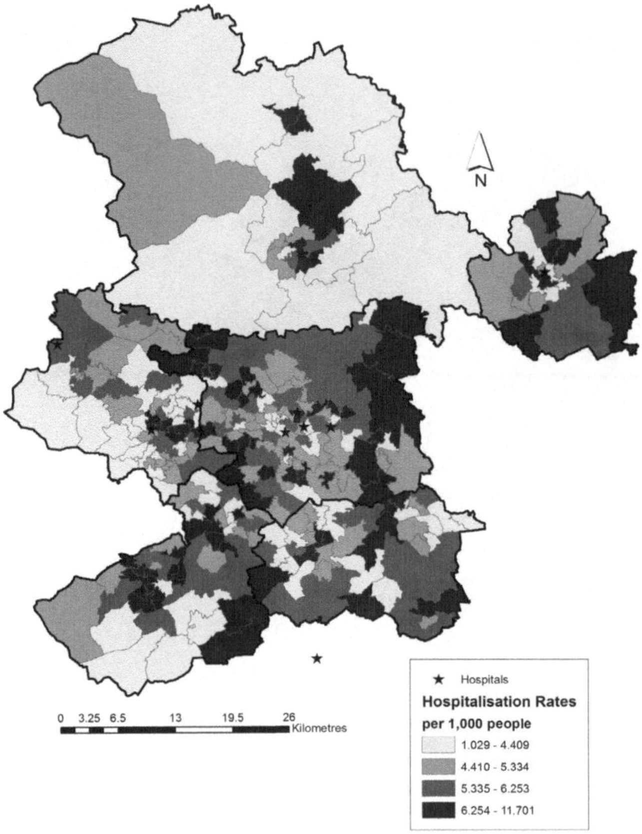


*Figure 13.3b* Hospitalisation rate of cataract treatment for anyone over 16 years of age per 1,000 people by MSOA, 2006–2007

individuals (Jenkins and Campbell, 1996). It may be more difficult for patients of an older age to make the journey to the hospital compared to younger persons. If services were located closer to a high proportion of the population estimated to have cataracts, then perhaps more persons with health care needs could be treated more effectively.

## 13.5 Modelling patient flows to hospitals

In order to examine the impact of changing accessibility to treatment centres for cataracts we first need to build a flow model for existing hospitalisation rates; then it is possible to make changes and explore what-if type scenarios. Spatial interaction models (SIMs) have been used frequently to study the flow of people, goods or services between an origin and a destination. Clarke and Wilson (1985) built a patient flow model for West Yorkshire in the mid-1980s as part of their work on creating a framework of spatial planning for the regional NHS. A patient flow model was incorporated as part of their 'allocation sub-model' for patients to hospital flows. Cowper and Kushman (1987) studied patient flows to primary care providers in Northern California. They explored how the entrance of a new GP to the market affected existing market share and the changes in service areas. As Cowper and Kushman (1987, p. 614) state: 'The cost-efficiency principle of spatial interaction, which describes interaction at an aggregate level, stems from the micro-level behavioural hypothesis that, all else being equal, individual consumers are more apt to make less costly trips than more costly trips.' Other useful applications can be found in Mayhew and Leonardi (1982), Lowe and Sen (1996) and Congdon (2001).

The formula for a spatial interaction model (SIM) can be given as:

$$S_{ij} = A_i O_i W_j^{\alpha} exp^{-\beta C_{ij}} \tag{13.1}$$

$S_{ij}$ is the flows of patients between the MSOA and any hospital. $O_i$ is the demand at each MSOA (the number of people estimated to have cataracts or the actual number of people treated at hospital). $W_j$ is the attractiveness of the hospital. $\alpha$ represents scale economies: when it is above one, then the greater the attraction of the larger hospitals to patients. $exp^{-\beta C_{ij}}$ is the distance decay factor, incorporating $\beta$, which measures patients 'willingness to travel' to a destination, and $C_{ij}$, which is the distance between the centroid of each MSOA and each hospital. $A_i$ is a balancing factor to ensure that all demand is allocated between the hospitals:

$$A_i = \frac{1}{\sum_{ij} W_j^{\alpha} exp^{-\beta C_{ij}}} \tag{13.2}$$

In previous studies, attractiveness has been often measured as the number of beds (attractiveness of size and capacity). However, increasingly, cataracts are performed as day cases and hence number of beds is not so useful as an attraction term. Thus we substitute number of beds for waiting times (a higher waiting time clearly makes a hospital less attractive). Table 13.5 summarises this information for hospitals in our region.

Three different spatial interaction models were run. The first attempted to reproduce the existing flows between residences of patients and hospitals. The SIM was validated based on the existing HES data. For the cataract spatial interaction model, $\beta$ (the distance decay parameter) was calibrated to 0.5 and $\alpha$ (scale economies parameter) to 1. Table 13.6 shows the initial model results.

Overall, there were very small percentage differences seen for the SIM when compared to the actual patient intake as obtained from HES for 2006/2007. More people were treated at St James University Hospital for cataracts than were estimated by the model. However, this 5% difference was small: Birkin et al. (2002, p. 151) state: 'We would expect a good model to be able to reproduce existing flows and revenue totals within 10% of reality.'

*Table 13.5* Descriptive information for each hospital as gathered from Dr Foster Intelligence (2012) and NHS Choices (2013)

| *Hospital* | *Number of Beds* | *Waiting times (in days) for Cataract Treatment* |
|---|---|---|
| Airedale NHS Trust | 334 | 140 |
| Bradford NHS Trust | 874 | 56 |
| Castle Hill Hospital (Hull) | 631 | 55 |
| Harrogate District Hospital | 339 | 98 |
| Leeds General Infirmary | 789 | 98 |
| St James University Hospital | 1041 | 98 |
| Wharfedale Hospital | Unknown | 98 |
| York NHS Trust | 592 | 91 |

The next version of the model replaces existing hospital patients with those estimated to have cataracts in the microsimulation model (which might be a better representation of actual or latent need). This demand layer was calculated based on who would potentially be in need of cataract treatment at any point in time, not just for one given year. The only change with this model was thus the new pattern of simulated demand, which was inputted as the $O_i$ term.

In this run, demand is significantly higher than for the existing situation. Thus, all hospital workloads are over-estimated. Clearly, the present system could not cope if all individuals estimated to have cataracts required treatment at the same time. The model results show that areas with low hospitalisation rates at present, such as Bradford, place particular strains on the local hospitals if all likely demand is to be satisfied.

In order to address the issue of under-capacity, the third model run introduces five new ambulatory care centres (or small community hospital facilities) located throughout the study area (see Figure 13.4). These locations were chosen based on the results shown in Table 13.7 and the pressure on existing facilities in particular.

Figure 13.4 details the location of these five proposed new locations. Facilities 1, 3, and 5 were located within the city of Leeds, where a significant proportion

*Table 13.6* Percentage of patient intake from SIM versus HES data 2006/2007 for cataract treatment

| *Hospital* | *Patient Intake Estimates from the SIM* | *Patient Intake Estimates from HES* |
|---|---|---|
| Airedale NHS Trust | 246 | 215 |
| Bradford NHS Trust | 1,467 | 1,418 |
| Harrogate District Hospital | 598 | 704 |
| Leeds General Infirmary | 1,244 | 1,166 |
| St James University Hospital | 1,228 | 1,524 |
| Wharfedale Hospital | 434 | 190 |
| York NHS Trust | 777 | 777 |

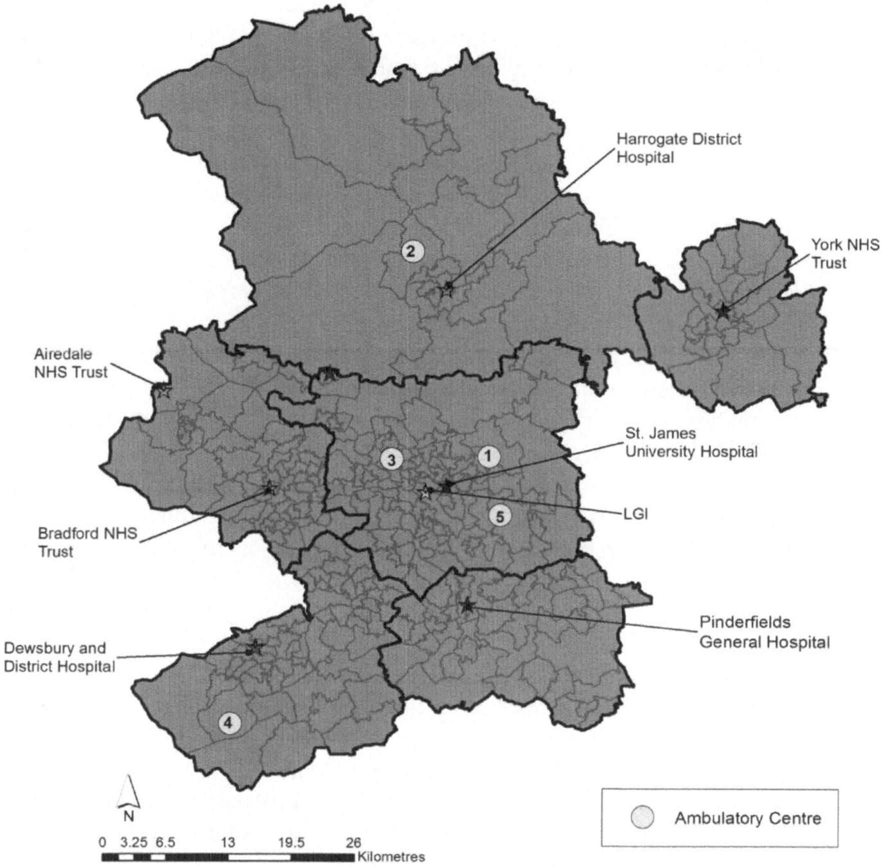


*Figure 13.4* Location of proposed ambulatory care centres in the study area

*Table 13.7* Percentage of patient intake for each hospital for cataract treatment with demand from the microsimulation model

| *Hospital* | *Estimated Patient Intake* | *Patient Intake Estimates from HES* |
|---|---|---|
| Airedale NHS Trust | 747 | 215 |
| Bradford NHS Trust | 6,525 | 1,418 |
| Harrogate District Hospital | 1,618 | 704 |
| Leeds General Infirmary | 5,429 | 1,166 |
| St James University Hospital | 5,261 | 1,524 |
| Wharfedale Hospital | 1,080 | 190 |
| York NHS Trust | 1,896 | 777 |

of the population was estimated to be at risk of cataracts. Location 4 was located in the south of the study area in Kirklees, in the Huddersfield region. This new facility has a marked impact and was estimated to treat people from many areas in the southwest of the study region. Location 2 was located near Harrogate District Hospital, but also close to the more elderly population to the north of the town.

The attractiveness of these facilities – waiting times – was set at three different arbitrary values of 5, 10 and 25 days. Such low waiting times for these centres made them all very attractive to nearby residents. The final results are shown below for the 25-day scenario. The existing Dewsbury District Hospital and Pinderfields General Hospital (in Kirklees and Wakefield, respectively) were also included in this version of the model, providing better access to existing urban hospitals in the south of the region. Demand was the estimated demand as derived from the spatial microsimulation model that was used in the previous spatial interaction models and the $\beta$ and $\alpha$ remained consistent with previous models.

Due to the three locations of ambulatory care centres included within the city of Leeds, the burden placed on the Leeds General Infirmary (LGI) and St James University Hospital was dramatically decreased for cataract treatment, considerably easing the workload on both hospitals. Harrogate District Hospital, Wharfedale Hospital and Bradford NHS Trust also experienced a decrease in patient intake but not to the same extent as the LGI and St James University Hospital. Airedale Hospital, even with the inclusion of these facilities, still has pressure to treat more patients (a possible location for a sixth community facility?)

Distance was a significant contributing factor determining where people were able to travel for cataract treatment; however, the inclusion of these five new facilities with dramatically lower waiting times did have a significant impact on the patient flows of the region. As the primary focus of these ambulatory care centres were for cataract treatment (hence the lower waiting time), it was

*Table 13.8* Percentage intake for each hospital with the inclusion of ambulatory centres at the 25-day proposed waiting time for cataract treatment

| *Hospital* | *Estimated Patient Intake* | *Patient Intake from HES* |
|---|---|---|
| Airedale NHS Trust | 739 | 215 |
| Bradford NHS Trust | 2,932 | 1,418 |
| Harrogate District Hospital | 451 | 704 |
| Leeds General Infirmary | 947 | 1,166 |
| St James University Hospital | 721 | 1,524 |
| Wharfedale Hospital | 541 | 190 |
| York NHS Trust | 1,849 | 777 |
| Dewsbury District Hospital | 1,962 | – |
| Pinderfields General Hospital | 2,706 | – |
| New community hospital 1 | 1,353 | – |
| New community hospital 2 | 1,218 | – |
| New community hospital 3 | 3,067 | – |
| New community hospital 4 | 1,172 | – |
| New community hospital 5 | 2,050 | – |

expected that these new facilities would lead to decreased intakes for the hospitals in the region, which allowed them to lower waiting times for other surgeries.

## 13.6 Conclusions

UK health planners are currently debating the building of new community hospitals within urban areas to alleviate pressure on large central hospitals and to increase access to more vulnerable sections of the population. In order to test the potential impacts of such new facilities, we have undertaken two sets of analysis: first, the construction of a spatial microsimulation model to estimate morbidity for cataracts (and hence potential or latent need). The microsimulation model for cataracts was based on the risk factors of age, gender, ethnicity, and self-assessed general health. In effect, the microsimulation model estimated the population who would be at risk of cataracts by linking variables in the Health Survey of England 2006/2007 and the 2001 UK Census. The microsimulation model was based on a simulated annealing methodology, which produced new estimates of morbidity based on the combination of the above factors: rates much higher than existing hospitalisation rates.

Certain MSOAs demonstrated a much higher estimated need than was seen with the hospitalisation rates for 2006/2007, suggesting that these areas have a higher demand for services than is currently being met. Harrogate demonstrated this issue very clearly as there are large MSOAs in the north of the region where people were estimated to have cataracts but this was not reflected in the hospitalisation rates. This could be due to accessibility; perhaps this area would be better served with smaller treatment facilities to meet this potential demand.

Another reason that this estimated need was not reflected in the hospitalisation rates could be because GPs are not referring patients onwards for treatment. Cataracts affect an older population, which could have more accessibility issues than other age brackets (i.e. no access to transportation) and thus might be less likely to be referred onwards to secondary care. If facilities were placed in these locations, a greater demand of the population could be met.

The second analysis involved the creation of a production constrained spatial interaction model to model the flows of patients between their home MSOA and the associated hospitals in the study area. Three different spatial interaction models were run. The first spatial interaction model was based on actual HES data for 2006/2007. It found that the estimated percentage of people treated at each facility was within 10% of actual flows to those hospitals and in some cases much lower. This first SIM was used for calibration purposes and was validated to within 10% of reality. Spatially, the estimation of flows of patients to hospitals was also similar to the actual flows from HES data. If smaller hospitals began to offer day-surgery options, then waiting times could be lowered for most hospitals and a greater demand could be met (as is seen SIM 3 with the introduction of ambulatory care centres).

Spatial interaction models helped us to better understand the demand and supply of health care in the study region. The inclusion of new community-style hospitals in key areas of the study region could help to mitigate the stress placed on hospital surgical units, especially with the increased waiting times of up to 140 days of some hospitals for cataract surgeries. If these care centres performed cataract day surgery at a much lowered waiting time of 25 days, patients could receive treatment faster and easier than travelling to the existing hospitals. Ambulatory care centres could be built to provide these services, so a greater demand of patients could be provided for, especially if they were located in areas of severe need or that currently have the lowest levels of accessibility to existing hospitals.

## References

Birkin, M., Clarke, G. P. & Clarke, M. (2002) *Retail geography and intelligent network planning*. Wiley, Chichester.

Carlisle, D. (2014) A new vision for cataract surgery. *The Health Service Journal*, 123(6391), 30–33.

Clarke, M. & Spowage, M. (1984) Integrated models for public policy analysis: an example of the practical use of simulation models in health care planning. *Papers of the Regional Science Association*, 55, 25–45.

Clarke, M. & Wilson, A. G. (1985) A model-based approach to planning in the National Health Service. *Environment and Planning B: Planning and Design*, 12, 16.

Congdon, P. (2001) The development of gravity models for hospital patient flows under system change: a Bayesian modelling approach. *Health Care Management Science*, 4(4), 289–304.

Cowper, P. & Kushman, J. (1987) A spatial analysis of primary health care markets in rural areas. *American Journal of Agricultural Economics*, 69(3), 613–625.

Cullinan, J., Hynes, S. & O'Donoghue, C. (2011) Using spatial microsimulation to account for demographic and spatial factors in environmental benefit transfer. *Ecological Economics*, 70(4), 813–824.

Dr Foster Intelligence (2012) Available at: www.drfosterhealth.co.uk.

Edwards, K. & Clarke, G. P. (2009) The design and validation of a spatial microsimuation mode of obesogenic environments for children in Leeds, UK: SimObesity. *Social Science & Medicine*, 69(7), 1127–1134.

Harland, K. (2013). *Microsimulation model user guide (flexible modelling framework)*. Working Paper, School of Geography, University of Leeds, UK.

Harland, K., Heppenstall, A., Smith, D. & Birkin, M. (2012) Creating realistic synthetic populations at varying spatial scales: a comparative critique of population synthesis techniques. *Journal of Artificial Societies and Social Simulation*, 15(1), 1–15.

Hermes, K. & Poulsen, M. (2012a) A review of current methods to generate synthetic spatial microdata using reweighting and future directions. *Computers, Environment and Urban Systems*, 36(4), 281–290.

Hermes, K. & Poulsen, M. (2012b) Small area estimates of smoking prevalence in London: testing the effect of input data. *Health & Place*, 18(3), 630–638.

Hodge, W. G., Whitcher, J. P. & Satariano, W. (1994) Risk factors for age-related cataracts. *Epidemiologic Reviews*, 17(2), 336–346.

Hynes, S., Farrelly, N., Murphy, E. & O'Donoghue, C. (2008) Modelling habitat conservation and participation in agri-environmental schemes: a spatial microsimulation approach. *Ecological Economics*, 66(2), 258–269.

Hynes, S., Morrissey, K. & O'Donoghue, C. (2009a) Building a static farm level spatial microsimulation model: statistically matching the Irish national farm survey to the Irish census of agriculture. *International Journal of Agricultural Resources, Governance and Ecology*, 8(2-4), 282–299.

Hynes, S., Morrissey, K., O'Donoghue, C. & Clarke, G. (2009b) A spatial micro-simulation analysis of methane emissions from Irish agriculture. *Ecological Complexity*, 6(2), 135–146.

Jenkins, C. & Campbell, J. (1996) Catchment areas in general practice and their relation to size and quality of practice and deprivation: a descriptive study in one London borough. *British Medical Journal*, 313(7066), 1189–1192.

Klein, B. E., Klein, R., Lee, K. E. & Meuer, S. M. (2003) Socioeconomic and lifestyle factors and the 10-year incidence of age-related cataracts. *American Journal of Ophthalmology*, 136, 506–512.

Leske, M., Wu, S., Hennis, A., Connell, A., Hyman, L. & Schachat, A. (1999) Diabetes, hypertension, and central obesity as cataract risk factors in a black poulation: the Barbados eye study. *Opthalmology*, 106(1), 35–41.

Lowe, J. M. & Sen, A. (1996) Gravity model applications in health planning: analysis of an urban hospital market. *Journal of Regional Science*, 36, 437–461.

Mayhew, L. D. & Leonardi, G. (1982) Equity, efficiency, and accessibility in urban and regional health-care systems. *Environment and Planning A*, 14(11), 1479–1507.

Mayo Clinic (2008) *Cataracts definition* [Online]. Available at: www.mayoclinic.com/health/cataracts/DS0050 [accessed 10/02/2009].

McCarty, C., Mukesh, B. N., Fu, C. & Taylor, H. (1999) The epidemiology of cataract in australia. *American Journal of Ophthmalogy*, 128(4), 446–465.

Morrissey, K., Clarke, G. & O'Donoghue, C. (2012) Linking static spatial microsimulation modelling to macro models: the relationship between access to GP services & long-term illness. In: Edwards, K. and Tanton, R. (eds), *Microsimulation methods and models*. Springer, London, 127–144.

Morrissey, K., Hynes, S., Clarke, G. & O'Donoghue, C. (2010) Examining the factors associated with depression at the small area level in Ireland using spatial microsimulation techniques. *Irish Geography*, 43, 1–22.

NHS Choices (2009) Cataracts, age-related treatment [Online]. Available at: www.nhs.uk/Conditions/Cataracts-age-related/Pages/Treatment.aspx.

Resnikoff, S., Pascolini, D., Etya'ale, D., Kocur, I., Pararajasegaram, R., Pokharel, G. P. & Mariotti, S. P. (2004) Global data on visual impairment in the year 2002. *Bulletin of the World Health Organization*, 82, 844–851.

Seddon, J., Fong, D., West, S. K. & Valnadrid, C. T. (1995) Epidemiology of risk factors for age-related cataract. *Survey of Ophthalmology*, 39, 323–334.

Smith, D. M., Pearce, J. R. & Harland, K. (2011) Can a deterministic spatial microsimulation model provide reliable small-area estimates of health behaviours? An example of smoking prevalence in New Zealand. *Health & Place*, 17(2), 618–624.

The Eye Digest (2009) *What is a cataract?* [Online]. Available at: www.agingeye.net/cataract/cataractinformation.php [accessed 10/02/2009].

Tomintz, M. N., Clarke, G. P. & Rigby, J. E. (2008) The geography of smoking in Leeds: estimating individual smoking rates and the implications for the location of stop smoking services. *Area*, 40, 341–353.

Waudby, C., Berg, R., Linneman, J., Rasmussen, L., Peissig, P., Chen, L. & McCarty, C. (2011) Cataract research using electronic health records. *BMC Ophthalmology*, 11(1), 1.

# 14 SimSALUD – towards a health decision support system for regional planning

*Melanie N. Tomintz and*
*Victor M. Garcia-Barrios*

## 14.1 Introduction

The Austrian Federal Ministry of Health aims to improve the health of all people living in Austria and to decrease health and social inequalities. The expenditure for health care costs is increasing and resources are getting scarce. Health problems related to a poor lifestyle are especially problematic, e.g. obesity, smoking, type 2 diabetes. Austria is recognised as having one of the best health care systems among the European countries. The Austrian population has a high life expectancy, above the average in Europe (OECD, 2011). Women have a life expectancy of 83 years and men have a life expectancy of 78 years. All this sounds good, but in fact Austria is among the worst performers when it comes to living a healthy life, especially in later life (Bundesministerium für Gesundheit, 2012). In Austria, cardiovascular diseases account for the highest rate of deaths, with 37.1% for men and 47.1% for women, followed by cancer (28.9% males, 23.6% females). The consumption of tobacco products is responsible for many diseases and can even be the major cause of death. Looking at the EU27 countries, there has been a 16% decrease in the smoking rate between 2000 and 2010. The highest decrease is seen for Norway, with an incredible 41% reduction, followed by Iceland with a decrease of 38% and Denmark and Latvia with 34%. Austria achieved a decrease in smoking rates of only 5% and has a current smoking rate of 23.2%, which is 0.2% higher than the EU27 average (23%) (OECD, 2012a). The percentage of male smokers is higher in comparison to women smokers; however, the gap between men and women is closing, especially for younger age groups. For example, the smoking prevalence for people between 15 and 29 years is 29% for women and 32% for men (Statistik Austria, 2012). When it comes to young smokers, Austria is top ranked. A study conducted across most of the OECD countries (in terms of smoking and alcohol consumption in 2005/2006) allows a comparison of the smoking behaviour among 15-year-old children. It shows that Austria is top for young smokers (30% girls, 24% boys) in comparison to all other OECD countries, followed by the Czech Republic, Germany and Finland. A big issue here is that smoking is addictive and is often taken forward into adult life. At the worldwide level, the lowest 15-year-old smoking rates are found in the USA, followed by Sweden, Canada and Norway (OECD, 2012b). This is of serious concern in Austria because smoking is responsible for many types of disease and is still the most preventable cause of death.

Sensitive local planning is necessary to meet people's needs and to distribute scarce health care resources best to meet government aims. This requires good quality health data, which is often not available for smaller geographical areas. For example, in Austria, the National Health Survey is collected only once a decade. To overcome such obstacles, we can make use of small-area simulation methods; here we use *spatial microsimulation* modelling to model this 'missing' data. One of the main advantages of this method is the simulation of specific target groups, e.g. divorced women aged 30 to 44 without a job. The main research centres of spatial microsimulation modelling are found in the UK, the US and Australia. However, the advantages are getting increased recognition worldwide. Drawbacks for using the modelling method include the lack of available software and expert knowledge in the area of spatial microsimulation modelling.

In Austria, this methodology is not yet widespread, especially not in the area of health care and in the usage of spatial microsimulation models. Therefore, within the research project SALUD – funded by the Federal Ministry for Transport, Innovation and Technology and the Austrian Science Fund – we aim to develop a framework for spatial microsimulation modelling, called *simSALUD*, as to date no user-friendly open-source Web framework exists internationally. Spatial microsimulation provides the capability to combine survey and census data for modelling small-area health issues based on individuals or households where no data exists, so-called missing data. Further, the framework aims to implement different microsimulation algorithms, thus allowing an easy way of comparing methods for specific models and data sets. This will also allow interested researchers to get started more easily. Further, different combination of constraints (data variables such as age or income, for example) and sub-constraints (e.g. age groups) can easily be simulated and compared to find the optimal model. Attention is also given to an easy way of validating the simulation results and visualising them in the form of maps. The latter will help to interpret the results, especially for non-experts.

Applied geographers represent a main user group of simSALUD, as they build a bridge to policy decision makers by being experts in geographical analysis. This means applied geographers have expertise in working with spatial data as well as knowing how to identify and show results in an understandable and effective manner to third parties. Applied geographers do not need to be experts in spatial microsimulation modelling itself, but they will understand the purpose behind the methodology and will be able to build models for creating synthesised data for their areas of interest. To give a practical example, they can use simSALUD for simulating specific population groups at high risk of developing cardiovascular diseases for small areas and map the results to identify high-risk areas (see also the discussion in Chapter 13 of this volume). On the one hand, this provides information for policy makers about where more preventive activities are required (e.g. sending out information material related to cardiovascular diseases) and, on the other hand, the simulated data can be combined with other analysis methods to build more sophisticated models for regional planning purposes. Further, simSALUD can be used to simulate different kinds of output data, including diseases, health problems related to lifestyle, chronic diseases, etc., as long as the appropriate input data is available (see section 14.4).

The aim of this chapter is to introduce the framework simSALUD for spatial microsimulation modelling and to highlight its applicability for applied geographers in the area of health policy. Section 14.2 describes the framework with its components. Section 14.3 shows the usage of each component with the framework illustrated by the case study 'simulation of smokers'. Section 14.4 leads to a discussion and outlook. Section 14.5 concludes with a summary.

## 14.2 Spatial microsimulation framework 'simSALUD'

SimSALUD is a flexible Web-based spatial microsimulation modelling package for health decision support with an easy-to-use (wizard-based) graphical user interface for experts, academics and non-experts to produce, map and analyse microdata without programming skills. Health policy makers are one of the major target users of our solution approach. A number of microsimulation models focusing on health care demand have already been built (for example, Clarke and Spowage, 1984; Falkingham and Hills, 1995; Abello and Brown, 2007; Morrissey et al., 2013; Tomintz et al., 2013). However, many of these models are not flexible enough regarding user handling (e.g. changing model input parameters, etc.) and are not accessible to everyone. Often, they are implemented for a specific project purpose only or are implemented as proprietary approaches (e.g. Microsoft Excel, Access, etc.). Therefore, they are either user-unfriendly or simply unable to run for relatively big data. Recently, some community-oriented frameworks have been published but these are often hard to use and very inflexible in terms of data handling. Moreover, special domain knowledge and software programming skills are required; hence they are mainly operated by experts or those that have undergone extensive training. Examples include Life-Cycle Income Analysis Model 2 (LIAM2) (see De Menten et al., 2014), Java MicroSIMulation (JAMSIM) combining the statistical package R and the Java-based Ascape (see Mannion et al., 2012), a spatial microsimulation (sms)-library for parallel simulation using the software language R (see Kavroudakis, 2014), a JAVA-desktop Flexible Modeling Framework (FMF) (see Harland, 2013) or Microsimulation Modelling and Predictive Policy Analysis System (Micro-MaPPAS) for predictive policy analysis (see Ballas et al., 2007). All these frameworks work, of course, but in terms of usability attributes they present a very long learning time until the user acquires expert knowledge in programming and statistical software languages such as R. Further, the user interfaces are not designed for novice users. The import of data is rather inflexible (i.e. they use proprietary data structures) and some of the software show a lack of functionality for data and method validation as well as for visual spatial mapping.

Within the area of spatial microsimulation, our freely available Web application called simSALUD is in current development. Our design and development

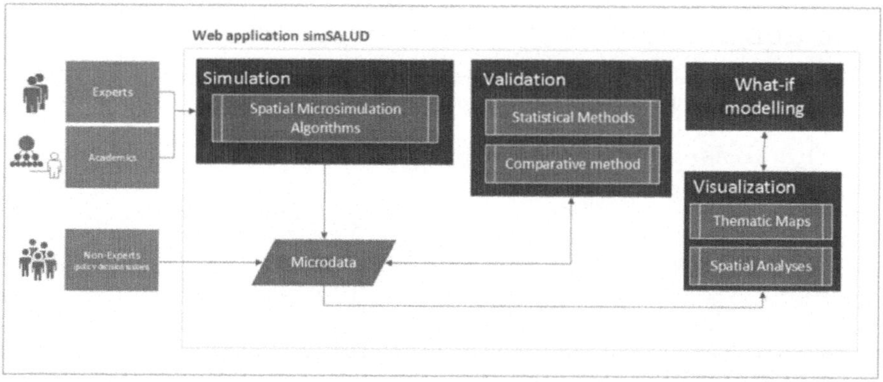


*Figure 14.1* A simplified conceptual overview of the simSALUD Web application

efforts aim at overcoming the aforementioned drawbacks of currently available frameworks. A simplified conceptual overview of the Web application is given in Figure 14.1. The application targets three user groups in particular:

- *Experts* in the area of spatial microsimulation, who know how spatial microsimulation works and how to prepare data so that it can be used in the Web application.
- *Academics*, meaning the framework can be used for lecturers (undergraduates and postgraduates) with accompanying tutorials and test data.
- *Non-experts* (e.g. health policy decision makers (HPDM)), who are mainly interested in using validated microdata, which are produced by experts, to visualise the results in the form of maps. The map results allow the HPDM to show hot spots of areas with highest rates for the simulated data for different population groups.

The framework has been developed in JAVA using the open source Web application framework Struts2. Currently, it is accessible online anywhere and anytime (see www.simsalud.org/). Functionally, the architecture of simSALUD consists of three modules (see Figure 14.2): (i) Spatial Microsimulation Framework, (ii) User Interaction Wizard and (iii) Web Mapping Framework:

- The Spatial Microsimulation Framework (SMF) represents the main server module and is responsible for the data analysis within the system. On this server the (resource- and performance-consuming) algorithms are applied, so the client's software (e.g. the notebook or desktop computer of the users) will not block users in their work.
- The User Interaction Wizard (UIW) represents the graphical user interface of the package. The wizard allows the user to upload all required data sets as well as match the survey and the small-area data together using a drag

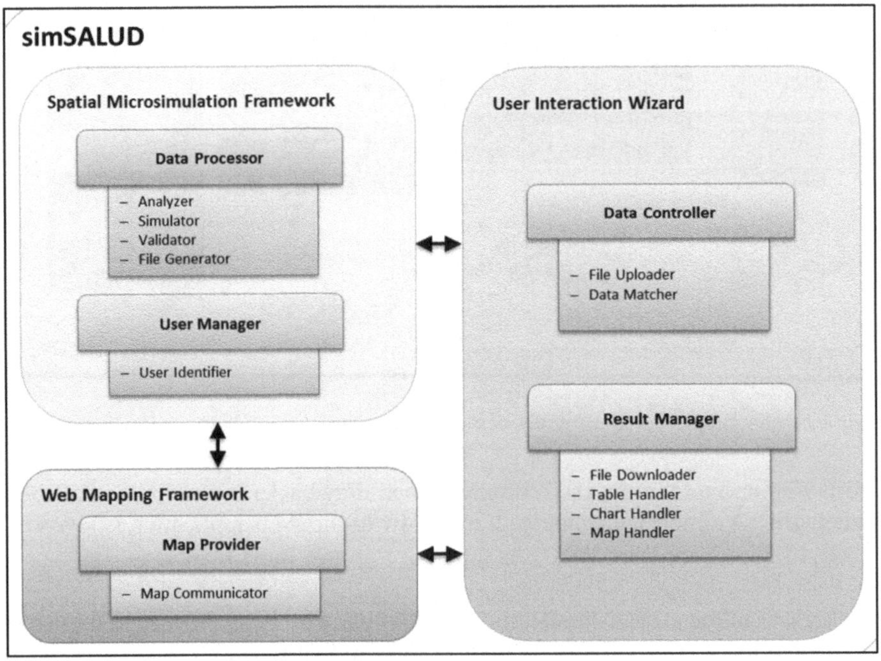


*Figure 14.2* System architecture of the simSALUD framework

and drop technique. The UIW also holds the results module, where the user can download the simulated data file and see all the results as tables, charts and/or map visualisations. The visualisation area is an interactive map with standard functionalities, including zoom and pan.

- The Web Mapping Framework (WMF) module comprises a set of geoprocessing tools and allows the spatial analysis of the data to answer what-if scenarios. The Map Provider with its Map Communicator allows users to process and edit geospatial data.

The application is implemented in the form of a wizard in order to better guide the user through the application and catch possible errors in the data files in advance. The workflow is shown in Figure 14.3. At first, the user of the simSALUD framework has to select their prepared survey file on their local computer and upload the file to the server. The next step is to upload all prepared single constraint files (small-area data), which the experts have specified as strong predictors of the variable being estimated. During this process, the framework picks up errors such as empty data cells, wrong data formats (e.g. there are letters in columns where only numbers are allowed) or the wrong order of the data columns (e.g. the survey file needs to have the weighting in the second column, followed by the variable to simulate and the constraint variables). After successfully uploading the data to the

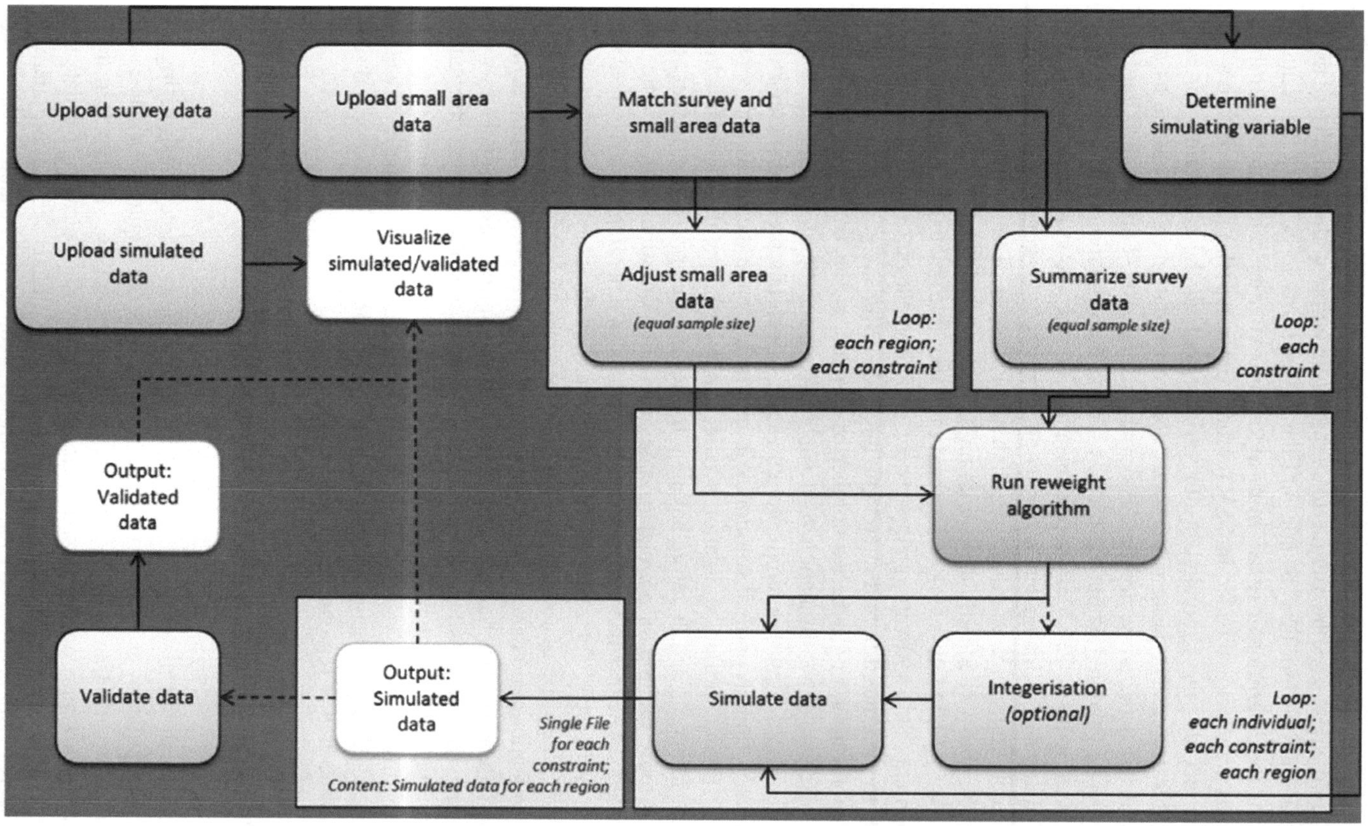


*Figure 14.3* Workflow of the deterministic reweighting approach within the simSALUD framework

server, the user has to match the survey data with the small-area data. This means the user has to match each survey constraint using a simple drag and drop technique. For the next step, the user has to start the simulation, with the choice of using the 'optimisation algorithm' or not, and the server starts to estimate the required data using the deterministic reweighting algorithm. After the calculation is finished, the results can be downloaded in CSV (comma separated file) format within a ZIP file that includes the simulated health variable and all the constraints for each geographical area. The latter is important for the validation of the model, as now the simulated data can be compared to real existing data. Further, the simulated results can be validated within the framework and the validation results can be also downloaded within a ZIP file. The visualisation module allows a direct mapping of the simulated and validated data for either constraints or sub-constraints. This offers a first quick impression of areas with high and low rates of the simulated data, as well as the ability to identify areas with high and low variations.

The simSALUD framework is designed to be reusable and expandable for future development. This software quality aspect is highly relevant because one future focus is to integrate different microsimulation algorithms, e.g. deterministic iterative proportional fitting, where a number of simulation runs can be set to achieve better matches (more simulation runs can improve the fit of the model: until a certain threshold is met) or simulated annealing, which is a probabilistic approach that seeks to find an optimal solution. All approaches have their pros and cons in terms of simulation time, simulation accuracy, required input data, model calibration and validation, and therefore having a framework that provides different algorithms is more suitable for a broader range of research questions with specific user requirements. The provision of such comparisons is an emerging need (Hermes and Poulson, 2012), e.g. for benchmarking test runs for the research community in particular. Thus, one of the main advantages of simSALUD is the possibility to build synthetic microdata without programming an application from scratch. Further, HPDM can use the application and perform various spatial analyses without needing specialist skills in spatial microsimulation modelling or geoinformation science. The first prototype of simSALUD has already been introduced to our project partners, the University of Leeds and the University of Sheffield, as well as Austrian health departments in order to evaluate its practical usefulness and relevance in the area of regional health planning.

## 14.3 SimSALUD case study: estimating the distribution of smokers

This section shows the usage of simSALUD using the example of estimating the smoking population at the municipality level in Austria. Figure 14.4 shows the modelling process that is applied to find the best model fit. The first step deals with data processing, accessing spatial and non-spatial data, cleaning the data sets and checking for possible data errors (e.g. missing values). Second, pre-analysis of the survey data is required to identify the best constraint variables that predict whether a person is more likely to smoke or not. This includes statistical analyses

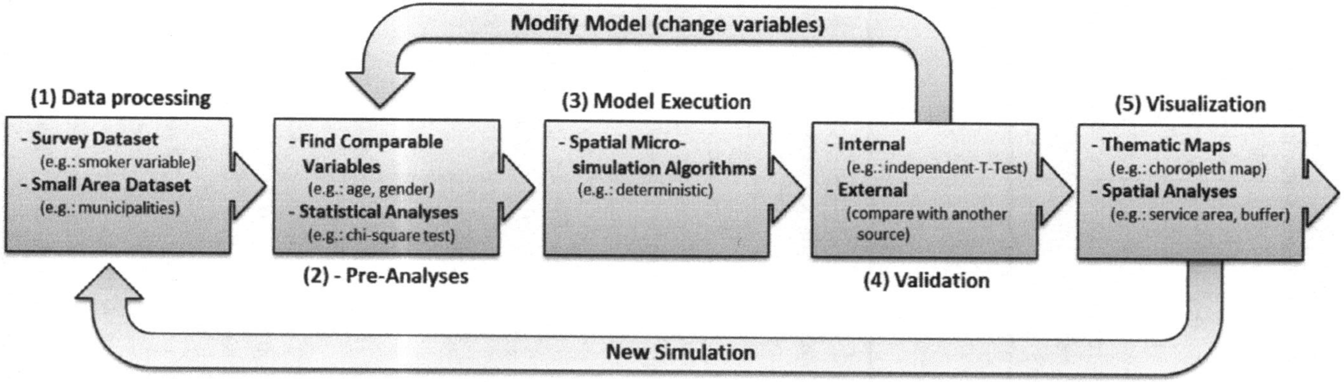


*Figure 14.4* Modelling workflow for finding the best simulation model fit

Source: Kosar and Tomintz (2014)

such as regression and chi-square analysis. Further, the data needs to be prepared in a structure that is obligatory for the simSALUD framework. Third, the data can be uploaded to the system and the simulation can be run. The fourth step includes the validation of the simulation results, using statistical methods (e.g. Standardised Absolute Error) to compare to any actual data that may exist at any spatial scale. Model results are often aggregated to higher spatial scales where data do exist. If the validation results are reasonable, the results can then be mapped using the integrated visualisation tool or the data saved to be visualised elsewhere, e.g. QGIS (previously known as 'Quantum GIS'), which is an open-source Geoinformation package. When the validation results are not satisfied, it is possible to go back to step two (pre-analysis) and change the input data, e.g. choice of constraints or different groupings of constraints. The process can be iterated until the best model is found.

## 14.4 Data and data processing

The simSALUD framework deals with two main data sources for the reweighting procedure, survey and census data. First, there is the individual or household data, which are mainly large-scale surveys holding a huge amount of information about persons or households, but not available for small-area geographies. Second, there are census-based data for small areas providing information about demographic and socio-economic status. The framework requires certain data structures to start the simulation. Possible data errors can be caught in advance when uploading the data files. The advantage of the package is to control each step of the modelling process and highlight errors to the user in advance to save time and increase the traceability of the system.

Thus, first, we need individual or household data that holds information on demographic, socio-economic and health-related issues; in our case, the smoking habits of individuals. Second, we need data for small areas holding demographic and socio-economic information for small areas, e.g. municipalities. A match of key variables (so-called constraint variables) is important for the modelling process. This means finding common variables within both data sets that best predict the variable of interest. To find these variables, statistical pre-analyses are needed, including chi-square analyses and regression analysis. The *Austrian Health Survey* (AHS) has been collected once a decade by the Statistics Austria since the 1970s. The last national survey on health was conducted in 2006/2007, where data on demographic, socio-economic and health issues were obtained from people aged 15 and above. This so-called microdata set can be accessed and used free of charge for academic purposes. The AHS 2006/2007 consists of 15,494 cases and 658 variables. The next AHS will be conducted in 2014/2015. This will be very interesting for us in order to explore trends over time. The *Austrian census data* is collected by the Statistics Austria, also every ten years, with 2001 being the last collection using the traditional paper format. Then, the decision was made to replace the traditional paper data collection with a digital data collection by obtaining data direct from the registers office. The first test collection was in 2006, followed by 2011, and now data is provided every two years. The advantage of this method of data collection is the saving of time and money. However, the data are no longer free for academic researchers to access. Data are provided in different packages and the costs depend on the area and resolution chosen. Costs can add up to well over €10,000. However, to access data for municipalities is more affordable as a one-year license costs €800 and the data can be accessed and downloaded *ad hoc* from an online Web platform. However, the volume of data that can be downloaded is limited and does not include all data categories. Some of them need to be requested and paid for separately.

For the modelling process, common variables between the AHS and the registered-based census need to be identified as these are required as constraints for the model. Having identified these variables, statistical analysis using the Pearsons chi-square test and Cramers V, as well as regression analysis and independent t-tests, are applied to test the independency between the variable 'smokers' and the possible explanatory variables for this variable (e.g. age, sex, marital status). Further, the strength of independence can be explained by the Cramers V test. In total, many variables were tested to act as potential constraints in the model, i.e. sex, age, marital status, education level, citizenship, country of birth, occupational status and household size. Following the statistical tests, four variables were selected as input for the model (sex, age, marital status and education level). Further, the variables were grouped, for example marital status was grouped into single, married, widowed, divorced. In total, we ended up with 19 groups that are consistent between the spatial (census data) and the survey data set (AHS).

The user of the simSALUD framework needs to be sure that the data and data format are valid according to the requirements of the framework. The data format is CSV (comma separated file) and a template on how the data needs to be prepared is available through a help menu within the framework. For the simulation to run without errors, it is important that both data sets (health survey and small-area census data sets) are prepared according to the template. The spatial microsimulation algorithm requires the match of the constraint variables and both data sets on which the simulation is based. But a detailed step-by-step help and tutorial system is available online to guide the user through the steps in case they are not self-explanatory.

## 14.5 Static deterministic reweighting simulation

In the area of spatial microsimulation, different algorithms can be used depending on the available input data (Rahman et al., 2010). The simSALUD framework uses two algorithms. First, a deterministic reweighting algorithm is used to allocate people from the survey to the small geographic areas, depending on the constraint variables. This activity can also be seen as a 'cloning' exercise, whereby each person from the survey is matched to a small area based on the constraints and the health variable from the survey is therefore assigned to an area. Each constraint data file needs to have the same number of people (total population) in an area. This can vary when receiving the original census data. If there are deviations (a different total number of persons in an area for one or more constraints), an adjustment is applied to ensure a consistent data set, where, in the end, all constraints have the same total number of persons in an area (Edwards et al., 2010). The first reweighting algorithm integrated into the framework is based on Edwards and Clarke (2013), Morrissey et al. (2008) and Tomintz et al. (2013). The equations can be written as follows:

$$nw_{ij} = w_i * a_{ij} / s_{ij} \quad (14.1)$$

$$y_{ij} = w_i * \text{sum}(a_j) / \text{sum}(w_i) \quad (14.2)$$

where $nw_{ij}$ is the new weight of individual *i*, $w_i$ is the original weight of individual *i*, $a_{ij}$ is element *ij* of the small-area statistics table (e.g. municipality), and $s_{ij}$ is element *ij* of the survey data table. $Y_{ij}$ is the new weight aggregated with the previous constraints, $w_i$ is the new weight from equation 1 ($nw_{ij}$), sum ($a_j$) is the total population for one constraint, and sum ($w_i$) is the sum of the new weights for all persons in an area from equation 1.

Equation 14.1 simulates the likelihood that an individual is living in a specific area and the result shows the final weight for each individual. This has to be multiplied with each constraint to get the total number of simulated people for each area. Equation 14.2 (integerisation) is then applied to allocate only 'whole' people to the final data set. The algorithm uses an iterative methodology to convert the decimal values from the results of the first algorithm to integers, which should optimise the results (Edwards and Clarke, 2013). Within the framework, this is called the 'optimisation algorithm'.

In this example, the simulation was executed four times using different settings, as follows:

- Model 1 uses the original weights of the survey data file and no optimisation algorithm
- Model 2 uses '1' as input weight and no optimisation algorithm
- Model 3 uses '1' as input weight and the optimisation algorithm
- Model 4 uses the original weight of the survey data file and the optimisation algorithm

## 14.6 Validation within the framework

As noted above, for the spatial microsimulation model four constraints divided into 19 sub-constraints are used to predict the smoking population for Austria. Four different types of models are prepared using the same constraints as input data. The differences lie in terms of the weighting for the input data and if the optimisation algorithm is used. Within the simSALUD framework seven statistical analysis tests are implemented for calibration purposes, namely Total Absolute Error (TAE), the personalised percentage of regions meeting a certain threshold of the TAE, Standardised Absolute Error (SAE), Percentage Error (PSAE), T-test,

Pearson Correlation and Linear Regression ($R^2$). This means that immediately after the simulation has finished, the user can select one or more validation analysis options. The framework stores the last simulation results and uses them directly as input data. A short description about each statistical method is given directly on the website, where an appropriate method can be chosen. After the validation is finished, the user can download the data or visualise the results immediately as a map within the framework.

Table 14.1 shows the results of four statistical validation tests that are analysed for each model separately and across all four models to find the best one. First, the TAE is calculated based on the total population of an area, where the simulated sub-constraints are compared with the census sub-constraints (real numbers). A tight threshold was set where the constraint variables should not have more than 10% TAE in 90% of the small areas (i.e. municipalities in this example) (see column TAE 10%) and, second, the percentage of TAE with more than 20% in 80% of the small areas (see column TAE 20%). In total there are 2,379 areas. The PSAE is calculated using the Standardised Absolute Error times 100 and the $R^2$ shows how well the data points fit the line or curve. Comparing all four models, then, model 2 is the one with the lowest simulation errors in comparison to the other three models. Using the integerisation within the framework results in slightly higher errors, but they are marginal. However, a big influence is the selection of the input weights, where it can be seen that the input weight of 1 (models 2 and 3) produces better simulation results than using the original survey weights. Further, the very young population group and the widowed persons have the highest simulation errors. For example, when looking at the sub-constraint 'Single', it can be seen that there is a decrease of error using the weight 1 in the input data set (models 2 and 3) and, further, it can be seen that the models without the integerisation process perform better (models 1 and 2). The simulated data have a better fit using weight 1, as is seen in the column $R^2$. Looking closer at model 2 and model 3, then model 2 performs better as the TAE10% and TAE20% was reduced. The next phase is adjusting the input data to achieve better results or certain sub-constraints. Nevertheless, this shows the advantage of using the framework for comparing models by adjusting the parameters.

As shown in this section, the simSALUD framework allows an easy and relatively fast comparison of the simulation results for all sub-constraints choosing one or more statistical methods; and this can be performed straight after the simulation model was run. The results can be downloaded or directly visualised within the framework, e.g. to highlight and detect areas with high errors. The visualisation works the same as visualising the simulation results and this is shown in the next section.

## 14.7 Visualisation within the framework

A visual interpretation of the results is helpful in understanding those results without looking at many thousands of rows of numbers in tables. An online visualisation of the simulation and validation results is therefore a useful tool to

*Table 14.1* Comparison of four types of models using the same constraints as input variables

| *Constraint* | *Sub-constraint* | *Model 1* | | | | *Model 2* | | | | *Model 3* | | | | *Model 4* | | | |
|---|---|---|---|---|---|---|---|---|---|---|---|---|---|---|---|---|---|
| | | *TAE 10%* | *TAE 20%* | *PSAE* | $R^2$ | *TAE 10%* | *TAE 20%* | *PSAE* | $R^2$ | *TAE 10%* | *TAE 20%* | *PSAE* | $R^2$ | *TAE 10%* | *TAE 20%* | *PSAE* | $R^2$ |
| Age | Age 15–19 | 55.4 | 17.4 | 0.7 | 0.9931 | 67.7 | 28.9 | 0.9 | 0.9945 | 68.0 | 28.2 | 0.9 | 0.9945 | 56.4 | 20.6 | 0.7 | 0.9931 |
| | Age 20–24 | 68.5 | 13.0 | 1.4 | 0.9909 | 31.0 | 3.3 | 0.9 | 0.9857 | 31.7 | 3.9 | 0.9 | 0.9857 | 67.2 | 20.7 | 1.5 | 0.9909 |
| | Age 25–29 | 97.7 | 65.3 | 3.1 | 0.9857 | 24.2 | 3.7 | 1.3 | 0.9793 | 26.2 | 4.9 | 1.3 | 0.9792 | 92.6 | 63.5 | 3.1 | 0.9856 |
| | Age 30–34 | 95.2 | 2.8 | 1.4 | 0.9986 | 4.0 | 0.2 | 0.4 | 0.9971 | 7.6 | 0.7 | 0.4 | 0.9971 | 76.8 | 15.9 | 1.4 | 0.9986 |
| | Age 35–39 | 39.9 | 1.7 | 1.1 | 0.9987 | 2.8 | 0.1 | 0.3 | 0.9991 | 6.9 | 1.1 | 0.3 | 0.9991 | 53.6 | 7.8 | 1.1 | 0.9987 |
| | Age 40–44 | 4.0 | 0.1 | 0.4 | 0.9981 | 3.6 | 0.1 | 0.3 | 0.9987 | 6.8 | 0.6 | 0.3 | 0.9987 | 20.0 | 4.9 | 0.4 | 0.9981 |
| | Age 45–49 | 6.9 | 0.2 | 0.4 | 0.9975 | 3.7 | 0.2 | 0.3 | 0.9980 | 6.3 | 0.5 | 0.3 | 0.9980 | 22.2 | 5.2 | 0.5 | 0.9975 |
| | Age 50–54 | 3.5 | 0.5 | 0.4 | 0.9923 | 7.2 | 0.7 | 0.4 | 0.9932 | 9.8 | 1.4 | 0.4 | 0.9932 | 19.4 | 4.5 | 0.5 | 0.9923 |
| | Age 55–59 | 74.0 | 11.0 | 1.0 | 0.9900 | 5.0 | 0.8 | 0.4 | 0.9913 | 9.5 | 1.4 | 0.4 | 0.9913 | 66.5 | 24.1 | 1.0 | 0.9898 |
| | Age 60–64 | 90.4 | 35.1 | 1.2 | 0.9877 | 8.9 | 1.0 | 0.5 | 0.9906 | 14.9 | 2.4 | 0.5 | 0.9906 | 78.1 | 42.9 | 1.3 | 0.9875 |
| | Age 65 above | 88.4 | 60.8 | 4.9 | 0.9809 | 30.9 | 5.8 | 2.5 | 0.9841 | 31.9 | 6.3 | 2.5 | 0.9841 | 85.3 | 59.8 | 4.9 | 0.9809 |
| Education | University degree | 99.9 | 98.4 | 3.7 | 0.9978 | 2.9 | 0.1 | 0.5 | 0.9995 | 12.6 | 2.3 | 0.5 | 0.9995 | 97.5 | 87.1 | 3.7 | 0.9978 |
| | A-level | 95.3 | 55.6 | 3.4 | 0.9972 | 29.1 | 1.1 | 1.4 | 0.9955 | 29.3 | 1.9 | 1.4 | 0.9955 | 90.8 | 52.8 | 3.4 | 0.9972 |
| | Without A-level | 13.2 | 0.8 | 7.1 | 0.9966 | 0.4 | 0.1 | 1.9 | 0.9988 | 0.4 | 0.1 | 1.9 | 0.9988 | 16.2 | 0.8 | 7.1 | 0.9966 |
| Marital Status | Single | 27.6 | 1.1 | 3.8 | 0.9963 | 18.4 | 2.9 | 2.6 | 0.9953 | 18.9 | 3.2 | 2.6 | 0.9953 | 31.8 | 1.9 | 3.8 | 0.9963 |
| | Married | 2.2 | 0.2 | 2.1 | 0.9972 | 1.0 | 0.1 | 1.4 | 0.9974 | 1.1 | 0.1 | 1.4 | 0.9974 | 4.2 | 0.4 | 2.1 | 0.9972 |
| | Widowed | 87.1 | 71.6 | 2.2 | 0.9706 | 64.3 | 35.7 | 1.3 | 0.9762 | 64.6 | 36.2 | 1.3 | 0.9762 | 84.9 | 68.9 | 2.2 | 0.9705 |
| | Divorced | 26.6 | 0.1 | 0.7 | 0.9956 | 2.5 | 0.2 | 0.3 | 0.9971 | 13.1 | 3.4 | 0.3 | 0.9971 | 46.2 | 12.2 | 0.7 | 0.9956 |
| Sex | Male | 0.3 | 0.0 | 3.4 | 0.9999 | 0.1 | 0.0 | 0.9 | 0.9998 | 0.2 | 0.0 | 0.9 | 0.9998 | 7.6 | 0.1 | 3.4 | 0.9999 |
| | Female | 15.7 | 0.0 | 3.4 | 0.9999 | 0.0 | 0.0 | 0.9 | 0.9998 | 0.2 | 0.0 | 0.9 | 0.9998 | 22.7 | 0.8 | 3.4 | 0.9999 |

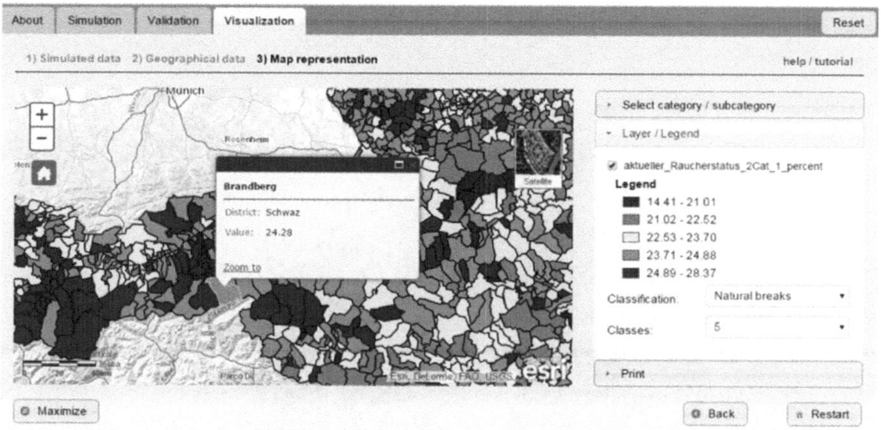


*Figure 14.5* Visualisation mask within the simSALUD framework

immediately identify hot spots. For non-experienced users of Geoinformation software, this tool is of special relevance. As mentioned above, the variables age, education level, marital status and sex are selected as constraint variables for the model. The simulated constraints as well as the simulated variable of interest, in this case the smoking population, can be mapped, as well as all sub-constraints. The same applies for visualising the validation results. Problem areas can therefore be identified immediately. Further, within the simulation model, the classification of the data classes can be changed between natural breaks, equal interval and quintile, and the number of classes (3, 5 or 7 classes). If the user is happy with the map, he or she can also personalise it with a title and description and print or save the map on the computer. In addition, it is possible to choose to print or save the map or with street or satellite view for easier orientation. The map is an interactive map which means that the user can click on an area in a map and a popup appears with related information to this area; for instance, the municipality and district name and the proportion of the simulated smoking population in this specific areas (see, for example, Figure 14.5).

The smoking population for Austria simulated at municipality level is shown in Figure 14.6. The smoking prevalence ranges from 14.4% to 28.4%. It can be seen that there is a north-west divide, with a lower smoking population in the north and a higher one in the west. Further, pockets of high and low prevalence are apparent within districts. These results need to be further analysed along with health policy planners, as the high proportion of smokers in the West is interesting and important to investigate further.

Running the simulation models using simSALUD allows an immediate visualisation of the results within this framework. In comparison to other model approaches, the advantage here is the possibility to visually explore spatial changes of the distribution of smokers immediately after the simulation and validation, respectively.

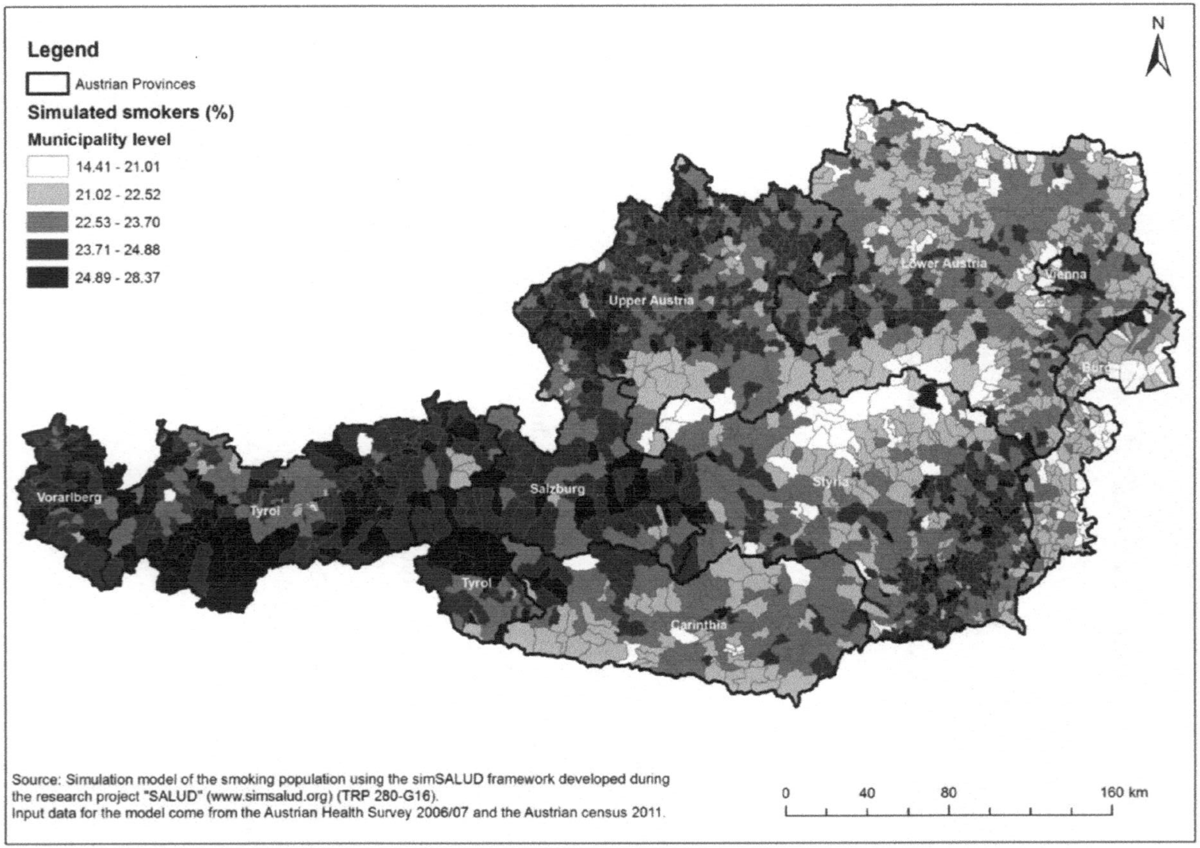


*Figure 14.6* Simulated smoking population at municipality level in Austria

Further, with simSALUD it is possible to change the model parameters, run the model again and visualise the new results. As it is a deterministic model, changes can be explored on an *ad hoc* basis.

Different model results are currently being discussed with the Carinthian health departments to explore reasons for the variations in certain districts and to receive detailed feedback about smoking cessation measures in Carinthia. Based on these results, it will be interesting to model specific population groups, for instance single people aged 25–34, to help target the population group highest on the policy agenda. Further, a new smoking cessation action plan is in development in the district of Carinthia, where such results are highly valuable. Moreover, the results can be linked with other spatial models for answering specific questions for future policy planning, i.e. where to distribute more persons that educate people about the harm of smoking and the possibilities of smoking cessation support.

## 14.8 Discussion and future outlook

SimSALUD provides a well-established framework for building and calibrating static deterministic spatial microsimulation models. The framework in its current version is the first attempt to build an easy-to-use Web application that does not require any programming skills to be operated and adapted. Integrating the algorithms into a single framework helps non-programming domain experts to test their data. The software architecture of simSALUD has been designed to ensure the reusability, flexibility and expandability of the system, and thus, being also an open-source project, will hopefully support and motivate developers to contribute to our research programme. Currently, the framework is accessible without any registration process so that anyone can use it at any time. In a future version, the system will also provide a login area for registered users in order to better manage personal data and processes, i.e. the system will provide a user with a personal simulation area with predefined settings, stored data sets and maps.

The framework shows currently some limitations in respect to visualisation as people can visualise their simulation results only with predefined geographical areas available within the framework. The geographical areas with a region ID are necessary to link (join) the simulation results with the areas to be drawn on the online map service. For example, when we simulate the smokers for all municipalities, then these geographical regions need to be available to select in the visualisation tab of the framework. But the geographical regions will be continuously expanded. To also allow usage for international users, they will be able in the future to send their geographical boundaries to the simSALUD team in order to be integrated into the framework. Currently, the user is able to map the simulation results and change the number of classes and classification method. However, the simSALUD team wants to integrate further visualisation possibilities and processing, for instance to draw buffers or overlay two or more variables.

In future, one of our main goals is to simulate more health-related variables for Austria, including diabetes and obesity. Further, Statistics Austria plans to publish a new health survey by the end of 2014/beginning of 2015. This will provide more up-to-date results and changes between the smoking populations of 2006/2007, whereby the latest one can be explored related to geographical variations. Further, cross-national spatial simulations are in preparation (with English colleagues) to identify changes when modelling the smoking population, with the focus on differences in data availability and structure, different model compositions and differences in spatial scales.

Lastly, the vision of simSALUD is to build a spatial decision support system where different parameters can be changed within the framework (so-called what-if analysis). For example, it would be interesting to model the impacts of an increase in the elderly population of 10%, to explore the impact on the smoking population spatially (for example, where would the smoking population live when accounting the elderly population in 2030). Another example might include finding the optimal locations of smoking cessation services based on existing quit statistics. Placing such services in high prevalence areas can help to show how the targets for stop smoking might be reached in the future. These are just simple examples, but they show in a comprehensible manner that changing some framework parameters lead to changes of the smoking population or the changes of people's health status.

This chapter has introduced our first attempt to build a spatial microsimulation Web framework that allows modelling small-area issues to simulate missing data, i.e. data not accessible or not surveyed. A main focus of the framework (called simSALUD) is to provide a very good user experience and usability in terms of interaction and user-friendliness, as the application can be used by novices and expert users.

Besides specific health outcomes, simSALUD can also be used to simulate other data sets related to other health areas. Examples include: (a) the simulation of economic data, for instance areas with a high proportion of unemployed people or deprived areas, which often leads to pure health outcomes; (b) the simulation of the distribution of current health expenditure to expand the analysis and thus to identify areas more in need – in order to reduce inequalities; (c) the simulation of areas with acute hospital demand; and (d) the geographical simulation of people with certain health conditions and their accessibility to the nearest general practitioners. Another very important advantage of spatial microsimulation modelling using the simSALUD application platform is the possibility of targeting specific population groups that are high on the policy agenda to simulate spatial changes when adjusting certain parameters to support future policy implications, i.e. for example 'what is the impact of the increase of diabetes on the health budget in five years' time and which areas might be most affected?' As shown, there is an increasing demand and huge potential in applying small-area modelling and GIS in areas of health and health care.

## Acknowledgments

This research project SALUD is funded by the Federal Ministry for Transport, Innovation and Technology (bmvit) and the Austrian Science Fund (FWF) (project number TRP280-G16). The census data is provided by 'STATcube – Statistical

Database of STATISTICS AUSTRIA' and the Austrian Health Survey is provided by Statistics Austria. We specially and gratefully acknowledge the implementation efforts of Bernhard Kosar to bring simSALUD to its present stage and for producing some of the graphics for this chapter. For requests, please contact info@simsalud.org.

## References

Abello, A. and Brown, L. (2007) Model 18: MediSim (static microsimulation model of the Australian pharmaceutical benefits scheme), in Gupta, A. and Harding, A. (eds), *Modelling our future: population ageing, health and aged care* (International Symposia in Economic Theory and Econometrics), 16, 533–539.

Ballas, D., Kingston, R., Stillwell, J. and Jin, J. (2007) Building a spatial microsimulation-based planning support system for local policy making, *Environment and Planning A*, 39(10), 2482–2499.

Bundesministerium für Gesundheit (HSg.) (2012) *Rahmen-Gesundheitsziele. Richtungsweisende Vorschläge für ein gesünderes Österreich*. Vienna: Langfassung.

Clarke, M. and Spowage, M.E. (1984) Integrated models for public policy analysis: an example of the practical use of simulation models in health care planning, *Papers in Regional Science*, 55, 25–45.

De Menten, G., Dekkers, G., Bryon, G., Liégeois, P. and O'Donoghue, C. (2014) LIAM2: a new open source development tool for discrete-time dynamic microsimulation models, *Journal of Artificial Societies and Social Simulation*, 17(3) (http://jasss.soc.surrey.ac.uk/17/3/9/9.pdf).

Edwards, K.L. and Clarke, G.P. (2013) SimObesity: combinatorial optimisation (deterministic) model, in Tanton, R. and Edwards, K.L. (eds), *Spatial microsimulation: a reference guide for users*. Dordrecht: Springer, pp. 69–85.

Edwards, K.L., Clarke, G.P., Thomas, J. and Forman, D. (2010) Internal and external validation of spatial microsimulation models: small area estimates of adult obesity, *Applied Spatial Analyses and Policy*, 4, 281–300.

Falkingham, J. and Hills, J. (1995) *The dynamic of welfare: the welfare state and the life cycle*. New York: Prentice-Hall.

Harland, K. (2013) *Microsimulation model user guide (flexible modelling framework)*, Working Paper, School of Geography, University of Leeds, UK.

Hermes, K. and Poulson, M. (2012) A review of current methods to generate synthetic spatial microdata using reweighting and future directions, *Computers, Environment and Urban Systems*, 36(4), 281–290.

Kavroudakis, D. (2014) SMS is an R library for spatial microsimulation [Online], http://cran.r-project.org/web/packages/sms/ [last accessed 31 August 2015].

Kosar, B. and Tomintz, M.N. (2014) simSALUD – a Web-based spatial microsimulation to model the health status for small areas using the example of smokers in Austria, in Vogler, R., Car, A., Strobl, J. and Griesebner, G. (eds), *Geospatial innovation for society*. GI-Forum 2014, Wichmann Verlag, pp. 207–216.

Mannion, O., Lay-Yee, R., Wrapson, W., Davis, P. and Pearson, J. (2012) JAMSIM: a microsimulation modelling policy tool, *Journal of Artificial Societies and Social Simulation*, 15(1) (http://jasss.soc.surrey.ac.uk/15/1/8.html).

Morrissey, K., Clarke, G.P., Ballas, D., Hynes, S. and O'Donoghue, C. (2008) Examining access to GP services in rural Ireland using microsimulation analysis, *Area*, 40, 354–364.

Morrissey, K., O'Donoghue, C., Clarke, G.P. and Li, J. (2013) Using simulated data to examine the determinants of acute hospital demand at the small area level, *Geographical Analysis*, 45, 49–76.

OECD (2011) *Health at a glance 2011: OECD indicators*. Paris: OECD Publishing (http://dx.doi.org/10.1787/health_glance-2011-en).

OECD (2012a) Smoking among adults. In *Health at a glance: Europe 2012*. Paris: OECD Publishing, pp. 58–59 (http://dx.doi.org/10.1787/9789264183896-graph60-en).

OECD (2012b) Smoking and alcohol consumption among children. In *Health at a glance: Europe 2012*. Paris: OECD Publishing, pp. 50–51 (http://dx.doi.org/10.1787/9789264183896-20-en).

Rahman, A., Harding, A., Tanton, R. and Liu, S. (2010) Methodological issues in spatial microsimulation modelling for small area estimation, *International Journal of Microsimulation*, 3(2), 3–22.

Statistik Austria (2012) *Jahrbuch der Gesundheitsstatistik 2011*. Vienna: Statistik Austria.

Tomintz, M.N., Clarke, G.P., Rigby, J.E. and Green J.M. (2013) Optimising the location of antenatal classes, *Midwifery*, 29, 33–43.

# 15 Small-scale agent-based modelling of infectious disease transmission

## An example in a primary school

*Mike Bithell*

## 15.1 Introduction

Spatial information is a crucial component of many types of social model. Transmission of infectious disease, for example, depends on how people come into contact with each other, and therefore on the way in which social behaviours lead to interpersonal interactions. As pointed out by Koopman (2004), knowing who contacts whom is central to modelling infection transmission, and making use of data collected about disease transmission requires individual-level models that take into account the real-world places where people meet, and the systematic and stochastic variability in those meetings. Conversely, Bian (2004) stressed the heterogeneity of both people and their interaction environment as factors that imply the need for individual-based approaches. Auchincloss and Diez Roux (2008) further point out that both the individual and environment change and adapt to each other over time, with consequences not just for disease spread, but for more general health-related issues. So, in order to disentangle the effects of social processes and spatial constraints from physiological mechanisms of disease spread, we need modelling of disease transmission at the scale of individuals and their day-to-day activities, including the effects of real-world spatial geometries.

Agent-based models, which are able to represent directly the behaviour and locations of individuals, are ideally placed to address these kinds of questions. Early examples were very abstract (Epstein and Axtell 1996; Wilensky 1998), but the range of different specific diseases that have now been considered using multi-agent models has increased over recent years, although a major focus has been influenza (Bian 2004; Ferguson et al. 2005; Ferguson et al. 2006; Germann et al. 2006; Stroud et al. 2007; Amouroux et al. 2008; Cooley et al. 2008; Halloran et al. 2008; Lee et al. 2008; Yang and Atkinson 2008; Epstein 2009; Brouwers et al. 2009; Rao et al. 2009; Ajelli et al. 2010; Bian et al. 2012; Duan et al. 2013), particularly with regard to the possibilities of a pandemic outbreak. Similar levels of attention have been given to smallpox (Chen et al. 2004; Eidelson and Lustick 2004; Epstein et al. 2004; Eubank et al. 2004; Burke et al. 2006; Brouwers et al. 2010), where an accidental or terrorist engineered epidemic might have large-scale consequences (Halloran et al. 2002; Longini et al. 2007).

Other specific examples include anthrax (Chen et al. 2006), cholera (Augustijn-Beckers 2011; Hailegiorgis and Crooks 2012), dengue fever (Jacintho et al. 2010; Lourenço and Recker 2013; Rodríguez 2013), foot and mouth (Dion et al. 2011), hepatitis-A (Ajelli and Merler 2009), HIV (Teweldemedhin et al. 2004), lassa fever (Dunham 2005), malaria (Linard et al. 2009), measles (Perez and Dragicevic 2009), mumps (Simoes 2012), respiratory syncytial virus (Dunham 2005), and tuberculosis (Patlolla et al. 2006). Such models even extend to direct modelling of the pathogen as individual agents (Segovia-Juarez et al. 2004), but here we are concerned with agents as people.

Coupling of agent models to real-world spatial data is becoming more routine, as evidenced by the increasing availability of GIS layers as components of such models (Gimblett 2002; Bian 2004; Brown et al. 2005; Perez and Dragicevic 2009; Amouroux et al. 2008; Simoes 2012). Tools such as RePast, AgentAnalyst, Netlogo, Mason and GAMA all include the capability to add GIS layers to agent-based models. However, even with increased spatial realism, typically most agent-based disease models have focused on relatively large scales, at the neighbourhood level or above, and on scaling up to country or global modelling, with rather abstract specifications for both movement and spatial structure, when the interaction between physical and social spaces may be quite complex, even at the level of a single building. A few examples (Patlolla et al. 2006; Duan et al. 2013) have looked at specific building layouts and the movement of people within and between rooms, but often, for example, individual interaction between agents and obstacles, and between moving agents, is not included as an effect, although this can change the way in which people gain access to spaces where they may become infected. Modelling at this very high level of detail presents a number of challenges that have yet to be thoroughly addressed in many agent-based models. In the first place, we have to deal with the semantic use of space (Bian 2004), rather than simply its geometry: the reasons why people go to particular places and the manner of their use have to be included in the model. In addition, the way in which generic descriptions of places are interpreted by agents as references to specific locations requires attention (e.g. an instruction to 'go home' has different meanings for different agents). Furthermore, the way in which social structures condition spatial patterns of movement and spatial aggregation need to be simulated, mediated by these interpretations of spatial position: one needs to deal with collective motions, where groups of people move in specific ways because of their group membership, while still maintaining their individual motivations for movement. Finally, model time stepping needs not only to include sufficient resolution to deal with individual physical interactions, but also with event-based scheduling, so as to govern the changes in movement patterns that occur as a result of routine daily activity.

In this chapter we focus on the construction of a model of a village primary school (in the UK this means children in the age range 4–11) with approximately 100 pupils, for which we have collected a range of observational data. Children are particularly susceptible to infectious diseases, having had little time to build up resistance to common complaints against which the adult population tends to be immune; in this respect children, particularly of pre-school and primary age, are thought to act critically in the transmission of some infectious diseases (Keeling et al. 1997; Brownstein et al. 2005; Glass et al. 2006; Merler et al. 2011; Gog et al. 2014). Furthermore, children spend much time in school in close contact with each other, and this might be expected to increase the rate of disease spread over that which would be expected through family contacts alone. Schools are in many ways a good focus for study, as they form a relatively closed system during the daytime, are of small spatial extent, have a relatively fixed level of occupancy, with highly constrained activities and relatively unchanging sets of social relationships during the course of a year. Nevertheless they have a very rich social structure that presents a number of modelling challenges, not least being able to observe the system unobtrusively without exciting undue comment and variation in behaviour on behalf of the pupils, although the highly constrained environment means that it may be possible to gather activity data effectively. During the year these children experience many ill-defined diseases, many of which have a considerable impact on the school day, leading to absences of many children from school at once, but which do not fall under the remit of the serious types of infection listed above, and may not be associated with a larger-scale epidemic. Nevertheless these absences are worthy of some study, as they hold out the prospect of improving the understanding of the way in which disease spreads through a population at a very fine scale, and so exploring ways in which the management of disease might be improved.

## 15.2 Model description

The model is described using the Overview, Design Concepts and Details + Decisions (ODD + D) framework (Müller et al. 2013), although re-ordered slightly to bring the 'theoretical and empirical background' up to the front with 'model purpose' as these seem to rest naturally together.

However, while ODD + D adds a considerable extra degree of detail to the original ODD protocol (Grimm et al. 2006), it still misses some components that are useful in re-creating models, namely testing and validation. These are rarely formalised, although in the software development world 'unit testing' is a key part of code design: writing formal test routines as part of the process of model building ensures that expected behaviours are followed in a way that can be reproduced. Changes or updates can be checked to ensure they do not break other parts of the system, and they also help to avoid having to repeat walk-throughs or code reviews when some changes are made, as re-running the tests can highlight any problems. Other users of the code can also use the test routines to check that they have a properly working version, and reproducing or re-engineering the code becomes easier as the tests can be used as an initial check that the model behaves as expected. In the original ODD protocol, the suggestion is made that testing be described separately for each sub-model – this is indeed a good idea, but overall testing of the full code also deserves a place, as interactions between sub-models may give rise to unexpected problems that do not show in tests of sub-models by themselves.

As far as validation goes, a set of test data sets needs to be added that can show how well the model is able to reproduce either the results of other model frameworks, or else measurements or observations from real systems. This suggests that some of the ideas from 'pattern oriented modelling' (POM – Grimm and Railsback 2012), in which the outputs of a model are tested against multiple different patterns of data, some of which might be qualitative, others quantitative, some observed and others maybe more theoretical, should be added to the ODD+D protocol as part of a formal model description. In this chapter we add a section to the end of the usual ODD description to show how well the current disease simulation is able to match the classic differential-equation based SIR (susceptible-infected-recovered) model in the well-mixed case (bearing in mind that this model is itself just an approximation to the behaviour of a real system), and how well its spatial behaviour matches to an idealised spatially propagating plane-wave of infection.

Finally, the ODD description does not include a section for description of how the model is run: whether Monte Carlo or other types of simulation were made, sensitivity studies carried out, parameter-space explorations undertaken or any other detail of the process by which outputs are produced. These again form a significant part of the effort of producing model output, and are potentially needed by anyone seeking to reproduce the code independently. This section should also give some indication of how long the model takes to run in a given programming environment and on a given hardware platform.

To avoid confusion in the following, wherever 'class' is referred to, we will mean one of the actual classes (which is to say sets of pupils of similar age) in the real school, rather than using the term in its programming sense. Where we need to refer to classes that are structures in model code, we will specify these as 'computational classes'.

### *15.2.1 Purpose*

The model is intended to represent transfer of infectious disease (i.e. not vector borne diseases such as malaria, although the distinction may not always be completely clear (Koopman 2004)) that are transmitted purely by spatial proximity, with little or no latency and of short-term duration, where an outbreak in the school lasts only a few weeks. We consider diseases for which there are generally very obvious symptoms (e.g. vomiting) that last only a few days and lead to a short-term absence. Nevertheless in a small school such diseases are not only common but can lead to classes missing significant numbers of pupils for several days. By constructing the model with a detailed representation of movement with sub-second time resolution and corresponding sub-metre scale spatial resolution, we can examine to what extent including collision detection and avoidance, navigation around walls and other obstacles might be significant in disease spread. With the inclusion of the school class structure and timetables, we can investigate where and when diseases are acquired and from whom, and thus detach the social and spatial structures from the physiology of the disease.

In principle, simplified models with, say, stochastic occupation of classrooms, can then be tested against the more detailed model to find out what can be left out without compromising the dynamics. We can also look at how to choose model parameters so as to accurately represent the duration and size of an outbreak, since we use realistic time and space scales throughout. The model is designed around a real village school so that we can compare model output with real-world data.

### *15.2.2 Theoretical and empirical background*

It has been argued that one should try to base agent-models in a well-specified theoretical framework rather than making *ad hoc* assumptions or appealing to 'reasonable estimates' for parameter values (Grune-Yanoff 2010). Unfortunately in many social situations we do not have well-attested theories in which to ground models (although we usually have plenty of available hypotheses). In such cases we may have to rely on guesses and reasonable assumptions, which we then test through the modelling process by challenging the model against data. In the present case, though, we can constrain the model by grounding it in observation (Glaser and Strauss 2012): the school timetable is largely followed throughout the day, the children in the school in question, in broad terms, follow the instructions given by teachers, with some degree of variation. Each teacher or responsible adult monitors the location of all children and ensures that they move between the required rooms in the school at the necessary times, but does not generally constrain movements beyond general instructions. There is a regular pattern of movement from room to room, regulated by lengths of lessons, duration of assembly, breaks and lunch, and bounded by the ends of the school day when children move freely in the playground. During movement agents generally wish to avoid collisions with each other and with walls and desks. The latter obstacles are assumed immoveable, but agents that are currently stationary may move to one side to allow the passage of others on the way to their destination. All of these phenomena have been subject to observation, as mention below.

As far as disease goes, we assume the usual compartmental categories of susceptible, infected and recovered. Recovered agents cannot re-acquire the disease. Disease carrying agents infect susceptible with equal probability per unit time per capita inside a spatial distance, D. This rather sharp cut-off model may not in practice be realistic, but in fact we are unlikely to have any good information about the variation with distance for an arbitrary infection, since in the usual case the data will not be available: rates of spread typically convolve infection per unit time and spatial distance to give a value for the average number of people infected by one disease carrying agent (the 'basic reproductive ratio' $R_o$). This is clearly not a property of the disease alone, but depends on movement patterns. The way in which the transfer of disease falls off with distance can be a critical factor in infection spread (Keeling and Rohani 2008), but we argue below that the cases we are investigating here have a rather short range, for which a sharp cut-off would seem to be reasonable first approximation.

## 15.3 Entities, state variables and scales

The main entities in the model are the spatial layout of the school, and the teachers and pupils, who are each explicitly represented as individual agents, with their own internal data (which includes whether they are infected by a disease).

### *15.3.1 Spatial structure*

The model represents the school buildings and parts of the school grounds, most notably the playground. Conceptually, we set up the building as a container for rooms, and rooms contain walls, doors and desks. Each of these is given appearance data, allowing the model to be visualised while running, if desired. Since this is a rather small area, we can just use a simple planar co-ordinate system, with linear scales measured in metres. It is important for the spatial units to be correct here, as we will need both to get the relative sizes of children and the school buildings right, but also speeds of movement depend on having correct relative spatial and temporal scales, and the timetable requires that we use the real-world time-units so that it can be matched to the actual school day, as detailed below. The overall size of the school buildings is approximately 50m × 25m, although the grounds extend over a region roughly 100m × 100m. We use a domain of size 1000m × 1000m so as to allow children to spread out into the surrounding area during the time they spend at home. At present we do not represent homes explicitly, but simply make sure that children are spread out far enough that infections are not transmitted out of school (unless coincidentally they happen to fall within the infection distance D). In order to get an accurate representation of the school buildings and surrounds, we measured the entire structure inside and out with a Leica HDS 3000 laser scanner during the course of one weekend during half-term. This ensured we could get data on the physical environment while the school was empty, and all furniture tidily stored to give best possible access for the scans. This device both takes photographs of the scene being scanned and uses a laser to obtain full three-dimensional information about the location of objects in its field of view. Scans from different positions can be tied together to give a three-dimensional model with typical matching errors for scans taken from different places of less than 6mm. This allowed us to obtain a full three-dimensional data set with resolution as good as 5cm, and generally no worse than 1.5m. In the interior we were able to get the locations of cupboards, desks, whiteboards, coat-hooks, and other furniture, even down to the location of books and pencils. The advantage of having this very detailed information is that we can at any stage revisit the virtual school buildings in order recover new data, and we can section and view the rooms and exterior grounds from any required viewpoint. In practice, at the current stage of the model we use only a small fraction of this very rich data set, extracting a very simple ground-plan that includes the most salient features relevant to the school day, namely the playground, classrooms, school hall and the locations of desks. Examples from the raw laser scanner data are shown in Plate 9. The plan view extracted for modelling is given in Plate 10.

The model has two background grid structures that are used for search processes. Both grids are set with 0.5m resolution (rather smaller than a school desk, and about the size for occupancy by one child), although this can be varied independently for the two. The first grid holds a set of stationary obstacles that impede agent movement – namely walls and desks. To make it simple for agents to detect these, we digitise the boundaries of each into sets of points at resolution better than 0.25m and use the grid cells as containers to hold these points. In the second grid, cells act as containers for moving objects, in this case the teachers and pupils. This second set of cell data is updated on every model timestep to ensure that these moving entities are always located in the correct cell. Agents can query the cell containing their own location, and surrounding cells within a given range, in order to find either sets of obstacles or other agents in the immediate locality. Note that the position of obstacles and agents is not dictated by the grid resolution – they can be located in any position down to the accuracy of machine double-precision numbers, and search can be made out to any desired distance. The grid resolution just determines the finest possible grain with which the location can be searched – this has a bearing on the efficiency of the model (as the time taken to run will depend on the square of the number of neighbours returned by any search process), but the necessary physical search distances depend on agent movement and the physical range over which disease is assumed to be able to spread. The grid structures allow for either periodic wrapping or hard edges to be included as part of their specification, but in practice, for this model we do not make use of either, as the agents never reach the boundaries of the domain.

The classroom names that appear on the plan in Plate 10 denote the way in which each of the spatial locations would be referred to in the school, and are used explicitly in the model to identify areas by name. Within these, each agent stores a set of named 'points of interest' that are mapped to sets of two-dimensional co-ordinates within the domain. Some of these have a common definition for all agents – for example, the location 'startOfLibrary' always denotes the same two-dimensional co-ordinate point. However, others vary by individual – so 'classSeat' is a two-dimensional point denoting where a given child sits in the classroom, and therefore is different for every child. However, by using a single name, a teacher can indicate a location to all children (effectively issuing the message 'go to your seat in the classroom') and each child agent will then target its own individual location as a destination. The sets of points of interest are grouped into hard-coded pathways for navigation through the school. For example, the path 'classSeattoAssembly' is defined as a sequence of named points of interest that vary depending on the originating classroom. So for class3 the sequence is

'classSeat → classDoor → startOfLibrary →
endOfLibrary → startOfTeachingCorridor → assemblyPointb →
assemblyRow3 → Assembly'

where the initial and final points ('classSeat' and 'Assembly') differ for each child. Effectively, these sets of paths form a graph that allows agents to find their way around the school geometry, but by hard-coding them, indexed by the end-points, we avoid having to have the agents compute the route from one place to another continually as they change the main location for their current activity. Since the overall school pattern of behaviour is very regimented, and the domain is small, this is not a problem in the current implementation: the loss of flexibility is made up for by a gain in run-time. However, it does mean that if we wished to include learning behaviour attached to navigation, or spontaneous activity that did not include the known paths, then the model would need to be modified. However, variation over time of the location denoted by each named point can be included: we make use of this in the playground, where, after arrival, the pupils use a location called 'playgRandom', which is a random destination point chosen once they reach the playground area, and which then allows for free movement within this region.

### *15.3.2 Teachers and pupils*

Agents are defined as generic entities with a unique ID number, an age (which determines which class they are in, or, for teachers, which class they teach), location, height and facing direction (for setting the appearance), a set of known paths, the current path they are following and the next point of interest on the path, a destination point and a velocity. Pupils also keep a pointer to their class teacher. For tracking disease, each agent also keeps a set of boolean variables that mark it as at the point of infection, infected and recovered. Other than this the agents act as containers for lists of processes, which is to say time-dependent activities that the agent can undertake. Teachers and pupils are distinguished by having separate processes that dictate their daily activities. Both teachers and pupils have a movement process, and may acquire a disease process by reason of contact with other infected agents. These processes are detailed further below in the section on sub-models.

The school in question has around 100 pupils, with a mixture of adult presence that includes teachers, class assistants, assistant staff and others, for example parents, at various times of the day. In the current case, we assume that the adults are mostly not material in spreading childhood diseases (as they may largely have built immunity, but they are also many fewer in number than the children), and just include one teacher for each class. Each teacher process holds a timetable which lists the times and locations at which events occur throughout the day. Teachers may change the times of some these events depending on their class. In addition, each teacher holds a lesson plan that allows them to customise pupil locations depending on in-classroom activity. For example, some of the younger pupils may spend some time sitting on a mat in front of a whiteboard, rather than sitting at desks. A timing process keeps track of the real-world time in date:hours:minutes:seconds, so that the timing of events on the timetable will be followed correctly even though the model timestep might be changed.

### *15.3.3 Model structure*

The model reads in configuration data such as timestep, initial and final dates, random seed, input and output file names from a parameter file, and holds these in a singleton data structure (Shalloway and Trott 2002) which allows instances of computational classes to be defined and accessed from arbitrary places in the code without the need for global variables. Singletons are also used for the model itself, defined as an independent computational class, and the list of known places. The model allows for the definition and addition of dynamic layers, which are used both for drawing during model visualisation and for timestepping model components. All the agents are held in a list in a single layer that can be used to track properties of the whole cohort (e.g. the total number of infections).

## 15.4 Process overview and scheduling

### *15.4.1 Timestepping*

The model has a fixed timestep, measured in seconds: all agents and the grid that holds the agent locations are updated synchronously each timestep. While it has been argued that asynchronous timestepping is required for individual-based models (Huberman and Glance 1993; Brown et al. 2005), this really only applies for event-based simulations in which the time of occurrence of a particular event might vary from place to place, and only one essentially instantaneous event type is being considered: even so, if synchronicity of *decision making* is not enforced, a synchronous model timestep can still be appropriate – in practice, all agents are doing *something* at any moment, even if only working through some sort of cognitive process. Where multi-stage activities are concerned, or multiple different activities are possible, or the initial state is not spatially symmetric, or agents can choose not to act on every timestep, this will generally lead to spatial asynchronicity of action even with synchronous updating, particularly when locations are not confined to a coarse grid structure.

Where realistic simulations of spatial movements are being considered for spatially interacting agents, however, synchronous timesteps are absolutely required: if agents interact physically, there will be a pair of agents that will need to update their states most urgently (e.g. they are about to collide, or are in contact). This determines a shortest time that can be allowed for a step, dependent on the actual physical interaction process. During this time the pair of agents may move by some small amount. However, all other agents who are not stationary also have to move during this time: otherwise inconsistencies arise in agent positions, and in particular things like contact or social forces may not be correctly updated. The fact that a pair of agents changes position, therefore, forces all other agents whose motion needs tracking to have their states adjusted with the same timestep. For simplicity, it is easiest to fix this step to a sufficiently small value at the start of a run and leave it unchanged: typically, for a model that incorporates forces this means that agents moving at their maximum speed should not be able to move more than a fraction of their own length in a single step, as then they will not be able to easily undertake unphysical actions such as moving through walls or other agents. However, to be safe, one can make a sensitivity study, reducing the timestep until results from the model converge to a consistent value, within some bounds of error (if the model does not converge, then that implies that there is some problem with the representation of the dynamics! Numerical schemes that are convergent as the timestep is reduced have been extensively discussed for models in the physical sciences (Morton and Meyers 2005), although this issue is rarely if ever raised in the agent modelling literature beyond a mention of synchronous/asynchronous concerns).

However, while these movement processes need synchronicity, it can be the case that some agents or some non-spatial processes do not need to participate in every update – an example in the current case might be parents, who might only need to be present at the school during certain hours, and might arrive and leave at different times, and therefore behave essentially asynchronously. Care has to be taken in such cases to be sure that agents that enter and leave an area where motion is being tracked do so in a way consistent with the other agents' timestepping.

One might also envisage a system in which transfers of diseases between agents happened on a separate timing and scheduling scheme from agent movement. In cases where more than one dynamical system is being simulated simultaneously, though, care has to be taken that the time resolution of events is capable of adequately representing the temporal behaviour of all systems under consideration adequately (Bithell and Brasington 2009). In the current model, we will assume that diseases are transferred by agents remaining in sufficiently close spatial proximity for long enough to become infected. The spatial movement processes and disease transfer are therefore not independent of each other. Construction of a probabilistic event schedule for disease transfer cannot be done ahead of time on a timescale larger than a movement timestep, as agents positions are not known in advance, and they may move out of disease-transfer range in the course of just one step. We therefore update the disease processes on the same timestep as movement: we can justify this on the basis that the timescale for the spread of disease through the school population is observed to be many days. Our movement timestep (typically ~0.25s) therefore defines a timescale much shorter than the typical disease transfer time.

In addition to the movement timestep, we also have two types of event schedule, namely the school timetable and teacher lesson plans. These constitute lists of activities with start and end times (e.g. assembly, morning break, in-class activity) with a time resolution of 1 second. Teachers are able to vary these schedules, also with a 1 second time resolution. For example, at the end of break time the children have to line up in the playground prior to entering the school. Children are not allowed to enter school until permission is given by their teacher. Although the school timetable has an event at 9:00am mandating that all children go and find their seat in the classroom, the individual teachers modify this so as to stagger the entry of each class by up to a few minutes. Schedules for activities within a classroom can be completely independent. Despite the underlying fixed timestep, the event schedules are often asynchronous. Again, this is absolutely required in order to match the way in which events are observed to take place in the school under study. This would seem to be typical of the way in which social systems act – some types of event need to have updates that are properly synchronous in order to get the correct, consistent, interaction between agent-agent and agent-environment physical properties. On the other hand, event schedules, and differences between the processes that agents are undertaking in different places, will typically lead to discrete activities taking place asynchronously. Provided the model is set up with a sufficiently small time grain, the two need not be in conflict, and indeed are probably generally needed: not a case of *either* asynchronous *or* synchronous timings, but both together.

### *15.4.2 Processes*

The main pattern of the model is one of movement between target locations governed by the daily school routine, except overnight when pupils and teachers go home. The length of disease outbreaks is such that multi-day runs are needed. The daily routine is controlled by the teachers: they continually check the general school timetable to see whether a new activity is indicated, and set their target location accordingly. An example timetable is as follows:

| | |
|---|---|
| 08:49:50 | leave home and go to playground |
| 08:58 | end of time in playground: line up ready to go into school |
| 09:00 | go into school when indicated by teacher: enter classroom for registration |
| 09:10 | end of registration: go to assembly |
| 09:30 | assembly ends: go to classroom for first class of the day |
| 10:45 | breaktime: go to playground |
| 10:58 | line up at end of break |
| 11:00 | go into class for second class of the day |
| 12:00 | start of lunch time (in the hall where assembly is held) |
| 12:20 | end of lunch: go to play in playground |
| 12:58 | line up for end of lunch break |
| 13:00 | go into class for final class of the day |
| 15:25 | school ends: go to playground |
| 15:30 | wait for parents in order to go home |

At present outdoor activities such as sports, or after-school clubs are not included. 'Homes' are random, well-separated locations, so that pupils arrive at school at slightly different times in the morning.

If the location is currently the classroom, teachers use their personal activity plan to set target locations. Pupils check with their class teacher and update their location target to match the teacher's. If their current location and target location are not matched, then agents select a path that leads from the current to the target location and begin to try to follow it. Otherwise they remain in their current area: this may still entail some degree of movement, particularly in the playground, where random motion anywhere within the playing area is assumed to be permitted. All movements are governed by obstacle and other-agent collision detection and avoidance using a social force model (Helbing and Molnár 1995). This means that although agents may attempt to follow a path, from, say, classroom to assembly, the track that they follow is adjusted to move around desks and walls, and go around other moving agents. In practice, this can mean that they have to wait for others to clear doorways, or that they may effectively push and shove other agents out of the way in order to get to their seat. Sometimes this can lead to the agents becoming 'lost' when they are pushed around the edge of a doorway, for example, and find themselves on the wrong side of a wall without a suitable path available to recover their route to their destination. Eventually these agents tend to find their way again when the timetable event changes, as this changes the path required in a way that typically allows them to navigate around walls that interfered with previous destinations. Correcting this behaviour would require adding some ability to search a local spatial graph to agents: while this is desirable, and the fact that agents lose their way is not an intended part of the model design, it happens relatively infrequently, and mimics quite well the case in which small children lose track of where they are meant to be without needing to add extra structure to the model. In practice, there would be many more adults in the school than we currently represent, and this means that lost children would be able to get back to their class more easily than they are able to in the present model version, but we leave this to future development. While the agents are moving around the school, if they carry the disease process, this checks to see whether they are still infective: if so, it finds all other agents currently in range, and then attempts to infect each one with a given probability per unit time. If this is successful, the disease gets added to the list of processes being executed by the target agent.

## 15.5 Individual decision making, learning and individual prediction

In the current model, the agents do not have any real decisions to take or any learning behaviours. The behaviour of children is entirely determined by the teachers, and that of the teachers by the timetable. This is consistent with the current goal of the model to see how spread of disease is related to routine activities and the resultant spatial patterns. In a more sophisticated model, some extra variation could be added: for example, at present the children make no use of toilets, and this could be a key location for the transmission of some types of disease. Decision making in this case would require children to have a method for deciding when they needed to go and a mechanism for asking the teacher for permission to do so. Children also at present have no mechanisms for preferentially choosing to sit with friends when the situation permits, or to interact preferentially with friends over lunch or in the playground. However, this would imply needing some knowledge of children's social networks, which we did not have the scope to investigate in the current project, and on which, at primary level, the literature is remarkably sparse (Moreno 1934; Rapoport 1957). At present, prediction by individual agents is not envisaged as part of the necessary model structure: this might become relevant if one were to include, for example, changes in weather and their link to sporting activities, or anticipation of timetable changes to account for school absences, but at present they are beyond the scope of the current model.

## 15.6 Individual sensing and interaction

The main sensing mechanisms are those involved in tracking motion of neighbours as part of the social force model, and sensing obstacles to avoid. These take place mediated by the grid structures that allow agents to find objects within a given spatial range. Disease processes also 'sense' local agents as targets for transmission. Pupils also interact with teachers by querying them to obtain information about the next target destination.

## 15.7 Collectives

Collectives exist in the sense that school classes represent meaningful spatial categories that relate to their use as social spaces. For example, in Plate 10 the area denoted 'Class1', first, denotes a room within the school, and is thus physically bounded by its walls. The physical objects within the room – desks, chairs, cupboards, whiteboard, mat, computers, etc. – are all related by their use during the teaching day. However, 'Class1' also represents a social category – a set of children in a given age range and their teacher. Children in 'Class1' make use of the physical space and objects within their own classroom as privileged objects compared to other locations within the school – the grouping of which partially determines the scope and type of activities that can take place within the room (e.g. the mat for sitting, or desks as alternative places to sit). Children have particular knowledge of items within the room – their place at a desk, for instance. At the start of the day they specifically 'go to Class1' in order to hang up coats, place lunch boxes in drawers, and begin the process of working under teacher supervision. Other locations, such as the hall and playground, have a more shared collective use for specialist activities like assembly, but otherwise may remain unused for large parts of the day. However, when the children from 'Class1' go to assembly, they do so only as and when instructed collectively by their teacher, move as a group, and go to a set of rows in the hall that are designated for their use during assembly time. In the playground, while there is free use of space during play, when break ends, children from 'Class1' line up in their own space on the playground, separated from the other children. These collective activities lead to a consistent spatial separation from other children in the other classes, and therefore potentially to a lower between-class than within-class spread of disease. In model terms, these class relationships are implemented, first, by the children only updating their own target location data by referring to their own teacher and, second, by having a set of destinations and paths through the school that are dependent on their school class designation. Plate 11 shows a model snapshot illustrating these sensing and collective processes.

## 15.8 Heterogeneity

The heterogeneity of agents exists at several levels: teachers and pupils have different roles, classes have different timetables and activities, and individuals have different destination points and paths to destinations dependent on time of day and the class to which they belong. There is also a dynamic heterogeneity that results from the social forces, which do not act uniformly across agents, and act differently each time an agent takes a particular path, as the surrounding agents are in different positions and will typically not be the same ones as on a previous excursion. Finally, the transmission of disease varies from place to place and agent to agent.

## 15.9 Stochasticity

Stochastic processes arise in several places throughout a model run. In the initialisation the location of 'home', the places that pupils go in the playground, the motion while in transit between locations, at desks or in assembly all have a small random component. The initial agent with the disease, and the transmission of disease to susceptible agents, also depend on random values. The paths taken through the school by individuals also vary in practice through interaction with other agents, even though the nominal routes go between fixed end-points. Pseudo-random values are generated where needed using the Mersenne Twister algorithm provided as part of the GNU C++ compiler 2011 standard library.

## 15.10 Observation

In the first place, the model produces a visualisation that shows the positions and movements of all the teachers and pupils in the school, along with the time, and a marker to show which agents are infected, so that the following of paths, changes of locations with time and interaction between the agents can be observed (particularly to ensure that the social force model is operating so as to keep agents from moving through each other or though obstacles, and that interpenetration of agents does not happen in crowded spaces). Second, the agent layer keeps count of the totals of infected, susceptible and recovered as a function of time and writes these to an output file. Additional data regarding the location of infection is also made available by the disease process, which writes out the time, agent and position at which the infection takes place.

## 15.11 Initialisation

After model parameters are read (including whether the model will set up a GUI for vizualisation, or just run without) and the set of places initialised so as to establish school walls, desks, points of interest and other geometry, 124 agents are then created, and given ages between 5 and 9 (these being used solely for purposes of identifying to which class they will belong: we assume just one age in each class, although in real school itself there may be children in each class that differ in age by as much as two years), and heights, for visualisation purposes. The first 120 have a 'pupil' process added and the remaining four a 'teacher' process. One agent picked at random is selected to be infected with the disease, and all agents are given a movement process. Teachers are set up first by giving each a school class identifier and default timetable, to which they add the necessary variations to customise the times at which pupils leave the playground, leave classrooms to go to assembly, break and lunch, and leave the school to go home in the evening. Each teacher then obtains a unique class identifier and then selects the set of pupils that belong to their class by age. Pupils and teachers are then allocated their set of known paths and points of interest, and a randomly allocated home destination outside the school grounds, but well inside the model domain.

## 15.12 Input data

As previously described, the school geometry was captured by use of a laser scanner. In order to examine the behaviour and timing of events in the school, one of the class assistants spent a day with each class surveying the activities of the children. This had the advantage that the teaching assistant was familiar with the school routine and pupils, and allowed us to maximise the effectiveness of data collection and minimise disturbance to the children's behaviour. Each classroom, the hall and the playground were notionally divided up into a small number of sub-areas (i.e. children were not aware of the way in which location recording was being done, nor were there visible marks that they could observe delineating the sub-areas) and the location of each child in the class recorded relative to the area they occupied at five minute intervals throughout the day. In addition, we were given access to five years of anonymised school attendance data, which included records of absence by reason of illness for all pupils in the school, for morning and afternoon periods (i.e. with a time resolution of one half school day).

## 15.13 Sub-models

### *15.13.1 Movement*

All agents have a desired walking speed when moving from place to place of 0.5 m/s. This is smaller than the numbers quoted for adults (Pelechano et al. 2008) by about a factor of two to account for the much smaller size of primary school children. We set the maximum speed to be larger by 30%. Each agent also has a desired speed, which could be different from the walking speed (but is less than the maximum), and which is used to determine the forces calculated below. Paths are modelled as finite state machines, with the current target node along a path as the state. At the start of a journey between locations, the agents load up a path in its initial state and set the next node of the path as their target destination. Once they are within their own physical size plus one timestep of movement at their current speed, they update the path to its next state. In each step the agents update their speed according to the force exerted by neighbouring objects, and their distance to the next path node, and then update their location at the current speed. Both of these updates use the forward Euler method, where changes are just linearly added with a constant timestep, i.e.

$$\underline{v}(t+dt)=\underline{v}(t)+\delta\underline{v}\cdot dt$$
$$\underline{x}(t+dt)=\underline{x}(t)+\underline{v}\cdot dt$$

where $\underline{v}$ and $\underline{x}$ are 2D vectors stored with each agent, $dt$ is the timestep, and $\delta\underline{v}$ the change in velocity calculated from the social force model. For large $dt$ this can potentially lead to considerable errors in path evaluation, and a second-order method such as a Verlet scheme (which uses information from timestep $t$-$dt$ and the acceleration at time $t$ to make a path that can simulate quadratic curves accurately, rather than just linear ones) might be preferable, but for the present case the Euler scheme seems to give paths that are acceptable in the sense of routing agents between obstacles without undue collisions or overlap.

Values for $\delta \underline{v}$ are calculated by adding together four force components. First, the speed is relaxed towards the desired speed on a timescale $t_{relax}$, and some random motion is added:

$$\delta \underline{v} = \frac{\left(v_{desired}\underline{\hat{r}} - \underline{v}\right)}{t_{relax}} + \underline{\delta v}_r$$

Where we typically use $t_{relax}$~0.6s, $\underline{\hat{r}}$ is a unit vector pointing to the next node along a path, and $\underline{\delta v}_r$ is a random vector with maximum size generally 0.05m/s$^2$, although we allow this to increase to 1.5m/s$^2$ in the play area. $v_{desired}$ will usually be walking speed until the agent gets close to the final node in a path, when we allow it to decay exponentially once the agents are within one second of their destination – this leads to a smooth approach to the final destination point on a path.

Second, we add a component for to account for other agents and obstacles as:

$$\delta \underline{v} = \frac{-ke^{\left(-\frac{dr}{s}\right)}}{s} \cdot c\underline{\hat{n}}$$

where $k$ and $s$ are constants. For agent–agent interactions, $k=0.5$, $s=0.07$, and $dr$ depend on agents' positions and relative velocities: we set $dr$ to be constant on elliptical contours with foci centred around the position of the agent and the position $\underline{r} + \underline{v}dt$, where $\underline{r}$ is agent position and $\underline{v}$ is the relative velocity of the agent and the moving object it is trying to avoid.

$\underline{\hat{n}}$ is a unit vector joining the two agents. Factor $c$ is used to introduce an angular dependence – $c=1$ unless the angle between the target destination and $\underline{r}$ exceeds 200 degrees, when we set $c$ to 0.1.

For obstacles (which remain stationary) $\underline{\hat{n}}$ is a unit vector directed from the agent to the obstacle, $k=0.3$ and $s=0.2$, $dr$ is the agent–object distance and $c=1$.

Finally, the velocity is scaled to ensure the speed remains below the maximum value.

### *15.13.2 Infection*

We assume that all agents within range of an infected agent that are currently susceptible have a constant probability per unit time of infection. This implies that the probability of infection should be exponentially distributed over a finite timespan. For each agent in range D of an infectious agent, in each timestep we therefore choose a random number and infect the agent if

$$-\ln\left(\text{random}\left[0,1\right]\right) < \hat{\beta} \cdot dt$$

where $\hat{\beta}$ is the per-capita infection rate.

Similarly, we assume that the per-unit-time chance of recovery is constant, so that an agent with

$$-\ln(\text{random}[0,1]) < \gamma \cdot dt$$

recovers during that step, with recovery rate $\gamma$, although they first still get a chance to infect other agents. Agents that have recovered retain the disease process, and we use this as an indicator that they cannot be re-infected. Since the probabilities are set assuming constant per unit time infection rates, we do not allow agents to infect their neighbours in the same timestep in which they themselves acquire the infection. This makes little difference in the well-mixed case, but allowing agents to infect their neighbours inside a timestep can lead to over-rapid propagation of the infection when agents are spatially distributed.

## 15.14 Testing

Since the grids are crucial to the operation of the model, unit tests were written to check grid operations, such as addition and removal from a grid cell, correct operation at boundaries, correct re-allocation to grid cells on spatial translation of agents, and correct discovery of neighbours for idealised spatial arrangement of sets of agents. Unit tests were also written for the disease model to ensure that infections only spread to spatial neighbours within an infection radius, and that the rates of infection and recovery were indeed distributed as would be expected for constant per unit time rates.

To test the operation of the movement and daily timetables, the model visualisations were used. These show the motion of all children in the school along with the layout and a clock showing the time of day. Inspection of a whole day's worth of activity allowed the model to be checked for correct transitions between events at expected times, correct following of paths though the system, avoidance of obstacles and other agents, and required lack of interpenetration of agents when close, and checking that walls formed real barriers to motion in the expected fashion. This also allowed some simple checking of timestepping, since the above conditions tend to be violated for timesteps that become too large (>~0.4s) (in particular large timesteps not only allow agents to intersect significantly, and to pass through walls, but tend to lead to targets along paths being missed).

As a third test we looked for convergence of the time-series of disease transmission as a function of timestep. Two hundred model runs were carried out with a recovery time set to three days and infection rate $\hat{\beta}$=10 day$^{-1}$person$^{-1}$. The timestep was varied linearly from 0.1s to 1s in steps of 0.1s. In each case a mean time series of infection was calculated, and the absolute maximum relative difference from the 0.1s case calculated. This is plotted in Figure 15.1. As can be seen, with a timestep of 1s differences can be over 100% and a given point in the time series, but this drops to under 10% once the timestep drops below 0.3s – beyond this the change per unit seems to level off, with additional reductions in timestep producing relatively little gain. In the following we therefore use a timestep of 0.25s as a reasonable compromise between accuracy and acceptable runtime.

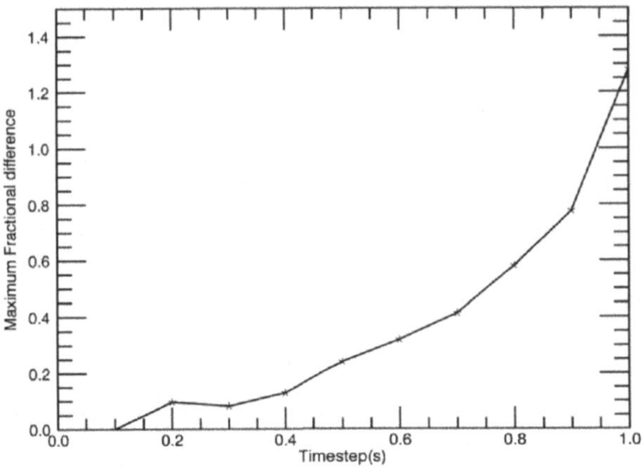


*Figure 15.1* Changes in model output as a function of timestep. Runs with timesteps > 0.1s were compared to the 0.1s case. For each set of 200 runs, the mean time series of infections was calculated. The relative differences from the 0.1s case were then found at each point of the time series, and the maximum value of this is plotted above

## 15.15 Validation

In the 'well-mixed' case, it is expected that the agent model should largely reproduce the results that would be obtained from the Kermack-McKendrick SIR formalism (Keeling and Rohani 2008). That is, it should reproduce the case where all infected agents are able to transmit the disease without regard to spatial location or behaviour, but simply have a probability per unit time to transmit the disease equally to all other agents in the domain. However, because the agent model is stochastic, and the usual SIR equations are deterministic, we would only expect the average over many agent model runs to match up. As pointed out by Chen et al. (2004), this is not true 'validation' as we are only comparing one model with another, rather than with data. However, it does give some confidence that model is acting in the expected manner, and that there are not global systematic errors being introduced by the model formulation. Note that this is a different exercise from tuning model parameters to match together an agent model and a SIR model (sometimes referred to as model alignment: Chen et al. 2004; Rahmandad and Sterman 2008). Here we force the agent model setup to be identical to that under which the SIR complete mixing model operates, anticipating that the two should then be in agreement. Unfortunately there is a further difficulty for small populations, since in many cases when there is only one initial agent infected the disease fails to get going even if the expected number of infections by a given agent with the disease ($R_0$) exceeds one. In these cases, in order to match the SIR results, those agent model runs in which the disease fails to spread to at least a few other agents from the initial case have to be discarded from the average. For a further discussion of the 'early burnout' and more detailed comparison between agent disease models and a-spatial network cases, see Rahmandad and Sterman (2008). For the parameters used here, when the number of initial infected agents is large enough (for a population of 100 agents this is when the initial number of infections is about 10), this ceases to be a problem, as the infection can generally spread. The SIR equations in the usual standard form are:

$$\frac{\partial S}{\partial t} = -\beta SI$$

$$\frac{\partial I}{\partial t} = \beta SI - \gamma I$$

$$\frac{\partial R}{\partial t} = \gamma I$$

where the proportion of the population susceptible to infection is *S*, infected is *I*, and recovered *R*. The ratio of infection to recovery rate $\beta/\gamma = R_o$ determines a threshold below which an infection dies out and above which it can spread to involve a significant fraction of the total population. We would expect that the agent model should be able both to demonstrate this threshold behaviour, as well as the time series of the evolution of the infection (a somewhat more stringent criterion, as the disease not only has to cross the threshold, but also get the proportions of all three variables to match at each timestep). Note that because these equations refer to population proportions, the infection rate is $\beta = \hat{\beta}$N, where N is the total number of agents in the simulation.

To make the comparison, the above equations were solved using a short python program that called the ordinary differential equation solver ('odeint') routine from the scientific python ('scipy') package (which solves first-order systems of ordinary differential equations (ODE) using ODEPACK (Hindmarsh 1983)), varying β and γ over the range 0.01 to 0.19 in steps of 0.01 so as to go from $R_o$ ~0.05 to $R_o$~20. The agent model was run with 100 agents all co-located and stationary at the model origin (so that all could infect each other), and with initial numbers of infected agents 1 and 10 (so initial fraction infected, I(0) set at 1% and 10%). One hundred runs were made for 1500 steps with a timestep of 0.2s. For the case with initially one agent infected, those runs for which the disease died out in the first 50 steps were discarded, to exclude those cases where small I(0) led to a lack of an epidemic despite high basic reproductive ratio, as mentioned above. Results for the case $\gamma = 0.03$ are shown in Figure 15.2, where the mean of the 100 runs are plotted with standard errors.

As can be seen from Figure 15.2a, while the general form of the agreement is good, when the initial number of infections is small, there are some systematic differences between the two models: the agent model has a slightly lower and later peak in fraction infected than the differential equation version. However, as Figure 15.2b shows, this difference gets systematically smaller as the fraction of initial infections increases, even though the total population of agents is quite small. In Figure 15.3 the final fraction of the population that gets infected is shown as a function of $R_o$, again for $\beta = 0.03$, and I(0)=1%. The threshold can be clearly seen in the region where $R_o$~1.

Once the condition of perfect mixing is relaxed, it becomes more difficult to test whether the model conforms to analytic models, as the appropriate form of analytic

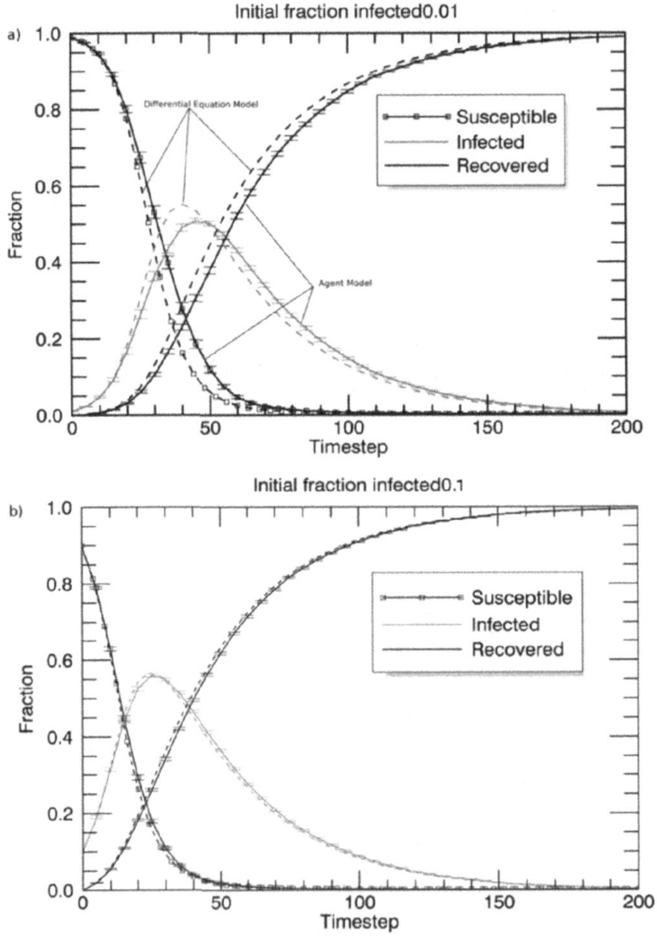


*Figure 15.2* Comparison between agent-based and differential equation models in the well-mixed case. Dotted lines show the differential equation solution, and lines with error bars show the mean of up to 200 runs of the agent model with 100 agents, where all agents are co-located and thus can infect each other. Error bars show the standard error in the mean of the 200 runs. Recovery rate y=0.03 and Ro=6. a) lnitial fraction infected=O.Ol b) initial fraction infected 0.1line symbols and styles as for a). Agreement between the two models improves as the initial infected fraction is increased

model becomes more difficult to identify. In particular for spatially distributed systems, the way in which the disease spreads may have a non-trivial dependence on the form of the spatial pattern of dispersal. However, we can set up a simple test of spatial propagation on a lattice in the case of a 'plane-wave'. Here we assume that agents are static and arranged on a regular square grid with some constant spacing $\delta$. We initially set a few columns of the grid along one edge to be infected. Since the density of the agents is uniform, and the distance out to which they can infect each other finite, we expect that provided $R_o>1$ the infection will spread at

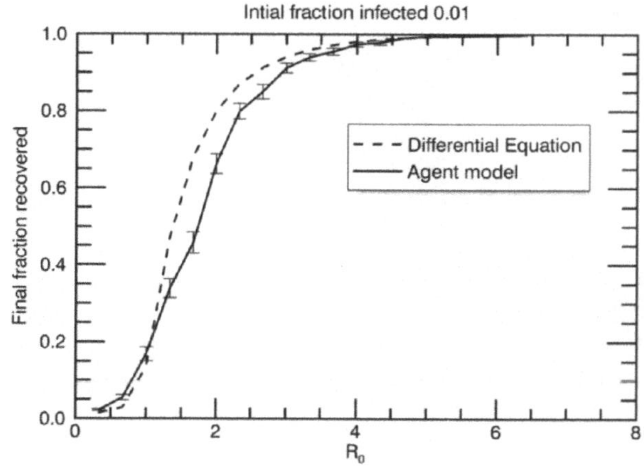


*Figure 15.3* Second comparison between agent-based and differential equation models in the well mixed case. Here the final fraction recovered gives the proportion of the population that are ever exposed to the disease and become infected over the whole course of the epidemic. The dotted line shows the expected threshold near to $R_0 = 1$ for the standard differential equation model. The heavy solid line shows the agent case, which clearly has the same kind of threshold, but slightly delayed. Error bars again show the standard error in the mean for 200 runs of 100 agents, recovery rate y = 0.03

uniform speed across the lattice, maintaining an essentially planar wave front (barring stochastic noise). At the edge of the infected region, approximately half the neighbourhood of any agent will be infected (those behind the advancing infection front), and these can infect only those agents up to infection distance D away. Since the timescale for this to take place is $1/\beta$, we expect the propagation speed to be about $0.5\ \beta\ D$, independent of the density of agents or the recovery rate. Agents stay infected for a constant time of order $1/\gamma$, so we expect the rear edge of the infected region to move at the same rate as the front, and since the velocity is $0.5\ \beta\ D$, the width should be of the order of $0.5\ \beta\ D/\gamma = 0.5\ D\ R_o$. In Figure 15.4 we show the location at which agents become infected as a function of time, along with the location for recovery – it can be seen that the infection front does indeed move uniformly at the expected average rate, and with the expected width. The spatial spread of the disease in the model would therefore seem to be valid in this idealised case.

## 15.16 Coding and running

The model was written in C++ using Qt widgets for visualisation. This allowed a pre-existing code-base to be exploited (Bithell and Brasington 2009). The model needed to be run for a sufficient time to allow a complete infection to propagate through the school population, this being about 30 days. With a timestep of 0.25 seconds, this implied runs of up to 10,368,000 steps. Since visualisation slowed the model by about a factor of 30, most runs were done without. Single-core runs on an intel core-i7 machine took approximately one hour 20 minutes to complete (although this is dependent on range of infection). The real

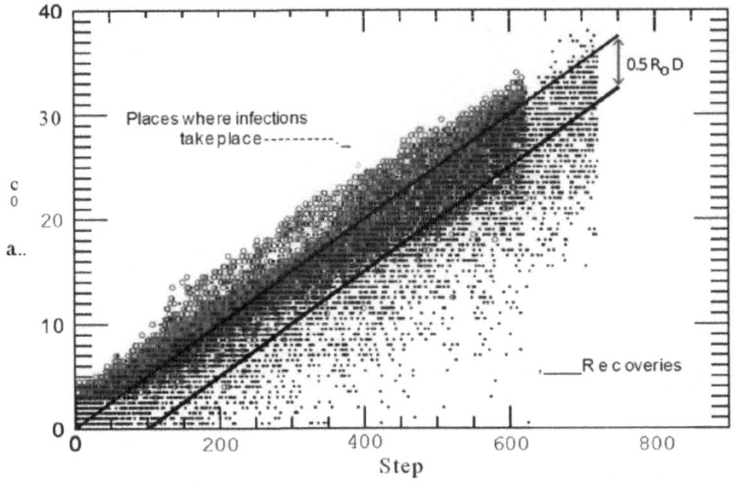


*Figure 15.4* Plane wave propagation on a 50 × 50 lattice. Circles show the locations in which agents become infected as a function of model timestep, and black dots the location and time of recovery. The upper heavy line shows the expected mean propagation rate with slope 0.5βD, and the lower line the expected recovery propagation, shifted down by $0.5R_0D$

time-per-step is quite variable, however, as agents that are not infected have less need to query nearby neighbours (this being the most expensive part of the model, as the infection distance can exceed the collision-detection distance), and, since we do not currently model siblings, or interaction with agents outside the school, night-times lead to faster time-steps as agents are isolated. However, time spent outside school is still required as this constitutes a vital recovery period.

## 15.17 Results

Examination of the five years of absence data shows that most epidemic histories involve low-level extended infection where only one or two pupils are absent at a time. In three cases, however, the total number of children absent at once exceeds 30, and in one case it exceeds 25: these have relatively high impact on the school day over periods of up to 30 days. We show the full time series for all children in Figure 15.5, and break these down by class for two large outbreaks in Figure 15.6.

Figure 15.6(a) seems to show some evidence that the disease infects the classes as distinct units, with high absence levels in each class separately, perhaps with some class-to-class propagation, and with one class having a couple of separated peaks. Figure 15.6(b), however, seems to show some co-ordination, with all classes having their peak on infection together. In both cases the disease onset is relatively slow, with rises to a peak taking several days. Typically for these sorts of infection, the recovery is not much more than three days for each pupil, and often they are only absent for a single day. This suggests that the per capita rate of infection is relatively small, given that the largest rates of infection seem to be 3–4

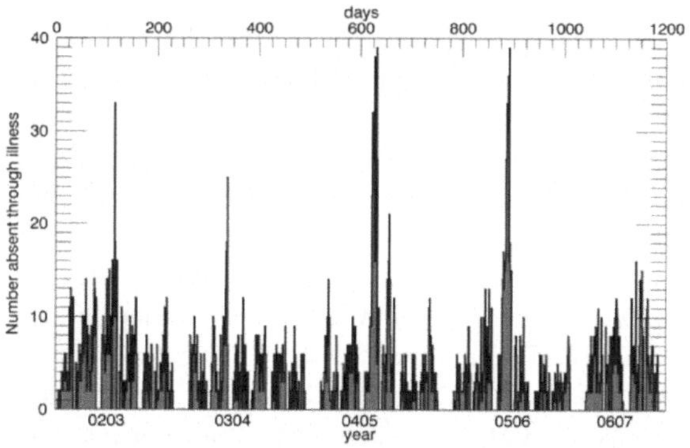


*Figure 15.5* Five annual records showing total absences from school between 2002 and 2007. Each record is 260 days from the start of the school year, except from 2006–2007 where we have only 146 days. Note this is not a continuous series but has gaps for the school holidays, which have been elided

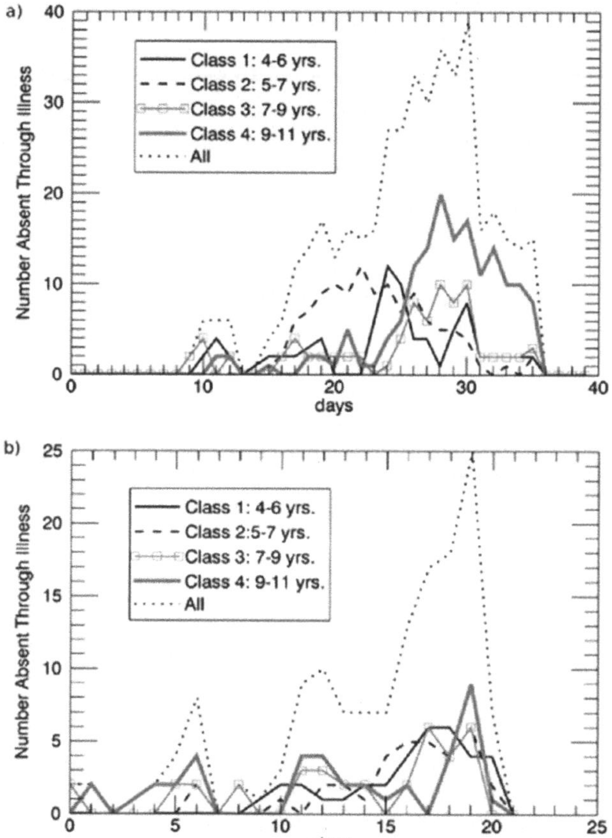


*Figure 15.6* Data from school records showing two separate outbreaks of disease that caused short-term absences from school, broken down into individual classes. Outbreak a) seems to show some evidence of class-to-class propagation, with different classes having peak infection at different times, whereas outbreak b) appears to show all classes developing the disease in tandem

pupils per day, and that the pupil density can be as high as 3 or 4 per square metre. Values for $R_o$ would therefore seem to be less than about 4. To try to match these characteristics we have three parameters, namely $\hat{\beta}$, $\gamma$ and D. The data suggest values of $\gamma > 0.3$ $day^{-1}$, but not too much larger, as a disease from which recovery happens in under 12 hours is unlikely to lead to a half-day of absence from school, unless there are very obvious symptoms (such as vomiting) for which a precautionary exclusion of at least a day will typically be applied. For the other parameters we have little information – D could be as small as 0.5m, or extend across the whole school, with values of $\hat{\beta}$ varying accordingly. However, in the cases where the infection seems to be peak separately in each class, this suggests that the range of infection is smaller than the size of a classroom, as children appear to be acquiring the disease from their immediate classmates, implying values of D less than a few metres, and values of $\hat{\beta}$ of order 1 $day^{-1}$ $person^{-1}$.

To test whether we could reproduce a large outbreak, the model was run 500 times varying D between 0.75m and 2.5m, $\hat{\beta}$ between 0.25 and 4 $day^{-1}$ $person^{-1}$, and $\gamma$ between 0.3 and 1 $day^{-1}$. Time series with outbreak patterns similar to those in Figure 15.6 could be observed provided that was $\hat{\beta}$ sufficiently small and $\gamma$ about one third $day^{-1}$ (Figure 15.7).

Note that because of the small numbers and stochastic nature of the model, we do not expect to see exact agreement. Spatial patterns of infection and recovery show that in such a case the typical infection locations are indeed in classrooms, or in the hall, with only small numbers in the playground (Plate 12). On the other hand, recoveries tend to take place at home, since this is where the majority of time is spent as a fraction of the whole day. However, although the infection time series in Figure 15.7 seem to match the observed patterns, the distribution of consecutive days sick is not well reproduced (Figure 15.8). In the first place, the

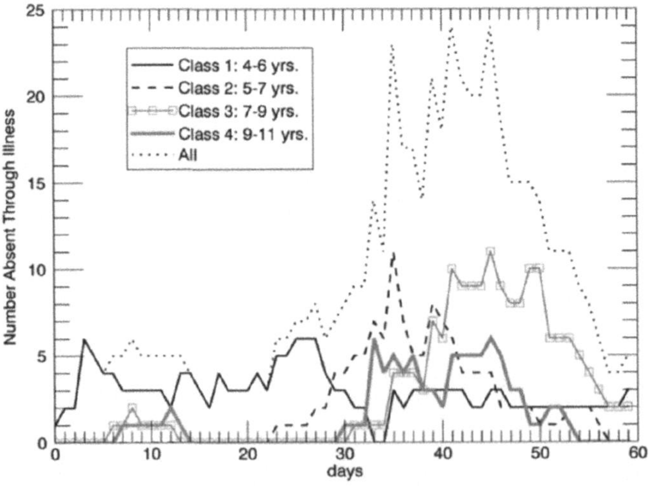


*Figure 15.7* Simulated disease outbreak using D = 2.0m, $\hat{\beta}$= 0.25, y = 3. Compare with Figure 15.6, particularly 15.6a. Outbreak timing is similar, with a long low-level lead-in for the overall total, and a magnitude of about the right size. Overall duration is also well captured

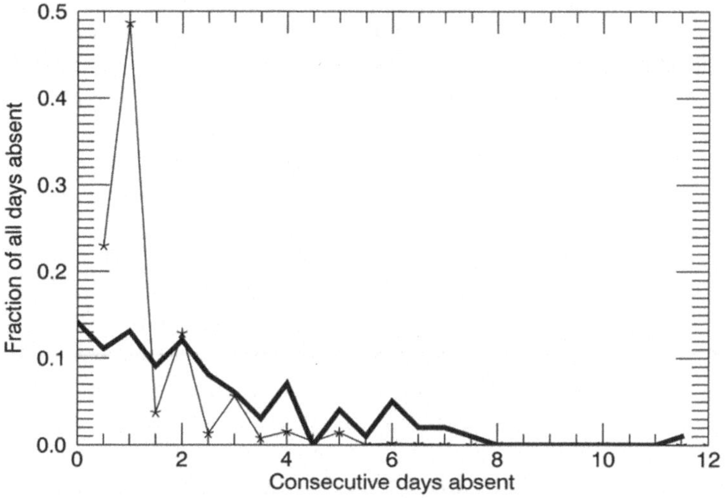


*Figure 15.8* Histogram of disease duration. The thin black line with stars shows the data from school registers. Half of the time children are only absent for a single day, although some may be absent for as many as 12 consecutive days. The heavy black line shows the number of consecutive days for which children are ill with the model parameters corresponding to Figure 15.7

data tend to have larger values on complete days, indicating that children are more likely to be off for a whole than a half-day, something that is not represented in the current model formulation. Second, the data show that the majority of children are only off for a single day – although there is a long tail out to as many as 12 days, this tail is less prominent than in the model, which tends to have fewer children sick for a short period, and more of them sick for more than one day.

## 15.18 Discussion

Qualitatively, the patterns of movement produced by the model inside the school building reproduce those observed very well, to an extent because these movements are highly constrained by the geometry and the school regime. This is less true in the playground, where the model spatial distributions are not as clustered as those in the real school. Since the model pupils currently have no social networks, they cannot use the presence of friends to determine their locations and interactions, and gather together during play: so transfer of disease in the playground is likely to be less rapid in the current model than might be the case in reality.

Furthermore, we do not implement any social distancing measures: in practice, if a child is physically sick, they are expected to remain away from school for a time, or indeed their parents may keep them at home. At present the model allows all children to remain in school at all times during the school day. This may partly account for the mismatch between the modelled and observed numbers of consecutive days sick. Similarly, we do not currently simulate weekends (although, as pointed out by Cooley et al. (2008), the expected effect of weekends is unclear, as there are many opportunities out of school for diseases to be exchanged), have little in-class behaviour (such as clustering around the sink after a painting session), no outside time (sports, for example) and no out-of-school activities or external reservoirs such as out-of-school clubs, or disease transfers at home through friends or siblings. The latter would benefit from embedding the school into a realistic village environment with the addition of GIS layers giving locations of houses and roads for routes to school. On the other hand, we should note that Figure 15.8 includes all outbreaks of disease across all five years, so a more careful breakdown of the data into individual outbreaks might also be instructive.

The lack of latency in the disease representation may also play a role here, as this could lead to changes in the rate at which pupils are perceived to be ill (and therefore need time off school). We could envisage, here, experimenting with specific real disease parameters rather than a generic infectious disease, including the detail that is known about rates of spread and proximity. Other factors, such as vector-borne spread, might also be important. We are limited in terms of the locations that can be visited or discovered by agents, since they are confined to fixed paths. Adding a simple A* algorithm to the graph structure would allow agents to find their own ways around – this might be useful for adding new pupils, or for adding slightly more independence for pupils, so that they could, for instance, visit the toilets (although this could also be achieved in the fixed-path formalism), and then some kinds of water-mediated infection might be included.

Finally, from the pure modelling point of view, we have yet to explore the full parameter space and map out all the conditions under which the types of infection pattern in the data might be matched by the model. The effect of the presence or absence of obstacles, or the social forces between agents still remains to be investigated in order to see whether these really have significant effects.

## 15.19 Conclusions

The micro-level activity of children moving through a small school can be quite nicely simulated by agent-based methods. The paths followed by children and the physical movement match remarkably well to the real-world case. Communication of infections can be simulated reasonably accurately, with recovery of some of the parameters that seem relevant, but in particular allowing the locations of infection to be tracked and the social factors to be separated from intrinsic disease properties. However, matching against multiple patterns shows that reproducing the overall time series of infection might *not* be sufficient to generate correct micro-scale disease dynamics: this only becomes clear when one has sufficiently fine-grained individual-level data that includes the times for which infections persist. Such small-scale detailed models are a useful check for more abstract large-scale simulation, as one can test the effects of neglecting certain aspects of the system and measure the way in which this might perturb the dynamics.

However, that fact that we can only partly match the data should sound a note of caution for the use of models that seek to advise on real-world practice. This comment applies not only to agent-based or other geographical models, but all models that attempt to represent aspects of social situations. There is often a temptation, or even a pressure, to jump from a relatively simple model to making recommendations for change. If the aim is to improve the way in which societies achieve their aims, though, then perhaps we should hold these models to as high a standard as those in other disciplines that affect people's lives. New drugs should not be brought to market without randomised controlled trials, and new engineering structures should not be built until thorough testing has shown them to be safe, and even then a considerable margin for error has to be included. Whether one should hold social-systems models to such rigorous standards perhaps depends on the subject matter in question. In some cases it may be ethical to hold randomised controlled trials for a particular intervention, for example – minor infections in primary schools might be a case where this could be tried – but in others we are limited by ethical or other considerations (such as long times or large spatial scales) to observation only, and in many cases obtaining time series of sufficient length and detail to really test a model can be difficult.

Even so, we need to be able to show that models reproducibly agree with data (to the extent that the actual data are available or believable) across multiple sets of situations and measurements, so that we can hope to be confident they agree *mechanistically* with the system under consideration (Grimm and Railsback 2012) – otherwise any suggestions for changes may push the system into a region which has no measurement, and if the mechanisms are not right then neither will be the advice. This applies equally to exploratory or participatory modelling, where models are used along with policy-makers, stakeholders or other participants in decision making, as a tool for stimulating discussion. If the mechanisms the models embody in such a case have not been shown to reproduce adequately the dynamics of the system in question, the detail of such simulations may prove misleading, particularly to those that are not expert in model interpretation. Recent financial crises have shown clearly the consequences of following models that have not been shown to represent sufficiently all of the relevant dynamics.

Real geographies are complex and difficult to understand. Given this complexity, it may be hard to be sure when a model reaches an adequate standard to represent a given problem (Spiegelhalter and Riesch 2011). On the other hand, the explosion of 'big data', and the progressively more detailed and realistic models that are being made at large scale (Parker and Epstein 2011; Barthelemy and Toint 2015) means that the prospects for repeatedly challenging models is improving rapidly, and this invites us to try to embrace the full complexity of the world in their creation. The more thoroughly and widely a model can be tested, the more we can hope that it is better than guesswork, and that any conclusions drawn from the model might be robust. It is possible that development of agreed standards for model specification (such as ODD + D) and *testing* may be helpful here, although this is itself not an area without problems (Collins et al. 2015). Without proper testing, though, we run the risk of doing no more than telling stories.

## Acknowledgments

Grateful acknowledgment goes to the head teacher, school governors and staff who gave permission for and facilitated the data collection exercise. Dr E.G. Bithell and Professor J. Brasington are also thanked for assistance with data gathering and subsequent discussions.

## References


Ajelli, M., B. Gonçalves, D. Balcan, V. Colizza, H. Hu, J. J. Ramasco, S. Merler, and A. Vespignani (2010) Comparing large-scale computational approaches to epidemic modeling: agent-based versus structured metapopulation models, *BMC Infectious Diseases*, 10, 190.

Ajelli, M. and S. Merler (2009) An individual-based model of hepatitis A transmission, *Journal of Theoretical Biology*, 259(3), 478–488.

Amouroux, E., S. Desvaux, and A. Drogoul (2008) Towards virtual epidemiology: an agent-based approach to the modeling of H5N1 propagation and persistence in North-Vietnam, in Q. T. H. T. D. Bui and T. V. Ho (eds), *Intelligent Agents and Multi-Agent Systems*, Volume 5357 of the series *Lecture Notes in Computer Science*, Springer, Berlin, 26–33.

Auchincloss, A. H., and A. V. Diez Roux (2008) A new tool for epidemiology: the usefulness of dynamic-agent models in understanding place effects on health, *American Journal of Epidemiology*, 168(1), 1–8.

Augustijn-Beckers, E. (2011) Simulation of cholera diffusion to compare transmission mechanisms, in *Proceedings of the 11th International Conference on Geocomputation*, no. 1, 39–42.

Barthelemy, J. and P. Toint (2015) A stochastic and flexible activity based model for a large population: application to Belgium, *Journal Artificial Societies and Social Simulation*, 18(3), 15.

Bian, L. (2004) A conceptual framework for an individual-based spatially explicit epidemiological model, *Environmental Planning: B Planning Design*, 31(3), 381–395.

Bian, L., Y. Huang, L. Mao, E. Lim, G. Lee, Y. Yang, M. Cohen, and D. Wilson (2012) Modeling individual vulnerability to communicable diseases: a framework and design, *Annals of the Association of American Geographers*, 102(5), 1016–1025.

Bithell, M. and J. Brasington (2009) Coupling agent-based models of subsistence farming with individual-based forest models and dynamic models of water distribution, *Environmental Modeling & Software*, 24(2), 173–190.

Brouwers, L., B. Cakici, M. Camitz, A. Tegnell, and M. Boman (2009) Economic consequences to society of pandemic H1N1 influenza 2009-preliminary results for Sweden, *Eurosurveillance*, 14(37), 1–7.

Brouwers, L., M. Boman, M. Camitz, K. Mäkilä, and A. Tegnell (2010) Micro-simulation of a smallpox outbreak using official register data, *Eurosurveillance*, 15(35), 1–8.

Brown, D. G., R. Riolo, D. T. Robinson, M. North, and W. Rand (2005) Spatial process and data models: toward integration of agent-based models and GIS, *Journal of Geographical Systems*, 7(1), 25–47.

Brownstein, J.S., K. P. Kleinman, and K. D. Mandl (2005) Identifying pediatric age groups for influenza vaccination using a real-time regional surveillance system, *American Journal of Epidemiology*, 162(7), 686–693.

Burke, D. S., J. M. Epstein, D. a T. Cummings, J. I. Parker, K. C. Cline, R. M. Singa, and S. Chakravarty (2006) Individual-based computational modeling of smallpox epidemic control strategies, *Academy of Emerging Medicine*, 13(11), 1142–1149.

Chen, L., B. Kaminsky, and T. Tummino (2004) Aligning simulation models of smallpox outbreaks, in *Intelligence and Security Informatics*, Volume 3073 of the series *Lecture Notes in Computer Science*, Springer, Berlin, 1–16.

Chen, L.-C., K. M. Carley, D. Fridsma, B. Kaminsky, and A. Yahja (2006) Model alignment of anthrax attack simulations, *Decision Support Systems*, 41(3), 654–668.

Collins, A., M. Petty, D. Vernon-Bido and S. Sherfey (2015) A call to arms: standards for agent-based modelling and simulation, *Journal Artificial Societies and Social Simulation*, 18(3), 12.

Cooley, P., L. Ganapathi, G. Ghneim, S. Holmberg, and W. Wheaton (2008) Using influenza-like illness data to reconstruct an influenza outbreak, *Mathematical and Computer Modelling*, 48 (5–6), 929–939.

Dion, E., L. VanSchalkwyk, and E. F. Lambin (2011) The landscape epidemiology of foot-and-mouth disease in South Africa: a spatially explicit multi-agent simulation, *Ecological Modelling*, 222(13), 2059–2072.

Duan, W., Z. Cao, Y. Wang, and B. Zhu (2013) An ACP approach to public health emergency management: using a campus outbreak of H1N1 influenza as a case study, *IEEE Transactions on Systems, Man, and Cybernetics: Systems*, 43(5), 1028–1041.

Dunham, J. (2005) An agent-based spatially explicit epidemiological model, *Journal of Artificial Societies and Social Simulation*, 9(1), 3.

Eidelson, B. M., and I. Lustick (2004) Vir-pox: an agent-based analysis of smallpox preparedness and response policy, *Journal of Artificial Societies and Social Simulation*, 7(3).

Epstein, J. M. (2009) Modelling to contain pandemics, *Nature*, 460(7256), 687.

Epstein, J. M. and R. Axtell (1996) *Growing Artificial Societies: Social Science from the Bottom Up*, Brookings Institution Press, Washington, DC.

Epstein, J. M., D. a. T. Cummings, S. Chakravarty, R. M. Singha, and D. S. Burke (2004) *Towards a Containment Strategy for Smallpox Bioterror*, Brookings Institution Press, Washington, DC.

Eubank, S., H. Guclu, V. S. A. Kumar, and M. V Marathe (2004) Modelling disease outbreaks in realistic urban social networks, *Nature*, 429 (May), 180–184.

Ferguson, N. M., D. a T. Cummings, S. Cauchemez, C. Fraser, S. Riley, A. Meeyai, S. Iamsirithaworn, and D. S. Burke (2005) Strategies for containing an emerging influenza pandemic in Southeast Asia, *Nature*, 437(7056), 209–214.

Ferguson, N. M., D. a T. Cummings, C. Fraser, J. C. Cajka, P. C. Cooley, and D. S. Burke (2006) Strategies for mitigating an influenza pandemic, *Nature*, 442(7101), 448–452.

Germann, T. C., K. Kadau, I. M. Longini, and C. a Macken (2006) Mitigation strategies for pandemic influenza in the United States, *Proceedings of the National Academy of Science USA*, 103(15), 5935–5940.

Gimblett, H. R. (ed.) (2002) *Integrating Geographic Information Systems and Agent-based Modeling Techniques for Simulating Social and Ecological Processes*, Oxford University Press, Santa Fe, NM.

Glaser, B. G. and A. L. Strauss (2012) *The Discovery of Grounded Theory: Strategies for Qualitative Research*. Aldine Transaction, New Brunswick, NJ.

Glass, R. J., L. M. Glass, W. E. Beyeler, and H. J. Min (2006) Targeted social distancing design for pandemic influenza, *Emerging Infectious Diseases*, 12(11), 1671–181.

Gog, J. R., S. Ballesteros, C. Viboud, L. Simonsen, O. N. Bjornstad, J. Shaman, D. L. Chao, F. Khan, and B. T. Grenfell (2014) Spatial transmission of 2009 pandemic influenza in the US, *PLoS Computational Biology*, 10(6), e1003635.

Grimm, V. and S. F. Railsback (2012) Pattern-oriented modelling: a 'multi-scope' for predictive systems ecology, *Philosophical Transactions of the Royal Society B: Biological Sciences*, 367(1586), 298–310.

Grimm, V., U. Berger, F. Bastiansen, S. Eliassen, V. Ginot, J. Giske, J. Goss-Custard, T. Grand, S. K. Heinz, and G. Huse (2006) A standard protocol for describing individual-based and agent-based models, *Ecological Modelling*, 198(1–2), 115–126.

Grune-Yanoff, T. (2010) Agent-based models as policy decision tools: the case of smallpox vaccination, *Simulation Gaming*, 42(2), 225–242.

Hailegiorgis, A. and A. T. Crooks (2012) Agent-based modeling for humanitarian issues: disease and refugee camps, paper delivered at the Computational Social Science Society of America Conference, Santa Fe, NM.

Halloran, M. E., I. M. Longini, A. Nizam, and Y. Yang (2002) Containing bioterrorist smallpox, *Science*, 298(5597), 1428–1432.

Halloran, M. E., N. M. Ferguson, S. Eubank, I. M. Longini, D. a T. Cummings, B. Lewis, S. Xu, C. Fraser, A. Vullikanti, T. C. Germann, D. Wagener, R. Beckman, K. Kadau, C. Barrett, C. a Macken, D. S. Burke, and P. Cooley (2008) Modeling targeted layered containment of an influenza pandemic in the United States, *Proceedings of the National Academy of Science USA*, 105(12), 4639–4644.

Helbing, D. and P. Molnár (1995) Social force model for pedestrian dynamics, *Physical Review E*, 51(5), 4282–4286.

Hindmarsh, A. C. (1983) ODEPACK, a systematized collection of ODE solvers, in *Scientific Computing*, Volume 1. Edited by R. S. Stepleman et al. North-Holland, Amsterdam, 55–64.

Huberman, B. A. and N. S. Glance (1993) Evolutionary games and computer simulations, *Proceedings of the National Academy of Science USA*, 90(16), 7716–7718.

Jacintho, L., A. Batista, T. Ruas, M. Marietto, and F. Silva (2010) An agent-based model for the spread of the dengue fever: a swarm platform simulation approach, in *Proceedings of the 2010 Spring Simulation Multiconference*, 1–8.

Keeling M. J. and P. Rohani (2008) *Modeling Infectious Diseases in Humans and Animals*. Princeton University Press, Princeton, NJ.

Keeling, M. J., D. A. Rand, and A. J. Morris (1997) Correlation models for childhood epidemics, *Proceedings of the Biological Sciences*, 264(1385), 1149–1156.

Koopman, J. (2004) Modeling infection transmission, *Annual Review of Public Health*, 25(1), 303–326.

Lee, B. Y., V. L. Bedford, M. S. Roberts, and K. M. Carley (2008) Virtual epidemic in a virtual city: simulating the spread of influenza in a US metropolitan area, *Translational Research*, 151(6), 275–287.

Linard, C., N. Ponçon, D. Fontenille, and E. F. Lambin (2009) A multi-agent simulation to assess the risk of malaria re-emergence in southern France, *Ecological Modelling*, 220(2), 160–174.

Longini, I. M., M. E. Halloran, A. Nizam, Y. Yang, S. Xu, D. S. Burke, D. a T. Cummings, and J. M. Epstein (2007) Containing a large bioterrorist smallpox attack: a computer simulation approach, *International Journal of Infectious Diseases*, 11(2), 98–108.

Lourenço, J. and M. Recker (2013) Natural, persistent oscillations in a spatial multi-strain disease system with application to dengue, *PLoS Computational Biology*, 9(10), e1003308.

Merler, S., M. Ajelli, A. Pugliese, and N. M. Ferguson (2011) Determinants of the spatiotemporal dynamics of the 2009 H1N1 pandemic in Europe: implications for real-time modelling, *PLoS Computational Biology*, 7(9), e1002205.

Moreno, J. L. (1934) *Who shall survive? A New Approach to the Problem of Human Interrelations*, Nervous and Mental Disease Monograph Series, No. 58. US Nervous and Mental Disease Publishing Co., Washington, DC.

Morton, K. W. and D. F. Meyers (2005) *Numerical Solution of Partial Differential Equations*. Cambridge University Press, Cambridge.

Müller, B., F. Bohn, G. Dreßler, J. Groeneveld, C. Klassert, R. Martin, M. Schlüter, J. Schulze, H. Weise, and N. Schwarz (2013) Describing human decisions in agent-based models – ODD + D, an extension of the ODD protocol, *Environmental Modeling & Software*, 48, 37–48.

Parker, J. and J. M. Epstein (2011) A distributed platform for global-scale agent-based models of disease transmission, *Philosophical Transactions of the Royal Society A*, 369(1956), 4730–4750.

Patlolla, P., V. Gunupudi, A. Mikler, and R. Jacob (2006) Agent-based simulation tools in computational epidemiology, in *Innovative Internet Community Systems*, Volume 3473 of the series *Lecture Notes in Computer Science*, Springer, Berlin, 212–223.

Pelechano, N., J. Allbeck, and N. Badler (2008) *Virtual Crowds: Methods, Simulation and Control*, Synthesis, Morgan and Claypool, San Rafael, CA.

Perez, L. and S. Dragicevic (2009) An agent-based approach for modeling dynamics of contagious disease spread, *International Journal of Health Geography*, 8, 50.

Rahmandad, H. and J. Sterman (2008) Heterogeneity and network structure in the dynamics of diffusion: comparing agent-based and differential equation models, *Management Science*, 54(5), 998–1014.

Rao, D. M., A. Chernyakhovsky, and V. Rao (2009) Modeling and analysis of global epidemiology of avian influenza, *Environmental Modeling & Software*, 24(1), 124–134.

Rapoport, A. (1957) Contribution to the theory of random and biased nets, *The Bulletin of Mathematical Biophysics*, 19(4), 257–277.

Rodríguez, R. (2013) Enhancing dengue fever modeling through a multi-scale analysis framework: a case study in the Central Valley of Costa Rica, Texas State University, San Marcos, TX.

Segovia-Juarez, J. L., S. Ganguli, and D. Kirschner (2004) Identifying control mechanisms of granuloma formation during M. tuberculosis infection using an agent-based model, *Journal of Theoretical Biology*, 231(3), 357–376.

Shalloway, A. and J. R. Trott (2002) *Design Patterns Explained: A New Perspective on Object-Oriented Design*. Addison-Wesley, Reading, MA.

Simoes, J. A. (2012) An agent-based/network approach to spatial epidemics, in A. J. Heppenstall, A. T., Crooks, L. M. See, and M. Batty (eds), *Agent-Based Models of Geographical Systems*, Springer, Amsterdam.

Spiegelhalter, D. J. and H. Riesch (2011) Don't know, can't know: embracing deeper uncertainties when analysing risks, *Philosophical Transactions of the Royal Society A*, 369(1956), 4730–4750.

Stroud, P., S. Del Valle, S. Sydoriak, J. Riese, and S. Mniszewski (2007) Spatial dynamics of pandemic influenza in a massive artificial society, *Journal of Artificial Societies and Social Simulation*, 10(4), 9.

Teweldemedhin, E., T. Marwala, and C. Mueller (2004) Agent-based modelling: a case study in HIV epidemic, in *Proceedings of the Fourth International Conference on Hybrid Intelligent Systems (HIS '04)*, IEEE Computer Society, Washington, DC, 154–159.

Wilensky, U. (1998) Netlogo Virus model, http://ccl.northwestern.edu/netlogo/models/Virus. Center for Connected Learning and Computer-Based Modeling, Northwestern University, Evanston, IL.

Yang, Y. and P. M. Atkinson (2008) Individual space–time activity-based model: a model for the simulation of airborne infectious-disease transmission by activity-bundle simulation, *Environmental Planning B: Planning Design*, 35(1), 80–99.

# 16 Exploring small-area geographies of obesity in the UK

## Evidence from the UK Women's Cohort Study

*Michelle A. Morris, Graham Clarke, Kimberley L. Edwards, Claire Hulme and Janet E. Cade*

### 16.1 Introduction

Obesity is a huge public health concern, which poses both personal and societal health issues and is an economic burden. Prevalence of obesity is at a record high in the UK, with 57% of women and 67% of men in England classified as overweight or obese in 2013 (Health and Social Care Information Centre, 2013a, 2015). Prevalence has been increasing rapidly since the 1970s. The Foresight report, in 2007, predicted that obesity was likely to continue to increase unless serious changes are made at both an individual and population level (Foresight, 2007). While the rate at which obesity is increasing is beginning to decelerate (Rokholm et al., 2010), obesity is an enormous problem which needs drastic action to reverse.

Obesity is defined by the World Health Organization (WHO) as: 'abnormal or excessive fat accumulation that may impair health' (World Health Organization, 2013). It can be measured using a number of techniques, of which Body Mass Index (BMI) is the most widely used (Hu, 2008). BMI is calculated by dividing an individual's weight (kg) by the square of their height (m), which accounts for differences in weight by height, but is unable to take into account the body composition of an individual. However, for large-scale epidemiological studies, the use of BMI is cost-effective and convenient (Hu, 2008). The WHO has published cut-off points to determine categories of BMI in order to facilitate consistent use of terms such as overweight and obese: underweight $<18.5$ kg/m$^2$, normal weight 18.5 kg/m$^2$ to 24.9 kg/m$^2$, overweight 25 kg/m$^2$ to 30 kg/m$^2$ and obese $>30$ kg/m$^2$ (World Health Organization, 2013).

Causes of obesity are a network of complex individual, social, cultural and environmental interactions which require multifaceted change in order to reverse the problem. Most simplistically, obesity is a result of being in a state of positive energy balance for a prolonged period of time, where energy consumed is greater than energy expended. Therefore, obesity is often associated with an energy dense diet and a sedentary lifestyle (Prentice and Jebb, 1995), although this can be to different extremes depending upon the individual. The dramatic increase in overweight and obesity in recent years cannot be due to genetic changes alone, due to too few generations having passed since the 'outbreak' of obesity. While ultimately an individual has a choice over what to eat and whether or not to exercise, there is a complex interplay between individual, social, cultural and environmental factors contributing to behaviour choice. Not everyone responds to the environment in such a way that they become obese. The reason for this is understood to be 'people's latent biological susceptibility interacting with a changing environment' (Foresight, 2007).

Swinburn et al. (2011) discuss an 'energy balance flipping point', at which time energy intake and expenditure were knocked out of balance around the early 1970s. It is not surprising, given the changes to the environment, that an increase in overweight and obesity is widespread. With the development of an infrastructure conducive of cars, but not necessarily of active transport, a decrease in manual labour employment and technology advances making television and computers available to everyone, the result has been an environment which promotes sedentary behaviours and energy dense food consumption (Hill and Peters, 1998). With many of these changes, time has also become a restricting factor in that people will often choose convenience foods which involve less preparation time (Jabs and Devine, 2006) and are often more energy dense.

Obesity occurs along a socio-economic gradient, with those of higher socio-economic status in general displaying less obesity. This is contrary to patterns we have seen historically, where only the most affluent had the money to purchase enough food to eat and become obese. Research has shown that once GDP reaches US$5,000 per head, there is no longer a positive linear relationship, but one which is flat (Swinburn et al., 2011). In fact the opposite is seen, where obesity is more prevalent in those of lower socio-economic status (Drewnowski and Specter, 2004; Wilkinson and Pickett, 2010; Offer et al., 2012). In developing countries, a problem of under-nutrition is also accompanied by over-nutrition, seeing instances of both malnutrition and obesity in the same population, and worldwide there are now more obese people than underweight (World Health Organization, 2000; Delpeuch et al., 2009).

Considering the diversity of residential neighbourhoods in the UK, and the fact that obesogenic environments are common, it is not surprising that obesity rates can vary according to where an individual lives. National statistics for obesity are published by the Government Office Region (GOR) in the UK (Health and Social Care Information Centre, 2011). These are updated annually with results from the Health Survey for England (HSE). HSE has collected measured height and weight data for adults annually since 1991. This is considered to be the most robust measure of adult obesity in the UK. However, the HSE data displays the caveat: 'Data not sufficiently robust to measure geographical boundaries smaller than SHA [Strategic Health Authority]' (Public Health England, 2013). The annual sample size in recent years has included approximately 8,000 adults, which are representative at the GOR level (Health and Social Care Information Centre, 2013b). For more granular geographic data, the National Obesity Observatory uses synthetic estimates from the 2006–2008 HSE statistics, simulated by combining with demographic data from the 2001 Census.

Following initiation of the National Child Measurement Programme in 2006 (Health and Social Care Information Centre, 2013c) detailed and more current data for child obesity rates, with geographical location, are also now available. However, such data are not available for adults. There are plans to use self-reported data from the Active People Survey (Sport England, 2013), but this is still subject to methodological testing to assess the reliability of self-reported height and weight estimates.

Often obesity is associated with other chronic diseases such as hypertension, type 2 diabetes, some cancers, cardiovascular disease, depression, sleep apnoea, osteoarthritis and others – through an accumulation of excess body fat which can lead to organ-specific pathological consequences through various mechanisms. Being overweight increases risk of such diseases, risk that continues to increase with increasing BMI (Crawford et al., 2010). Despite the risks and potential causes of obesity being well documented (Gibney, 2004; Foresight, 2007; Hu, 2008; Pearce and Witten, 2010; Gortmaker et al., 2011; Hall et al., 2011; Swinburn et al., 2011; Wang et al., 2011), obesity remains a disease with no quick fix. More interdisciplinary research is required in order to cover combined aspects of influence on obesity. To understand the determinants of (and possible interventions to reduce) the prevalence of obesity and how risk factors interact also needs to be better understood so that they can be more effectively addressed by policy makers.

Using the available data to present descriptive statistics means that some generalisation occurs due to the large geographical scale. This may average out potential differences in areas of varying socio-economic status and rurality. Synthetic estimates are a beneficial alternative where data are missing, but there could be better use made of data which are collected for large cohort studies, such as the UK Women's Cohort Study (UKWCS) set up at the University of Leeds in 1994 (www.ukwcs.leeds.ac.uk). The cohort collected dietary information and self-reported height and weight from approximately 35,000 women in the late 1990s. While such data are subject to reporting bias, the large sample and wide geographical spread of women provides some confidence in the sample. The UKWCS is an appropriate population to use to answer the following research question: does overweight/obesity vary according to where women live in the UK Women's Cohort? It could be hypothesised that there will be a spatial variation in overweight/obesity in the UKWCS, with those in the South, Rural, Countryside and City Living Areas less likely to be overweight/obese.

The novelty of using data from this cohort is the opportunity to use postcode-level spatial-scale overweight/obesity and demographic information at an individual level with a large sample size. Results are aggregated for analysis and presentation of results. Other work in this area has used national survey data aggregated to a higher geographic unit such as ward, Primary Care Trust (PCT) or region and then combined with census data for those units (Moon et al., 2007), which relies on assumptions that the survey sample shares characteristics of the demographic data. Other methods generating local-level estimates also rely on sophisticated matching of data from different sources to produce synthetic estimates through spatial microsimulation (Pearce and Witten, 2010). This work offers a new alternative perspective on small-area geographies of obesity.

The objectives for this chapter are to:

- Describe overweight/obesity in the UKWCS using Government office Region (GOR), Urban and Rural and Output area Classification (OAC) Supergroup.
- Investigate associations between overweight/obesity, diet cost and GOR, Urban and Rural and OAC in the UKWCS, using both a continuous measure of BMI and by dichotomising the sample as normal weight or overweight/obese.
- Estimate weight status for women living in a large UK city, Leeds, based on the OAC group analysis and results.

## 16.2 Methods

### *16.2.1 Weight status*

Body Mass Index (BMI) in the UKWCS is calculated from self-reported height (m) and weight (kg) in the baseline survey, collected between 1995 and 1998, using the formula:

$$\text{BMI (kg/m}^2\text{)} = \text{weight (kg)/height (m)}^2$$

This produces a continuous measure of BMI. The effect of BMI was investigated using BMI as a continuous measure and also using the WHO BMI categories.

The World Health Organization (WHO) publishes cut-off points for underweight, normal weight, overweight and obese to ensure consistent recording of these weight status (World Health Organization, 2006). Each of the women was assigned a BMI category. The WHO cut-off points are summarised in Table 16.1.

### *16.2.2 Spatial analysis*

Three spatial measures are used for this analysis:

1. GOR – includes the nine GORs of England plus Scotland and Wales
2. Urban/Rural living – includes England, Scotland and Wales

*Table 16.1* WHO BMI categories

| *Weight status* | *BMI range* |
|---|---|
| Underweight | <18.5 |
| Normal weight | 18.5–24.9 |
| Overweight | 25–29.9 |
| Obese | >=30 |

Source: World Health Organization (2006)

3. OAC geodemographic classification – Supergroup (top layer of the classification hierarchy) and Group (middle layer) – for England, Scotland and Wales

The Geoconvert application was used to complete this process (Census Dissemination Unit, 2009). Geoconvert is an online tool provided by the Census Dissemination Unit at the University of Manchester, which will match one type of geography to another and provide a report of the target geographies. These target geographies can then be merged back into the source dataset.

These three spatial measures were chosen in order to investigate elements of both geographical context and composition. A GOR is a large geographical unit which is home to between two and nine million individuals. This unit is used by government and is known to most people. To report dietary consumption at this level means that substantial generalisation is made, but it may be a useful unit in which to convey results to the public and policy makers. In the UK, research has shown that differences in health and mortality occur at an even larger geographic scale. Within the analysis of GOR data it is also possible to say something about the health divide between the North and the South of England. Economic differences also occur at this scale, with a higher cost of living in the South. Thus we shall also investigate if any differences in health can be attributed to differences in dietary patterns or diet cost across the North–South divide.

The Urban/Rural and OAC units can be pinpointed to a specific geographical context, but also include elements of geographic composition. The OAC does this to the greatest extent, incorporating 41 demographic variables from the census.

The odds of women within each OAC Group (the second layer of the OAC hierarchy) being overweight/obese, compared to normal weight were calculated for the UKWCS. Assuming that weight status by OAC Group in the cohort is typical for all women in these Groups throughout the UK, due to their similar demographic characteristics, weight status from the UKWCS were used to estimate weight status of women in a specific UK city, Leeds. This pattern could be applied to any area, but Leeds was selected as this is an 'average' UK city, which has been used in other studies relating to geography and health (Tomintz et al., 2008; Edwards et al., 2009).

### *16.2.3 Index of Multiple Deprivation*

Index of Multiple Deprivation (IMD) in a UK index of deprivation (Department for Communities and Local Government, 2012). This is used to highlight the social geography of Leeds in comparison to overweight/obesity in Leeds.

## 16.3 Statistical analysis

All statistical analyses have been completed using Stata statistical software: version 12 (StataCorp, 2009, 2012). Due to the multiple testing which occurs in this analysis, statistical significance is considered at $p<0.01$.

### *16.3.1 Descriptive statistics*

Descriptive statistics are presented by spatial measure for the percentage of women falling within the four BMI categories and mean BMI. Difference in variation by spatial measures was tested using chi-squared or a t-test, as appropriate. Descriptive statistics for OAC Group (the middle layer of OAC hierarchy) are presented in a separate table.

UKWCS obesity prevalence results are presented alongside statistics for all women in the UK for comparison purposes.

### *16.3.2 Spatial measures and overweight/obese*

The UKWCS women were categorised into two groups: (1) normal weight and (2) overweight/obese. Underweight women were excluded from the analysis (n=726). Daily diet cost was presented by spatial measure for these two categories and the difference between these values was tested using chi-squared or a t-test, as appropriate.

Logistic regression assessed whether the spatial measures predict relative risk of being overweight/obese compared to normal weight at each spatial measure.

Reference categories for each spatial measure were assigned to the area with the largest number of women residing within, with the exception of OAC Supergroup, where the Typical Traits category is used as reference as these have the most average demographic characteristics (see Table 16.2; note here that North South has been additionally included as a separate geographical layer of analysis).

The reference category for GOR analysis, the South East, has the largest population within the cohort and the South East Strategic Health Authority also has the lowest obesity prevalence in the UK.

There are three regression models for each spatial measure. Model 1 looks at how well the spatial measure predicts overweight/obesity with no adjustments. Model 2 adjusts for dietary patterns and daily cost of diet. Model 3 adjusts for energy intake, physical activity, dietary pattern, daily cost of diet, smoking, menopause status, education and social class. For models where the OAC Supergroup is the predictor, education and social class are not included as these variables are already accounted for in the OAC and could cause co-linearity problems.

*Table 16.2* Reference categories for regression models

| *Spatial measure* | *Reference category* |
|---|---|
| GOR | South East |
| North–South | South |
| Urban–Rural | Urban |
| OAC Supergroup | Typical Traits |

### 16.3.3 Spatial measures and BMI

In order to quantify whether a dose response relationship exists for spatial measures and BMI, linear regression was carried out with the same spatial measures as predictor variables (GOR, North/South, Urban/Rural and OAC) and a continuous measure of BMI as the outcome. Three models were produced with adjustments, as described above.

## 16.4 Results

### 16.4.1 UKWCS compared to UK women

Less overweight and obesity is observed in the UKWCS compared to women of the same age during the same time period for the whole of the UK. There was 9% less overweight and 6% less obesity in the UKWCS. Prevalence of underweight was the same, with a higher proportion of UKWCS women being of normal weight (Table 16.3). This is also the case when looking at variation in obesity by GOR (Table 16.4).

Results suggest that there is a spatial variation in prevalence of overweight and obesity, but less obesity variation exists in the UKWCS by GOR than compared to the whole UK, suggesting more homogeneity in UKWCS women.

*Table 16.3* BMI category variation in UK women compared to UKWCS

| *BMI Category* | *Weight status for UK women 1995–1998 (%) (Public Health England, 2013)* | *Weight status for UKWCS (%)* |
|---|---|---|
| Underweight | 2.1 | 2.2 |
| Normal weight | 45.9 | 60.7 |
| Overweight | 32.9 | 24.2 |
| Obese | 19.2 | 13.0 |

*Table 16.4* Obesity prevalence from Public Health England Adult Obesity Prevalence data for women

| *Government office region* | *Obesity prevalence for UK Women 1998–2000 (%)* | *Obesity prevalence for UKWCS (%)* |
|---|---|---|
| North East | 19.8 | 13.7 |
| North West | 21.0 | 13.9 |
| Yorkshire and the Humber | 20.5 | 13.8 |
| East Midlands | 24.3 | 13.6 |
| West Midlands | 23.5 | 14.0 |
| East of England | 19.6 | 12.3 |
| Greater London | 20.2 | 12.6 |
| South East | 17.7 | 12.3 |
| South West | 18.8 | 12.3 |

Source: Public Health England (2013)

### *16.4.2 Spatial measures and weight status*

Weight status by BMI category varies by spatial measure. Table 16.5 shows the percentage of women in each BMI category by four spatial measures: GOR, North South, Urban/Rural and OAC Supergroup. Narrow confidence intervals exist around the percentages in each BMI category providing assurance in the estimates.

Prevalence varies significantly (chi-squared test $p<0.01$) by GOR in all weight status categories, except obese (Table 16.5). The lowest percentage of overweight exists in Greater London (21.1%) and the highest in the East of England (26.6%). The prevalence of obesity is homogenous across GORs in the UKWCS, with a range only between 12.3% and 14% throughout GORs of England and Wales and Scotland in the UKWCS, which is statistically non-significant.

There is little variation seen for mean BMI by spatial measure, although analysis of variance tests show this is significant between groups. On closer examination, using ANOVA with Scheffe post hoc tests, significant differences appear between Southern and Northern GORs, suggestive of a North South divide with BMI being significantly ($p<0.01$) higher in the North East, North West, Yorkshire and the Humber, East Midlands, West Midlands and Scotland compared to Greater London. There is also a significantly higher BMI in the North West and Yorkshire and the Humber compared to the South East.

When the women are categorised as North or South, there is a significant difference, for each BMI category and also for mean BMI (Table 16.5).

There is a significant difference in BMI by Urban/Rural, with Urban areas having higher BMI (Table 16.5). Urban areas have a higher percentage of obese and a lower percentage of normal weight women. Underweight and overweight groups do not differ significantly.

Those in areas Constrained by Circumstance and of Blue Collar Communities have a higher percentage of overweight and obese than other types of area, with prevalence being lowest in City Living areas. When the OAC Supergroups are further subdivided into OAC Groups, it can be seen that the high percentage of overweight and obese comes mainly from the Terraced Blue Collar and Younger Blue Collar Groups, but the Older Blue Collar women are also above average for percentage of overweight and obese. All groups within the Constrained by Circumstance Supergroups have a high percentage of overweight/obese. Within the Prospering Suburbs Supergroup, the Prospering Semis Group have a much higher prevalence of both overweight and obese, a difference most prominent when compared to the Thriving Suburbs Group. In general, confidence intervals about the percentage of women in each BMI category are slightly wider in the granular OAC Groups than when looking at the Supergroup level (see Table 16.6). Mean BMI is lowest in Transient Communities (23.7 $kg/m^2$). The highest BMI can be seen in Younger Blue Collar and Public Housing (25.9 $kg/m^2$).

In logistic regression analysis (see Table 16.7) odds ratios (ORs) show a statistically significant increase in risk of being overweight/obese in all areas – except the North East, Greater London and the South West – compared to those

*Table 16.5* Percentage of UKWCS women in each of the WHO BMI categories by spatial measure

| | *Underweight % (95% CI)* | *Normal weight % (95% CI)* | *Overweight % (95% CI)* | *Obese % (95% CI)* | *Mean BMI (95% CI)* |
|---|---|---|---|---|---|
| Whole cohort (n=32784) | 2.2 (2.0 to 2.3) | 60.7 (60.2 to 61.3) | 24.2 (23.7 to 24.6) | 13.0 (12.6 to 13.3) | 24.4 (24.4 to 24.4) |
| **Government office region** | **$Chi^2$ p=0.002** | **$Chi^2$ p<0.001** | **$Chi^2$ p<0.001** | **$Chi^2$ p=0.202** | **$Chi^2$ p<0.001** |
| North East (n=974) | 1.8 (1.0 to 2.7) | 60.3 (57.2 to 63.3) | 24.2 (21.5 to 26.9) | 13.7 (11.5 to 15.8) | 24.8 (24.5 to 25.1) |
| North West (n=3038) | 2.0 (1.5 to 2.5) | 57.9 (56.2 to 59.7) | 26.2 (24.6 to 27.8) | 13.9 (12.6 to 15.1) | 24.6 (24.5 to 24.8) |
| Yorkshire and the Humber (n=2561) | 1.5 (1.0 to 2.0) | 59.3 (23.7 to 27.1) | 25.4 (23.7 to 27.1) | 13.8 (12.4 to 15.1) | 24.7 (24.5 to 24.9) |
| East Midlands (n=2405) | 1.6 (1.0 to 2.1) | 59.6 (57.6 to 61.5) | 25.2 (23.5 to 27.0) | 13.6 (12.2 to 15.0) | 24.6 (24.4 to 24.8) |
| West Midlands (n=2534) | 1.8 (1.3 to 2.3) | 59.3 (57.4 to 61.1) | 24.9 (23.2 to 26.5) | 14.0 (12.7 to 15.4) | 24.6 (24.4 to 24.7) |
| East of England (n=3001) | 2.6 (2.0 to 3.1) | 58.5 (56.8 to 60.3) | 26.6 (25.0 to 28.2) | 12.3 (11.1 to 13.5) | 24.4 (24.3 to 24.6) |
| Greater London (n=3709) | 2.9 (2.3 to 3.4) | 63.4 (61.9 to 65.0) | 21.1 (19.7 to 22.3) | 12.6 (11.5 to 13.7) | 24.1 (23.9 to 24.2) |
| South East (n=6789) | 2.4 (2.0 to 2.7) | 62.9 (61.7 to 64.0) | 22.5 (21.5 to 23.5) | 12.3 (11.5 to 13.1) | 24.2 (24.1 to 24.3) |
| South West (n=4155) | 2.4 (1.9 to 2.9) | 62.2 (60.7 to 63.7) | 23.0 (21.8 to 24.3) | 12.3 (11.3 to 13.3) | 24.2 (24.1 to 24.4) |
| Scotland (n=2199) | 1.7 (1.1 to 2.2) | 58.9 (56.9 to 61.0) | 26.3 (24.5 to 28.2) | 13.1 (11.6 to 14.5) | 24.6 (24.4 to 24.8) |
| Wales (n=1419) | 1.9 (1.2 to 2.6) | 59.2 (56.6 to 61.8) | 25.6 (23.3 to 27.9) | 13.3 (11.5 to 15.1) | 24.6 (24.4 to 24.8) |
| **North South** | **T test p<0.001** | **T test p<0.001** | **T test p<0.001** | **T test p<0.001** | **T test p<0.001** |
| The North (n=11512) | 1.8 (1.5 to 2.0) | 59.1 (58.2 to 60.0) | 25.4 (24.6 to 26.2) | 13.8 (13.2 to 14.4) | 24.6 (24.6 to 24.7) |
| The South (n=17654) | 2.5 (2.3 to 2.7) | 62.1 (61.4 to 62.8) | 23.0 (22.4 to 23.6) | 12.4 (11.9 to 12.9) | 24.2 (24.1 to 24.3) |
| **Urban Rural** | **T test p=0.014** | **T test p <0.001** | **T test p=0.567** | **T test p<0.001** | **T test p<0.001** |
| Urban (n=21866) | 2.3 (2.1 to 2.5) | 59.9 (59.2 to 60.5) | 24.3 (23.7 to 24.8) | 13.6 (13.1 to 14.0) | 24.5 (24.4 to 24.6) |
| Rural (n=10843) | 1.9 (1.6 to 2.1) | 62.4 (61.5 to 63.3) | 24.0 (23.2 to 24.8) | 11.7 (11.1 to 12.3) | 24.2 (24.1 to 24.3) |
| **OAC Supergroup** | **$Chi^2$ p=0.003** | **$Chi^2$ p<0.001** | **$Chi^2$ p<0.001** | **$Chi^2$ p<0.001** | **$Chi^2$ p<0.001** |
| Blue Collar Communities (n=2114) | 1.8 (1.2 to 2.4) | 52.1 (50.0 to 54.2) | 27.8 (25.9 to 29.7) | 18.3 (16.7 to 20.0) | 25.3 (25.1 to 25.6) |
| City Living (n=2216) | 3.0 (2.3 to 3.7) | 65.2 (63.2 to 67.2) | 20.4 (18.8 to 22.1) | 11.3 (10.0 to 12.6) | 23.8 (23.7 to 24.0) |
| Countryside (n=7518) | 2.0 (1.7 to 2.3) | 63.1 (62.0 to 64.2) | 23.1 (22.2 to 24.1) | 11.8 (11.0 to 12.5) | 24.2 (24.1 to 24.3) |
| Prospering Suburbs (n=11029) | 2.0 (1.7 to 2.2) | 61.8 (60.9 to 62.8) | 24.5 (23.7 to 25.3) | 11.7 (11.1 to 12.3) | 24.3 (24.3 to 24.4) |
| Constrained by Circumstance (n=1411) | 1.7 (1.0 to 2.4) | 49.9 (47.3 to 52.5) | 29.1 (26.8 to 31.5) | 19.3 (17.2 to 21.3) | 25.5 (25.2 to 25.8) |
| Typical Traits (n=6576) | 2.4 (2.1 to 2.8) | 60.0 (58.8 to 61.2) | 24.4 (23.3 to 25.4) | 13.2 (12.4 to 14.0) | 24.4 (24.3 to 24.5) |
| Multicultural (n=1981) | 2.8 (2.1 to 3.6) | 59.8 (57.6 to 61.9) | 22.2 (20.4 to 24.0) | 15.2 (13.6 to 16.8) | 24.4 (24.2 to 24.6) |

*Table 16.6* Percentage of UKWCS women in each of the WHO BMI categories by OAC Group

| *Group name* | *Underweight (%) (95% CI)* | *Normal weight (%) (95% CI)* | *Overweight (%) (95% CI)* | *Obese (%) (95% CI)* | *Mean BMI kg/m² (95% CI)* |
|---|---|---|---|---|---|
| 1A - Terraced Blue Collar (n=344) | 1.5 (0.2 to 2.7) | 50.0 (44.7 to 55.3) | 28.8 (24.0 to 33.6) | 19.8 (15.5 to 24.0) | 25.5 (25.0 to 26.0) |
| 1B - Younger Blue Collar (n=600) | 2.2 (1.0 to 0.3) | 47.5 (43.5 to 51.5) | 29.3 (25.7 to 33.0) | 21.0 (17.7 to 24.3) | 25.9 (25.3 to 26.4) |
| 1C - Older Blue Collar (n=1170) | 1.7 (1.0 to 2.5) | 55.0 (52.2 to 57.9) | 26.8 (24.2 to 29.3) | 16.5 (14.4 to 18.6) | 25.0 (24.7 to 25.3) |
| 2A - Transient Communities (n=544) | 4.6 (2.8 to 6.4) | 65.1 (61.1 to 69.1) | 17.1 (13.9 to 20.3) | 13.2 (10.4 to 16.1) | 23.7 (23.3 to 24.0) |
| 2B - Settled in City (n=1672) | 2.5 (1.8 to 3.3) | 65.3 (63.0 to 67.5) | 21.5 (19.6 to 23.5) | 10.7 (9.2 to 12.2) | 23.9 (23.7 to 24.1) |
| 3A - Village Life (n=2457) | 2.2 (1.6 to 2.8) | 61.1 (59.2 to 63.0) | 24.9 (23.2 to 26.6) | 11.8 (10.6 to 13.1) | 24.3 (24.1 to 24.5) |
| 3B - Agricultural (n=2136) | 1.7 (1.2 to 2.3) | 63.5 (61.5 to 65.6) | 22.8 (21.0 to 24.5) | 12.0 (10.6 to 13.4) | 24.1 (24.0 to 24.3) |
| 3C - Accessible Countryside (n=2925) | 2.0 (1.5 to 2.5) | 64.5 (62.8 to 66.3) | 21.9 (20.4 to 23.4) | 11.6 (10.4 to 12.7) | 24.1 (23.9 to 24.2) |
| 4A - Prospering Younger Families (n=1124) | 2.2 (1.4 to 3.1) | 60.9 (58.1 to 63.8) | 24.0 (21.5 to 26.5) | 12.8 (10.9 to 14.8) | 24.3 (24.0 to 24.5) |
| 4B - Prospering Older Families (n=4196) | 1.7 (1.3 to 2.1) | 63.6 (62.1 to 65.0) | 24.0 (22.8 to 25.3) | 10.6 (9.7 to 11.6) | 24.3 (24.1 to 24.4) |
| 4C - Prospering Semis (n=2632) | 1.8 (1.3 to 2.3) | 57.1 (55.2 to 59.0) | 27.0 (25.3 to 28.7) | 14.2 (12.8 to 15.5) | 24.8 (24.7 to 25.0) |
| 4D - Thriving Suburbs (n=3077) | 2.4 (1.8 to 2.9) | 63.9 (62.2 to 65.6) | 23.1 (21.6 to 25.6) | 10.6 (9.5 to 11.7) | 24.1 (23.9 to 24.2) |
| 5A - Senior Communities (n=151) | 2.6 (0.0 to 5.2) | 46.4 (38.3 to 54.4) | 31.8 (24.3 to 39.3) | 19.2 (12.8 to 25.5) | 25.8 (24.9 to 26.7) |
| 5B - Older Workers (n=1106) | 1.4 (0.7 to 2.0) | 50.7 (47.8 to 53.7) | 28.9 (26.3 to 31.6) | 19.0 (16.7 to 21.3) | 25.4 (25.1 to 25.7) |
| 5C - Public Housing (n=154) | 3.2 (0.4 to 6.0) | 47.4 (39.4 to 55.4) | 27.9 (20.8 to 35.1) | 21.4 (14.9 to 28.0) | 25.9 (25.0 to 26.7) |
| 6A - Settled Households (n=1593) | 2.0 (1.3 to 2.7) | 57.0 (54.6 to 59.4) | 27.3 (25.1 to 29.5) | 13.7 (12.0 to 15.4) | 24.7 (24.5 to 24.9) |
| 6B - Least Divergent (n=1994) | 2.2 (1.6 to 2.9) | 59.8 (57.6 to 61.9) | 24.6 (22.7 to 26.5) | 13.4 (11.9 to 14.9) | 24.5 (24.3 to 24.7) |
| 6C - Young Families in Terraced Homes (n=1060) | 2.7 (1.8 to 3.7) | 57.0 (54.0 to 60.0) | 26.3 (23.7 to 29.0) | 14.0 (11.9 to 16.0) | 24.6 (24.3 to 24.8) |
| 6D - Aspiring Households (n=1929) | 2.9 (2.2 to 3.7) | 64.3 (62.1 to 66.4) | 20.7 (18.9 to 22.5) | 12.1 (10.7 to 13.6) | 24.0 (23.9 to 24.2) |
| 7A - Asian Communities (n=1245) | 2.4 (1.6 to 3.3) | 59.3 (56.5 to 62.0) | 23.1 (20.7 to 25.4) | 15.3 (13.3 to 17.3) | 24.5 (24.3 to 24.8) |
| 7B - Afro-Caribbean (n=736) | 3.5 (2.2 to 4.9) | 60.6 (57.1 to 64.1) | 20.8 (17.8 to 23.7) | 15.1 (12.5 to 17.7) | 24.2 (23.9 to 24.6) |

*Table 16.7* Logistic regression investigating the odds of being overweight/obese compared to normal weight by spatial measure. Model 1 = unadjusted. Model 2 = adjusted for diet cost and dietary pattern. Model 3 = adjusted for Energy intake, energy expenditure, diet cost, dietary pattern, smoking, menopause, education and class. OR = Odds ratio; CI = Confidence Interval

| | *Model 1* | | *Model 2* | | *Model 3* | |
|---|---|---|---|---|---|---|
| | *OR (95% CI)* | *P value* | *OR (95% CI)* | *P value* | *OR (95% CI)* | *P value* |
| **Government office region** | Pseudo $R^2$= 0.002 | | Pseudo $R^2$=0.018 | | Pseudo $R^2$= 0.035 | |
| North East | 1.14 (0.99 to 1.31) | 0.074 | 1.11 (0.97 to 1.28) | 0.139 | 1.12 (0.96 to 1.31) | 0.139 |
| North West | **1.25 (1.14 to 1.37)** | **<0.001** | **1.23 (1.12 to 1.34)** | **<0.001** | **1.27 (1.15 to 1.40)** | **<0.001** |
| Yorkshire and the Humber | **1.19 (1.09 to 1.31)** | **<0.001** | **1.18 (1.07 to 1.29)** | **0.001** | **1.19 (1.07 to 1.32)** | **0.001** |
| East Midlands | **1.18 (1.07 to 1.30)** | **0.001** | **1.16 (1.05 to 1.28)** | **0.002** | **1.17 (1.05 to 1.30)** | **0.004** |
| West Midlands | **1.19 (1.08 to 1.30)** | **<0.001** | **1.17 (1.06 to 1.29)** | **0.001** | **1.20 (1.09 to 1.34)** | **<0.001** |
| East of England | **1.20 (1.10 to 1.31)** | **<0.001** | **1.19 (1.09 to 1.30)** | **<0.001** | **1.19 (1.08 to1.32)** | **<0.001** |
| Greater London | 0.96 (0.88 to 1.04) | 0.334 | 1.02 (0.94 to 1.11) | 0.651 | 1.05 (0.96 to 1.16) | 0.268 |
| South East | 1.00 | – | 1.00 | | 1.00 | |
| South West | 1.03 (0.95 to 1.11) | 0.533 | 1.03 (0.95 to 1.12) | 0.483 | 1.01 (0.92 to 1.11) | 0.829 |
| Scotland | **1.21 (1.09 to 1.33)** | **<0.001** | **1.18 (1.06 to 1.30)** | **0.002** | **1.27 (1.14 to 1.42)** | **<0.001** |
| Wales | **1.19 (1.05 to 1.34)** | **0.005** | **1.18 (1.05 to 1.34)** | **0.006** | **1.20 (1.05 to 1.37)** | **0.006** |
| **North South** | Pseudo $R^2$= 0.001 | | Pseudo $R^2$= 0.017 | | Pseudo $R^2$= 0.035 | |
| The North | **1.16 (1.11 to 1.22)** | **<0.001** | **1.13 (1.08 to 1.19)** | **<0.001** | **1.15 (1.09 to 1.22)** | **<0.001** |
| The South | 1 | – | 1 | – | 1 | – |
| **Urban Rural** | Pseudo $R^2$= 0.0004 | | Pseudo $R^2$= 0.017 | | Pseudo $R^2$= 0.036 | |
| Urban | 1 | – | 1 | – | | – |
| Rural | **0.91 (0.86 to 0.95)** | **<0.001** | **0.89 (0.85 to 0.94)** | **<0.001** | **0.90 (0.85 to 0.95)** | **<0.001** |
| **OAC Supergroup** | Pseudo $R^2$= 0.005 | | Pseudo $R^2$= 0.021 | | Pseudo $R^2$= 0.033 | |
| Blue Collar Communities | **1.41 (1.28 to 1.56)** | **<0.001** | **1.33 (1.20 to 1.47)** | **<0.001** | **1.36 (1.23 to 1.52)** | **<0.001** |
| City Living | **0.78 (0.70 to 0.86)** | **<0.001** | **0.82 (0.74 to 0.91)** | **<0.001** | **0.81 (0.73 to 0.90)** | **<0.001** |
| Countryside | **0.88 (0.82 to 0.95)** | **<0.001** | **0.86 (0.80 to 0.92)** | **<0.001** | **0.85 (0.79 to 0.92)** | **<0.001** |
| Prospering Suburbs | 0.93 (0.88 to 1.00) | 0.035 | **0.88 (0.83 to 0.94)** | **<0.001** | **0.86 (0.80 to 0.92)** | **<0.001** |
| Constrained by Circumstance | **1.55 (1.38 to 1.74)** | **<0.001** | **1.47 (1.31 to 1.66)** | **<0.001** | **1.47 (1.30 to 1.66)** | **<0.001** |
| Typical Traits | 1.00 | – | 1.00 | – | 1.00 | – |
| Multicultural | 1.00 (0.90 to 1.11) | 0.984 | 1.08 (0.97 to 1.20) | 0.176 | 1.09 (0.97 to 1.21) | 0.136 |

in the South East of England. The reference category (the South East) odds of being overweight/obese are 1, therefore all odds ratio values greater than 1 represent increased odds of being overweight/obese and values less than one decreased odds. When the women are dichotomised as living in the North or South, those in the North have a significantly increased risk of being overweight/obese compared to the South, a difference which remains in the adjusted models. Urban living is also associated with an increased risk of overweight/obese with those in rural areas having a 10% reduced risk of overweight/obesity compared to the urban reference group. For the OAC Supergroup analysis, ORs show a 42% increased risk of being overweight/obese in Blue Collar Communities compared to areas of Typical Traits and 56% increased risk of being overweight/obese in Constrained by Circumstance. When the model is adjusted for dietary pattern and diet cost, the explained variance in the model increases and the effects of OAC Supergroup on prevalence of overweight/obesity relationships remain although slightly attenuated. Including further possible confounders and mediators in the model further increases the explained variance, with little effect on the ORs.

Table 16.8 shows how overweight/obese vary according to the OAC Groups, a more granular unit of the output area classification. Using this we can investigate differences within the OAC Supergroups. The Terraced Blue Collar and Younger Blue Collar have higher odds of being overweight/obese than the Older Blue Collar, with increased risks of 54%, 65% and 24%, respectively. Within the Countryside Supergroup, showing reduced odds of being overweight/obese, this is driven by the Agricultural and Accessible Countryside Groups with the Village Life Group showing no reduced risk.

Linear regression analysis using a continuous measure of BMI concurs with the findings reported above, quantifying the unit increase in BMI, which in most cases is only a small, but statistically significant difference (see Table 16.9). The largest unit increase in BMI is 0.61 $kg/m^2$ (95% CI 0.31 to 0.90) observed in the North East compared to the South East in the fully adjusted model. The only areas where BMI does not vary significantly from the South East are the other Southern areas: East of England, Greater London and the South West.

The regression models (see Table 16.7, Table 16.8 and Table 16.9) all have relatively small pseudo $R^2$ and $R^2$ values, suggesting that they explain less than 5% of variation in overweight/obesity and BMI, implying that perhaps spatial unit is not the best predictor of weight status.

### *16.4.3 Estimated overweight/obese for Leeds*

Using the unadjusted odds ratios for OAC group reported in Table 16.8, it is possible to estimate the odds of being obese in Leeds. Figure 16.1 shows the estimated distribution of overweight/obese in Leeds. By comparing this map to Figure 16.2 it can be seen how overweight/obesity visually correlates with Index of Multiple Deprivation (IMD) in Leeds.

*Table 16.8* Logistic regression investigating the odds of being overweight/obese compared to normal weight by OAC group. Model 1 = unadjusted. Model 2 = adjusted for diet cost and dietary pattern. Model 3 = adjusted for Energy intake, energy expenditure, diet cost, dietary pattern, smoking, menopause, education and class. OR = Odds Ratio; CI = Confidence Interval

| | *Model 1* | | *Model 2* | | *Model 3* | |
|---|---|---|---|---|---|---|
| | *OR (95% CI)* | *P value* | *OR (95% CI)* | *P value* | *OR (95% CI)* | *P value* |
| **OAC Group** | Pseudo $R^2$= 0.007 | | Pseudo $R^2$=0.023 | | Pseudo $R^2$= 0.039 | |
| 1A - Terraced Blue Collar (n=344) | **1.54 (1.22 to 1.94)** | <0.001 | **1.43 (1.13 to 1.81)** | **0.002** | **1.40 (1.08 to 1.82)** | **0.011** |
| 1B - Younger Blue Collar (n=600) | **1.65 (1.37 to 1.99)** | <0.001 | **1.58 (1.32 to 1.91)** | <0.001 | **1.60 (1.30 to 1.96)** | <0.001 |
| 1C - Older Blue Collar (n=1170) | **1.24 (1.07 to 1.44)** | 0.004 | **1.18 (1.02 to 1.37)** | **0.025** | **1.18 (1.00 to 1.38)** | **0.050** |
| 2A - Transient Communities (n=544) | **0.71 (0.58 to 0.87)** | 0.001 | **0.78 (0.63 to 0.95)** | **0.016** | 0.83 (0.66 to 1.03) | 0.094 |
| 2B - Settled in City (n=1672) | **0.78 (0.68 to 0.89)** | <0.001 | **0.81 (0.71 to 0.94)** | 0.004 | 0.87 (0.75 to 1.01) | 0.071 |
| 3A - Village Life (n=2457) | 0.95 (0.84 to 1.07) | 0.372 | 0.93 (0.82 to 1.05) | 0.248 | 0.92 (0.81 to 1.06) | 0.250 |
| 3B - Agricultural (n=2136) | **0.87 (0.76 to 0.99)** | **0.029** | **0.85 (0.74 to 0.96)** | **0.011** | **0.85 (0.74 to 0.98)** | **0.023** |
| 3C - Accessible Countryside (n=2925) | **0.82 (0.73 to 0.92)** | 0.001 | **0.81 (0.71 to 0.91)** | <0.001 | **0.81 (0.71 to 0.92)** | 0.001 |
| 4A - Prospering Younger Families (n=1124) | 0.95 (0.82 to 1.11) | 0.513 | 0.91 (0.78 to 1.06) | 0.219 | 0.93 (0.79 to 1.10) | 0.417 |
| 4B - Prospering Older Families (n=4196) | **0.87 (0.77 to 0.97)** | **0.010** | **0.82 (0.74 to 0.92)** | 0.001 | **0.81 (0.72 to 0.91)** | 0.001 |
| 4C - Prospering Semis (n=2632) | 1.14 (1.01 to 1.28) | 0.031 | 1.07 (0.95 to 1.20) | 0.291 | 1.03 (0.90 to 1.18) | 0.655 |
| 4D - Thriving Suburbs (n=3077) | **0.83 (0.74 to 0.93)** | 0.002 | **0.80 (0.71 to 0.90)** | <0.001 | 0.81 (0.71 to 0.93) | 0.002 |
| 5A - Senior Communities (n=151) | **1.70 (1.22 to 2.36)** | 0.002 | **1.65 (1.18 to 2.31)** | 0.003 | 1.59 (1.09 to 2.33) | 0.017 |
| 5B - Older Workers (n=1106) | **1.50 (1.29 to 1.74)** | <0.001 | **1.46 (1.25 to 1.69)** | <0.001 | **1.36 (1.15 to 1.61)** | <0.001 |
| 5C - Public Housing (n=154) | **1.59 (1.14 to 2.21** | 0.006 | **1.44 (1.03 to 2.00)** | **0.031** | 1.36 (0.93 to 1.97) | 0.109 |
| 6A - Settled Households (n=1593) | 1.13 (0.99 to 1.30) | 0.070 | 1.12 (0.98 to 1.29) | 0.097 | 1.13 (0.98 to 1.32) | 0.097 |
| 6B - Least Divergent (n=1994) | 1.00 | – | 1.00 | – | 1.00 | |
| 6C - Young Families in Terraced Homes (n=1060) | 1.10 (0.94 to 1.28) | 0.221 | 1.12 (0.96 to 1.30) | 0.158 | 1.07 (0.91 to 1.27) | 0.408 |
| 6D - Aspiring Households (n=1929) | **0.80 (0.70 to 0.91)** | 0.001 | **0.82 (0.72 to 0.93)** | **0.003** | **0.83 (0.72 to 0.96)** | **0.014** |
| 7A - Asian Communities (n=1245) | 1.01 (0.88 to 1.17) | 0.865 | 1.08 (0.93 to 1.25) | 0.303 | 1.12 (0.95 to 1.32) | 0.161 |
| 7B - Afro-Caribbean (n=736) | 0.91 (0.77 to 1.09) | 0.304 | 1.00 (0.84 to 1.20) | 0.929 | 1.06 (0.87 to 1.28) | 0.573 |

*Table 16.9* Linear regression showing how spatial measures predict BMI, with coefficients quantifying the unit change in BMI. Model 1 = unadjusted. Model 2 = adjusted for diet cost and dietary pattern. Model 3= adjusted for Energy intake, energy expenditure, diet cost, dietary pattern, smoking, menopause, education and class. OR = Odds Ratio; CI = Confidence Interval

| | *Model 1* | | *Model 2* | | *Model 3* | |
|---|---|---|---|---|---|---|
| | *Coef (95% CI)* | *P value* | *Coef (95% CI)* | *P value* | *Coef (95% CI)* | *P value* |
| **Government office region** | $R^2$ = 0.0027 | | $R^2$ =0.0323 | | $R^2$ =0.0491 | |
| North East | **0.60 (0.30 to 0.89)** | **<0.001** | **0.53 (0.24 to 0.82)** | **<0.001** | **0.61 (0.31 to 0.90)** | **<0.001** |
| North West | **0.45 (0.26 to 0.64)** | **<0.001** | **0.39 (0.20 to 0.58)** | **<0.001** | **0.43 (0.24 to 0.62)** | **<0.001** |
| Yorkshire and the Humber | **0.52 (0.32 to 0.72)** | **<0.001** | **0.47 (0.27 to 0.67)** | **<0.001** | **0.48 (0.28 to 0.68)** | **<0.001** |
| East Midlands | **0.41 (0.20 to 0.61)** | **<0.001** | **0.36 (0.16 to 0.56)** | **0.001** | **0.38 (0.18 to 0.59)** | **<0.001** |
| West Midlands | **0.37 (0.17 to 0.57)** | **<0.001** | **0.33 (0.13 to 0.53)** | **0.001** | **0.33 (0.13 to 0.53)** | **0.001** |
| East of England | 0.22 (0.03 to 0.41) | 0.022 | 0.19 (0.00 to 0.38) | 0.047 | 0.16 (−0.02 to 0.35) | 0.089 |
| Greater London | −0.12 (−0.30 to 0.05) | 0.171 | 0.03 (−0.14 to 0.23) | 0.711 | 0.05 (−0.12 to 0.23) | 0.549 |
| South East | 0 | – | 0 | – | 0 | – |
| South West | 0.06 (−0.11 to 0.22) | 0.515 | 0.06 (−0.11 to 0.23) | 0.473 | 0.06 (−0.11 to 0.23) | 0.488 |
| Scotland | **0.40 (0.19 to 0.62)** | **<0.001** | **0.31 (0.10 to 0.52)** | **0.004** | **0.39 (0.18 to 0.61)** | **<0.001** |
| Wales | **0.42 (0.17 to 0.67)** | **0.001** | **0.40 (0.15 to 0.65)** | **0.002** | **0.44 (0.18 to 0.69)** | **0.001** |
| **North South** | $R^2$ = 0.0023 | | $R^2$ =0.0319 | | $R^2$ =0.0489 | |
| The North | **0.43 (0.32 to 0.53)** | **<0.001** | **0.35 (0.24 to 0.45)** | **<0.001** | **0.37 (0.27 to 0.48)** | **<0.001** |
| The South | 0 | – | 0 | – | 0 | – |
| **Urban Rural** | $R^2$= 0.0076 | | $R^2$= 0.0317 | | $R^2$= 0.0481 | |
| Urban | 0 | – | 0 | – | 0 | – |
| Rural | **−0.27 (−0.38 to −0.17)** | **<0.001** | **−0.30 (−0.40 to −0.20)** | **<0.001** | **−0.30 (−0.40 to −0.20)** | **<0.001** |
| **OAC Supergroup** | $R^2$= 0.0076 | | $R^2$= 0.0370 | | $R^2$= 0.0544 | |
| Blue Collar Communities | **0.90 (0.68 to 1.11)** | **<0.001** | **0.73 (0.52 to 0.95)** | **<0.001** | **0.80 (0.58 to 1.02)** | **<0.001** |
| City Living | **−0.58 (−0.79 to −0.67)** | **<0.001** | **−0.43 (−0.64 to −0.23)** | **<0.001** | **−0.44 (−0.65 to −0.23)** | **<0.001** |
| Countryside | **−0.26 (−0.41 to −0.12)** | **<0.001** | **−0.32 (−0.46 to −0.18)** | **<0.001** | **−0.35 (−0.50 to −0.21)** | **<0.001** |
| Prospering Suburbs | −0.07 (−0.21 to 0.06) | 0.280 | **−0.21 (−0.35 to −0.08)** | **0.002** | **−0.30 (−0.43 to −0.16)** | **<0.001** |
| Constrained by Circumstance | **1.09 (0.84 to 1.35)** | **<0.001** | **0.96 (0.71 to 1.21)** | **<0.001** | **0.94 (0.69 to 1.20)** | **<0.001** |
| Typical Traits | 0 | – | 0 | – | 0 | – |
| Multicultural | −0.11 (−0.23 to 0.21) | 0.922 | 0.16 (−0.06 to 0.38) | 0.147 | 0.18 (−0.05 to 0.40) | 0.119 |

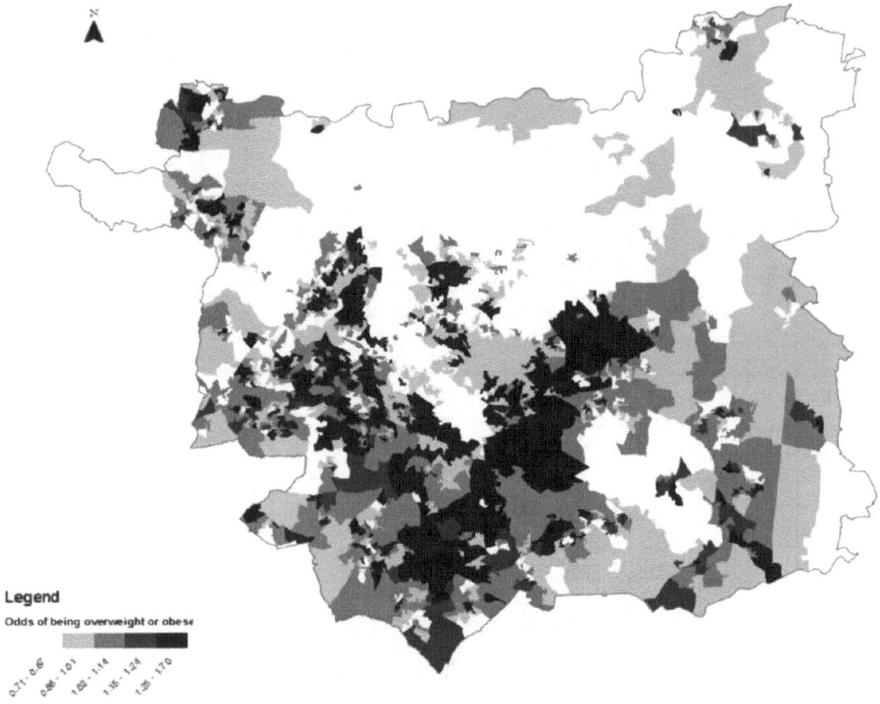


*Figure 16.1* Estimated odds of being overweight/obese in Leeds

## 16.5 Discussion

### *16.5.1 Spatial variation in obesity*

Differences exist despite less overweight and obesity prevalence in the cohort compared to the UK as a whole, which may reflect the above average social class of the women in the UKWCS (Cade et al., 2004). Overweight and obesity are significantly higher in the North than the South of the UK, which is as hypothesised. It is interesting to see that this difference remains even when controlling for all confounders and mediators – including social class and education – especially when variation was less evident when grouping by GOR. This is as hypothesised and in line with other spatial obesity research (Moon et al., 2007; Scarborough and Allender, 2008). An analysis of HSE data also showed increased odds of being overweight or obese in the North compared to the South for middle-aged women (Scarborough and Allender, 2008). It may be that obesity and related conditions contribute to the observed difference in mortality between North and South too (Shaw et al., 1998). Obesity varies significantly between Urban and Rural areas, with a higher percentage of obesity in the Urban environments,

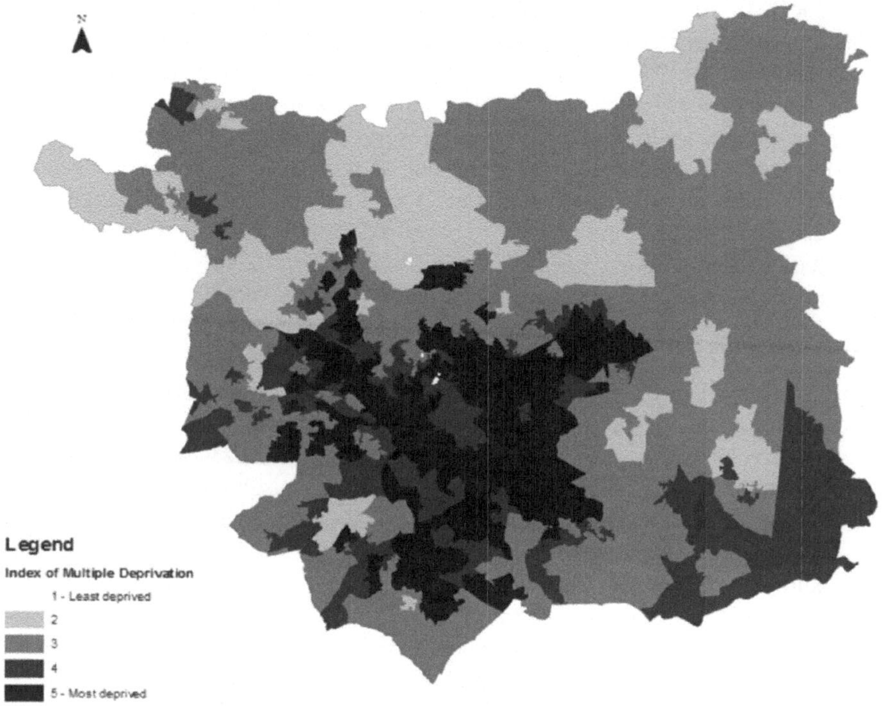


*Figure 16.2* Index of Multiple Deprivation for Leeds

consistent with existing literature in the UK (Riva et al., 2009), although no variation is seen for overweight women in the UKWCS. When the women are grouped by OAC Supergroup, there are large differences observed in percentages of both overweight and obese women, supporting evidence that inclusion of geodemographic characteristics aids the understanding of groups more or less likely to be overweight/obese.

When considering BMI as a continuous measure, rather than by category, more significant variation is observed by GOR, showing that while the women may not be classified as overweight or obese, they do have higher BMIs in Northern GORs, as seen when the sample is dichotomised to North or South. We still see that Rural areas are less likely to be overweight or obese, with a reduced BMI coefficient compared to the Urban reference group. These results show slightly different patterns compared to the categorical results as they take into account those who fall into the normal weight category but have higher BMI.

Use of either categorisation or BMI as a continuous measure shows a strong relationship that those in Blue Collar Communities and Constrained by Circumstance are significantly more likely to be overweight/obese than those in areas of Typical Traits. The City Living, Countryside and Prospering Suburb groups are significantly less likely to be overweight/obese with negative coefficients for BMI. No difference observed between the Typical Traits and Multicultural areas may be a result of the fact that while the UKWCS have women living in Multicultural areas, 99% of the UKWCS women report ethnicity as White. The ethnic characteristics which typify Multicultural areas do not exist in the UKWCS, which may explain why less difference is observed than would be expected.

Using data collected in large cohorts such as the UKWCS profiled by geodemographic categories is a novel method for estimating small-area geographic distributions in health – in this case overweight/obese – in areas of the UK. The case study example here is Leeds.

### *16.5.2 Strengths and limitations*

There is strength in the size of the sample and wide geographic spread of the women's locations, despite the sample not being designed to be geographically representative. The sample size far outweighs the sample sizes of other national surveys which collect height and weight data, such as the National Diet and Nutrition Survey (NDNS), which has data on ~3,000 adults and over three years of data collection (Department of Health and Food Standards Agency, 2012), and the Health Survey for England, which collects data on ~8,000 per year (Health and Social Care Information Centre, 2013a). No synthetic estimates were required to increase the sample size. However, using self-reported height and weight to calculate BMI – as the UKWCS does – can introduce systematic bias into data as heavier individuals are more likely to underestimate their weight and likewise short individuals are more likely to overestimate their height (Hu, 2008).

Underweight individuals were excluded from the logistic regression analysis and normal weight used as the reference group. These underweight women constitute only a small proportion (~2%) of the study population, but may provide important information if underweight and food insecurity were the research question of interest. In future work it would be interesting to look into higher prevalence of underweight observed in Transient Communities, Public Housing and Afro-Caribbean OAC Groups.

While results at each of the geographical scales add to understanding the issue of overweight and obesity in the UK, their use in application for public health policy needs some consideration. When planning how best to implement an intervention to reach those most in need in a cost-effective manner, a combination of approaches may be most beneficial. For example, based on the results of this research, to reach middle-aged women with the highest odds of being overweight or obese in the UK it would be most beneficial to target those in Constrained by Circumstance groups in the North West of England and in Scotland. Using the geodemographic profiles, it is possible to estimate the neighbourhoods within a given area which need most help.

The addition of spatial identifiers such as North/South, Urban/Rural and OAC to survey datasets, like the NDNS (Department of Health and Food Standards Agency, 2012) as standard could be useful for researchers and policy makers alike. The Family Spending Survey leads the way in this respect, presenting results using a range of spatial scales, including Urban/Rural, OAC, GOR and some reference to the North–South divide. If other surveys followed suit this could facilitate comprehensive research which in turn would benefit health care providers and policy makers and subsequently the UK population.

### *16.5.3 Continued research*

Obesity is a problem of epidemic proportions which has not yet shown signs of reprieve. Therefore, continued research is essential to combat the problem. Spatial modelling is a key element to this such that the effect of the environment and the spatial and social distribution of overweight and obesity can be better understood. With greater understanding comes more specific targeting of interventions and policies.

## References


Cade J. E., Burley, V. J., Greenwood, D. C. and Group, U.K.W.S.C.S.S. (2004) The UK Women's Cohort Study: comparison of vegetarians, fish-eaters and meat-eaters. *Public Health Nutrition*, 7, 871–878.

Census Dissemination Unit (2009) *GeoConvert* [Online]. Available at: http://geoconvert.mimas.ac.uk/ [accessed August 2010].

Crawford, D., Jeffery, R., Ball, K. and Brug, J. (eds) (2010) *Obesity Epidemiology*. New York: Oxford University Press.

Delpeuch, F., Maire, B., Monnier, E. and Holdsworth, M. (2009) *Globesity*. London: Earthscan.

Department for Communities and Local Government (2012) *English Indicies of Deprivation* [Online]. Available at: www.gov.uk/government/collections/english-indices-of-deprivation.

Department of Health and Food Standards Agency (2012) National Diet and Nutrition Survey: headline results from Years 1, 2 and 3 (combined) of the Rolling Programme (2008/2009– 2010/11). In: Beverley Bates, Alison Lennox, Ann Prentice, Chris Bates and Swan, G. (eds), *National Diet and Nutrition Survey*. London: Food Standards Agency.

Drewnowski, A. and Specter, S. E. (2004) Poverty and obesity: the role of energy density and energy costs, *American Journal of Clinical Nutrition*, 79, 6–16.

Edwards, K. L., Cade, J. E., Ransley, J. K. and Clarke, G. P. (2009) A cross-sectional study examining the pattern of childhood obesity in Leeds: affluence is not protective, *Archives of Disease in Childhood*, 95, 94–99.

Foresight (2007) *Tackling Obesities: Future Choices – Project Report*. London: Government Office of Science.

Gibney, M. (ed.) (2004) *Public Health Nutrition*. Oxford: Blackwell Science.

Gortmaker, S. L., Swinburn, B. A., Levy, D., Carter, R., Mabry, P. L., Finegood, D. T., Huang, T., Marsh, T. and Moodie, M. L. (2011) Changing the future of obesity: science, policy, and action, *Lancet*, 378, 838–847.

Hall, K. D., Sacks, G., Chandramohan, D., Chow, C. C., Wang, Y. C., Gortmaker, S. L. and Swinburn, B. A. (2011) Quantification of the effect of energy imbalance on bodyweight, *The Lancet*, 378, 826–837.

Health and Social Care Information Centre (2011) Health Survey for England. London: Health and Social Care Information Centre.

Health and Social Care Information Centre (2013a) *Statistics on Obesity, Physical Activity and Diet – England 2013*. London: Health and Social Care Information Centre.

Health and Social Care Information Centre (2013b) *Methodological Changes to Health Survey for England (HSE) 2013* [Online]. Available at: www.hscic.gov.uk/media/11361/Methodological-Changes-to-Health-Survey-for-England-HSE-2013-NS/pdf/MethChange201312_HSE_Sample_Design.pdf [accessed May 2016].

Health and Social Care Information Centre (2013c) *National Child Measurement Programme*. London: Health and Social Care Information Centre.

Hill, J. O. and Peters, J. C. (1998) Environmental contributions to the obesity epidemic. *Science*, 280, 1371–1374.

Hu, F. B. (2008) *Obesity Epidemiology*. New York: Oxford University Press.

Jabs, J. and Devine, C. M. (2006) Time scarcity and food choices: an overview, *Appetite*, 47, 196–204.

Moon, G., Quarendon, G., Barnard, S., Twigg, L. and Blyth, B. (2007) Fat nation: deciphering the distinctive geographies of obesity in England, *Social Science and Medicine*, 65, 20–31.

Offer, A., Pechey, R. and Ulijaszek, S. (eds) (2012) *Insecurity, Inequality, and Obesity in Affluent Societies*. New York: Oxford University Press.

Pearce, J. and Witten, K. (eds) (2010) *Geographies of Obesity*. Aldershot: Ashgate.

Prentice, A. M. and Jebb, S. A. (1995) Obesity in Britain: gluttony or sloth? *British Medical Journal*, 311, 437–439.

Public Health England (2013) *Health Survey for England* [Online]. Available at: www.noo.org.uk/data_sources/adult/health_survey_for_england [accessed September 2013].

Riva, M., Curtis, S., Gauvin, L. and Fagg, J. (2009) Unravelling the extent of inequalities in health across urban and rural areas: evidence from a national sample in England, *Social Science and Medicine*, 68, 654–663.

Rokholm, B., Baker, J. L. and Sorensen, T. I. (2010) The levelling off of the obesity epidemic since the year 1999: a review of evidence and perspectives, *Obesity Review*, 11, 835–46.

Scarborough, P. and Allender, S. (2008) The North–South gap in overweight and obesity in England, *British Journal of Nutrition*, 100, 677–684.

Shaw, M., Dorling, D. and Brimblecombe, N. (1998) Changing the map: health in Britain 1951–91, *Sociology of Health and Illness*, 20, 694–709.

Sport England (2013) *Active People Survey. National Obesity Observatory*.

StataCorp (2009) *Stata Statistical Software: Release 11*. College Station, TX: StataCorp LP.

StataCorp (2012) *Stata/IC 12.1*. College Station, TX: StataCorp LP.

Swinburn, B. A., Sacks, G., Hall, K. D., Mcpherson, K., Finegood, D. T., Moodie, M. L. and Gortmaker, S. L. (2011) The global obesity pandemic: shaped by global drivers and local environments, *The Lancet*, 378, 804–814.

Tomintz, M. N., Clarke, G. P. and Rigby, J. E. (2008) The geography of smoking in Leeds: estimating individual smoking rates and the implications for the location of stop smoking services, *Area*, 40, 341–353.

Wang, Y. C., Mcpherson, K., Marsh, T., Gortmaker, S. L. and Brown, M. (2011) Health and economic burden of the projected obesity trends in the USA and the UK, *The Lancet*, 378, 815–825.

Wilkinson, R. and Pickett, K. (2010) *The Spirit Level*. London: Penguin.

World Health Organization (2000) *Obesity: Preventing and Managing the Global Epidemic*. Report of a WHO consultation, WHO Technical Report Series 894 [Online]. Available at: http://libdoc.who.int/trs/WHO_TRS_894.pdf [accessed May 2016].

World Health Organization (2006) *BMI Classification* [Online]. Available at: http://apps.who.int/bmi/index.jsp?introPage=intro_3.html [accessed November 2013].

World Health Organization (2013) *Obesity and Overweight Fact Sheet (Number 311)* [Online]. Available at: www.who.int/mediacentre/factsheets/fs311/en/ [accessed September 2013].

# 17 Examining the socio-spatial determinants of depression in the UK

*Karyn Morrissey, Peter Kinderman, Eleanor Pontin, Sara Tai and Matthias Schwannauer*

## 17.1 Introduction

Individual risk factors for poor mental health status are well established at the national level and include female gender, low socio-economic status, and unemployment (Fone et al., 2007; Mallinson and Popay, 2007; Van de Velde et al., 2010). National-level analyses of health are beneficial, as they highlight the main determinants of health for a country and allow for cross-country analysis and comparisons. However, individuals live in households that reside within communities, which in turn are nested within regions, and countries (Riva and Smith, 2012). Research in health geography, epidemiology and public health has led to an increasing recognition among practitioners and policy makers that space and health are not mutually exclusive; people and their health are shaped by the places in which they live and inhabit on a regular basis (Stafford and Marmot, 2003; Smith and Easterlow, 2005; Gatrell and Elliott, 2009).

Residents of socially and economically deprived communities experience worse health outcomes on average than those living in more prosperous areas (Ellen et al., 2001). Research has shown that this is in part because people with similar characteristics, such as income level, age and employment status, cluster together (Congdon, 2006) and in part because individuals living in the same neighbourhood are subject to common contextual influences, such as higher crime rates and lower social cohesion (Walters et al., 2004; Merlo et al., 2005; Congdon, 2006). Studies to establish the impact of geographical variations on mental illnesses have found that once individual characteristics have been taken into account, the amount of variation attributed to the higher levels is very small (Stafford and Marmot, 2003; Walters et al., 2004; Skapinakis et al., 2005; Weich et al., 2005; Fone et al., 2007; Peterson et al., 2009).

Linking individual-level data on depression and area-level data, this study has three aims. First, for the population within the Stress Test, we wish to establish if residents of deprived areas are relatively more depressed. Second, using an intra-area income comparison, we wish to explore whether individuals within income categories and with depression live in areas with slightly higher rates of deprivation. Third, using a standardised logistic model we examine whether area-level deprivation is associated with depression, controlling for individual-level characteristics. The Stress Test is outlined in the following section.

## 17.2 Data

### *17.2.1 The Stress Test*

In June 2011 a web-based survey on the causes of mental distress, the Stress Test, was launched on *All in the Mind*, a BBC Radio 4 programme (Kinderman et al., 2011). The test's URL was publicised on radio and TV broadcasts, and was made available via BBC web pages and social media. It comprised 12 sections and included: demographic and socio-economic; familial mental health; social inclusion; negative life events (historic and recent); neurocognitive responses to negative feedback and positive and negative stimuli; psychological processes (response style and attribution style); depression and anxiety; and stress. Participants were also asked to write, using free text, their first and second biggest cause of stress. The optimal time to ensure response quality for online surveys is approximately 20 minutes (Cape, 2010). The Stress Test was piloted on BBC staff and took approximately 20 minutes to complete, and participants were required to complete the test in one sitting (Pontin, 2012). Standardising the experience of participation in internet research can be difficult (Nosek et al., 2002) because of the increased variability of the environments in which respondents participate. To minimise variation in the Stress Test, instructions were coherent and items completed in a fixed order. Answers were selected from a drop-down menu and some tasks were constrained within time limits. Once the test was completed, participants were not permitted to complete the test again (Pontin, 2012).

### *17.2.2 Depression*

Depression was measured within the Stress Test using the depression component of the Goldberg Anxiety and Depression Scale (GADs), which can detect levels of psychological disturbance in the general population and are widely used in psychological research (Goldberg et al., 1988). The GADs were developed from 36 items of the Psychiatric Assessment Schedule (Surtees et al., 1983) and are an 18-item self-report symptom inventory with 'yes' or 'no' responses to items asking how respondents had been feeling in the past few weeks. Nine items each comprise the anxiety and depression scales. Although the depression and anxiety scales are typically highly correlated, evidence has demonstrated them to be distinct entities (Christensen et al., 1999). Research has concluded the GADs to be a valid and acceptable method of detecting distress associated with depression and anxiety (Smith, 2004).

### *17.2.3 Individual and area-level data*

Over 32,000 individuals responded to the online Stress Test and 21,000 of these individuals were from England. The demographic and socio-economic variables included: age; gender; ethnic group; occupation; gross annual or weekly household earnings; highest level of formal schooling; occupational status; parents' income during childhood; relationship status; and number of children (Pontin, 2012). Ethnic groups were divided in concordance to the BBCs method of recording ethnicity data, which is comparable to 2001 Census data. Occupational status was indicated from eight categories and occupation from a list of 25. Gross annual or weekly household income was divided into nine ranges (from up to £9,999 per annum or £199 per week to £75,000 or more per annum or £1,450 or more per week). Parent's income during childhood was indicated by four categories that compared income to the rest of the population (e.g. 'much lower than others (bottom 25 percent of the population)') Highest level of formal schooling achieved was divided into six categories ranging from 'did not complete GCSE/CSE/O-level or equivalent' to 'postgraduate qualification'. Finally, relationship status was indicated from ten categories and number of children specified (from zero to six plus).

Of interest to the current analysis is that the survey also asked participants for their postcode. The postcode data released for the purpose of this research included only the first half of the UK postcode, for example, L1 for residents of Liverpool 1 postcode. The postcode in the Stress Test was linked to the National Statistics Postcode Lookup (NSPL) table produced by the Office of National Statistics (ONS) Geography (Office for National Statistics, 2012). The NSLP allows users to allocate data collected at postcode level to a range of higher geographies, including ward, Lower Super Output Area (LSOA) and Government Office Region (GOR). As only the first part of the postcode is available in the Stress Test, the full postcode assigned to each individual will not be unlikely to be his or her actual postcode. However, while this matching process lacks 100% precision, the spatial accuracy of the individual's approximate address is deemed to be high as the postcode is the smallest output geography available. Linking the data to the postcodes provided by the NSPL allowed our area-level variable, the Index of Multiple Deprivation (IMD) for England 2010, to be included in the original BBC dataset.

The Social Disadvantage Research Centre at the University of Oxford constructed the IMD on behalf of the Department of Communities and Local Government (DCLG) (2011). The IMD is partially based on Census data, but uses a combination of Census data and data derived from other sources, such as the Inland Revenue, the Department of Health and the Department of Transport. The purpose of the IMD is to measure deprivation at the small-area level to identify the most disadvantaged areas in England (ODPM, 2004). The IMD 2010 was constructed by combining seven general welfare domain scores weighted as followed: income (22.5%); employment (22.5%); housing and disability (13.5%); education, skills and training (13.5%); barriers to housing and services (9.3%); crime (9.3%); and living environment (9.3%). The IMD is based on data at the Lower Super Output Areas (LSOA) with each LSOA containing on average 1500 people (Westling et al., 2009). The IMD is available in two numerical forms: as a rank variable, which shows how an individual LSOA compares to other LSOAs in the country, and as an absolute score (ODPM, 2004). The LSOA ranked number one is the most deprived, with higher rankings indicating less deprived areas.

*Table 17.1* The regional population-based representativeness of the Stress Test data compared to the Health Survey of England 2011

| *Region* | *Stress Test* | *Health Survey of England* |
|---|---|---|
| North East | 3 | 5 |
| North West | 11 | 13 |
| Yorkshire & Humber | 8 | 10 |
| East Midlands | 7 | 9 |
| West Midlands | 8 | 10 |
| East England | 12 | 12 |
| Greater London | 16 | 15 |
| SE England | 22 | 16 |
| SW England | 13 | 10 |
| Chi-Squared Test | Not significantly different at the 90% level 0.69 | |

Given the aims of this research, it is important to examine if the data is spatially representative. A chi-squared test was undertaken to compare the proportional spatial representative of the Stress Test and the Health Survey for England (HSE), the largest representative Health Survey for England (2011). Table 17.1 presents the spatial rates at the regional level for the Stress Test and the HSE results. Running a chi-squared test, Table 17.1 found that there is no spatial difference in the respondent rates between the Stress Test and HSE (2011) at the 90% confidence level. Examining the spatial representativeness of the data at the sub-regional level, a simple count of the number of English wards represented in the Stress Test, 6,829 out of 8,576 wards (80%) were represented by 1 or more respondents.

## 17.3 Results

Examining the Stress Test dataset, Table 17.2 indicates that individuals with lower incomes live in areas with higher deprivation than individuals with higher incomes. Individuals earning less than £9,999 a year live in areas with an average deprivation score of 20.66. In contrast, individuals earning more than £75,000 a year live in areas with an average deprivation score of 16.28. Table 17.2 also highlights that the average deprivation score and the percentage of individuals scoring greater than 6 on the Goldberg Depression Scale decreases as the income categories increase. Indeed, 47% of individuals earning less than £9,999 score greater than 6 on the GDS and these individuals live in LSOAs with an average deprivation score of 21.27. In contrast, only 28% of individuals earning greater than £75,000 scored greater than 6 on the GDS and resided in areas with an average deprivation score of 16.93. A t-test was also employed to examine if there was a significant difference in the average residential IMD score for depressed and non-depressed individuals. The t-test confirmed that at the 95% level there is a significant (0.00) difference between the residential IMD of depressed and non-depressed individuals.

*Table 17.2* Average residential deprivation score and depression rates by income category

| | *Average residential deprivation score* | *Percentage of residents without depression & average LSOA deprivation rate* | *Percentage of residents with depression & average LSOA deprivation rate* |
|---|---|---|---|
| Income less than £9,9999 | 20.66 | 20.11 (53%) | 21.27 (47%) |
| Income £10,000–£19,999 | 20.38 | 19.84 (59%) | 21.17 (41%) |
| Income £20,000–£29,999 | 19.99 | 19.67 (64%) | 20.57 (36%) |
| Income £30,000–£39,999 | 19.11 | 18.84 (67%) | 19.65 (33%) |
| Income £40,000–£49,999 | 18.26 | 17.86 (67%) | 19.18 (33%) |
| Income £50,000–£74,999 | 17.14 | 17.28 (69%) | 16.85 (31%) |
| Income £75,000 plus | 16.28 | 16.02 (72%) | 16.93 (28%) |

Source: Kinderman (2013)

More interestingly, Table 17.2 also provides an intra income comparison between income categories. While, 53% of individuals earning less than £9,999 a year had scores less than 6 on the GDS, compared to individuals with depression in this income category, they resided in areas with a slightly lower deprivation score (20.11 versus 21.27). The same relationship is observed for all individuals scoring less than 6 on the GDS for each income category relative to individuals who scored greater than 6. Table 17.2 therefore indicates that individual's who earn low wages have higher rates of depression if they also live in relatively more deprived areas. This relationship is maintained across all income categories up until individuals earn greater than £50,000 a year. Individuals with depression who earn £50,000 to £74,999 and £75,000 plus a year reside in areas with slightly lower deprivation compared to those without depression in the same income categories.

### *17.3.1 Specification of the standardised logistic model*

Continuing, a logistic regression model was used to explore the individual and area level determinants of depression within the BBC dataset. The individual-level explanatory variables included: age; gender; British white ethnicity; income level (categorical variable: less than £9,999; £10,000–£19,999; £20,000–£29,999; £30,000–£39,999; £40,000–£49,999; £50,000–£74,999, £75,000 plus); and marital status. The spatial variables included the nine Government Office Region variables and the Index of Multiple Deprivation for England (Department of Communities and Local Government, 2011) at the Lower Super Output Area (LSOA). On constructing the model, the IMD score was included as a continuous variable and as a five-level categorical variable. Including the IMD as a categorical variable indicated that there were no non-linearities in the relationship between area deprivation and depression, and as such the IMD score as a continuous variable was used. Including the IMD as a continuous variable has important implications for the specification of the logistic model. When explanatory variables are measured in different units of measurement, for example the continuous IMD score versus the binary gender variable, standardised coefficients rather than odds ratios should be used (Menard, 2011). Finally, while employment status and education level were available within the BBC dataset, similar to recent research (McLennan et al., 2011), these two variables were highly correlated with the IMD, thus income alone was included as an individual level indicator of socio-economic status.

This section outlines the results of the logistic model. Examining age, respondents aged 25–34, 35–44, and 45–54 were 15, 17 and 16 times more likely to score 6 or above on the GDS relative to respondents aged 18–24 years old. Interestingly, age 55–64 did not reach significance and individuals aged 65 plus had a negative significant association (0.90) relative to individuals aged 18–24 years old. That is respondents aged 65 plus were 0.90 times less likely to score 6 or above on the GDS. Being married (0.92) relative to all other marital statuses had a negative association with scoring 6 or above on the GDS. Respondents who are married are 0.92 times less likely to score 6 or above on the GDS relative to all other relationship categories. Marriage would therefore seem to be a protector against depression for respondents to the BBC Stress Test. White British ethnicity (0.96) relative to all other ethnicity classifications was also negatively associated with depression. Thus, respondents of British white ethnicity are 0.96 less likely to score 6 or above on the GDS. As with being married, controlling for the other demographic and socio-economic variables included in the logistic model, white British ethnicity is also a protector from depression for respondents to the Stress Test. Interestingly, gender (female) failed to reach significance in the individual level model. Thus, gender was not significantly associated with depression for respondents to the Stress Test.

Examining the income variables, the six annual income categories were all negatively associated with depression relative to individuals on incomes less than £9,999. This means that individuals in all income categories greater than £0–£9,999 were less likely to score 6 or above on the GDS. From Table 17.3, it can be seen that this negative relationship is linear. Respondents earning £10,000 to £19,999 are 0.91 times likely to suffer from depression according to the Stress Test. This likelihood drops to 0.82 for respondents earning £20,000 to 29,999, 0.79 for respondents earning £30,000 to £39,999, 0.78 for respondents earning £40,000 to £49,999, 0.76 for respondents earning £50,000 to £74,999 and 0.74 for respondents earning £75,000 plus. Thus, analysing the Stress Test using a logistic modelling approach indicates that, similar to previous research on depression (Ellen et al., 2001; Weich et al., 2005; Morrissey et al., 2010), individuals with lower levels of income are statistically more likely to be diagnosed with depression on a clinical scale. Examining the regional variables, none of the eight regions (relative to the North East region) reached significance. More specifically, the logistic model did not find a statistically significant relationship between region of residence and depression. While insignificant, only the West Midlands (1.02), the East of England (1.04) and London (1.01) had a positive association with depression relative to the North East region.

Finally, the IMD score was included in the logistic model as a continuous, rather than rank, variable. Table 17.3 indicates that increases in area deprivation are

*Table 17.3* Logistic model of depression with standardised coefficients

| *Depression* | *Standardised Coefficient* | *P>z* | *95% Confidence Interval* | |
|---|---|---|---|---|
| Gender (Female, 1) | 1.00 | 0.82 | 0.93 | 1.06 |
| Age 25–34 (ref. 18–24) | 1.15 | 0.00 | 1.26 | 1.59 |
| Age 35–44 (ref. 18–24) | 1.17 | 0.00 | 1.29 | 1.64 |
| Age 45–54 (ref. 18–24) | 1.16 | 0.00 | 1.26 | 1.60 |
| Age 55–64 (ref. 18–24) | 1.03 | 0.27 | 0.95 | 1.22 |
| Age 65+ (ref. 18–24) | 0.90 | 0.00 | 0.51 | 0.73 |
| Married (1–Yes) | 0.92 | 0.00 | 0.78 | 0.89 |
| £10–19,999 (£0–9,000) | 0.91 | 0.00 | 0.69 | 0.88 |
| £20–29,999 (£0–9,000) | 0.82 | 0.00 | 0.54 | 0.68 |
| £30–39,999 (£0–9,000) | 0.79 | 0.00 | 0.46 | 0.59 |
| £40–49,999 (£0–9,000) | 0.78 | 0.00 | 0.46 | 0.59 |
| £50–74,999 (£0–9,000) | 0.76 | 0.00 | 0.42 | 0.54 |
| £75000+ (£0–9,000) | 0.74 | 0.00 | 0.36 | 0.47 |
| British White (1–Yes) | 0.96 | 0.01 | 0.78 | 0.96 |
| North West (reference North East) | 0.99 | 0.74 | 0.80 | 1.17 |
| Yorkshire and Humber (reference North East) | 0.99 | 0.84 | 0.80 | 1.19 |
| East Midlands (reference North East) | 0.97 | 0.26 | 0.73 | 1.09 |
| West Midlands (reference North East) | 1.02 | 0.39 | 0.89 | 1.33 |
| East (reference North East) | 1.04 | 0.26 | 0.92 | 1.35 |
| London (reference North East) | 1.01 | 0.88 | 0.84 | 1.22 |
| South East (reference North East) | 0.97 | 0.37 | 0.77 | 1.10 |
| South West (reference North East) | 0.95 | 0.15 | 0.72 | 1.05 |
| IMD score | 1.04 | 0.04 | 1.00 | 1.00 |

Source: Kinderman (2013)

associated with increased levels of depression. Respondents who lived in areas with higher levels of deprivation, controlling for the demographic and socio-economic variables included in the logistic model, were four times more likely to score 6 or above on the GDS than residents of less deprived areas. These findings are similar to previous research in the UK on the relationship between area-level deprivation and depression (Stafford and Marmot, 2003; Skapinokis et al., 2005; Weich et al., 2005). However, while the IMD score is significant, the standardised logistic model found that being aged 25–54 (relative to 18–24), married and increased income have a larger impact on being clinically diagnosed with depression. However, living in increasingly deprived areas has a greater significance and impact on having depression than gender, being aged 55 plus, ethnicity and regional residential location. It is also interesting to note that individual-level income was found to have a greater impact on depression than area-level deprivation.

## 17.4 Discussion

Including the Index for Multiple Deprivation for England (2010) (Department of Communities and Local Government, 2011) with the Stress Test, the aim of this study was to explore the association between depression and area-level deprivation, controlling for personal income and other demographic factors. Using descriptive statistics, it was found that individuals in lower income categories have higher rates of deprivation, and this difference is significant at the 95% level. Merging the LSOA defined IMD with the individual-level Stress Test also provides an intra-income comparison of depression. It was found that within income categories, individuals with depression resided in more deprived areas relative to those without depression. Interestingly, this intra-income relationship is maintained as income levels increase; that is, across all categories depressed individuals live in relatively more deprived areas than non-depressed individuals.

With regard to the individual-level variables included in the logistic regression model, increases in income level (Skapinakis et al., 2005; Fone et al., 2007; Morrissey et al., 2010) have a significant negative association with scoring 6 or above on the GDS. Being married is also found to be a protector from depression for both men and women. Increases in age were also found to be positively associated with being case prevalent in this survey. However, this association was only significant until age band 45–54. The model found that those aged 55–64 did not have a significant association with the depression variable and that individuals aged 65 years and older had a negative association with the depression variable. There is a general belief that depression is synonymous with ageing and that depression is in fact inevitable (Singh and Misra, 2009). However, recent studies have found that age isn't always significantly related to level of depression, and that the oldest of olds may even have better coping skills to deal with depression, making depressive symptoms more common but not as severe as in younger populations (Singh and Misra, 2009). Interestingly, European-wide research as part of the Survey of Health Ageing and Retirement in Europe (SHARE) project further found that the association between age and mental health is mediated by the health and living conditions of older persons; age by itself has no explanatory power (Buber and Engelhardt, 2011). Thus, the result of this study furthers the hypothesis that depression is not an inevitable outcome of ageing, particularly when individual income and area-level factors are accounted for in a model of depression.

Interestingly, the gender variable within this dataset is insignificant. The majority of international research has found that women are more likely to present symptoms of depression in a clinical setting and self-report higher levels of depression than men (Van de Velde et al., 2010). Ethnicity was also found to have a non-significant effect on depression. A number of studies have focused on the relationship between ethnicity and mental illness in the UK (O'Connor and Nazroo 2002; Sproston and Nazroo 2002). While some studies have pointed to differences in the rates of depression and anxiety and in the longevity of illnesses in different ethnic groups (Husain et al., 1997), others have pointed to rates being broadly similar once adjusted for socio-economic characteristics (Jenkins et al., 1997; Shaw et al., 1999). Thus, the relationship between ethnicity and depression remains unclear (Mallinson and Popay, 2007). With regard to the significant positive impact of white British ethnicity on depression found in this study, it must be noted that 90% of the respondents to the Stress Test recorded being of white ethic origin. The limited sample size for other ethnicities may therefore be a factor for this significant relationship.

## 17.5 Conclusions

Research over the last three decades has shown that geography is relevant to health outcomes (Smith and Easterlow, 2005; Gatrell and Elliott, 2009), including mental health (Walters et al., 2004). This study found that increases in area-level deprivation are associated with increased rates of scoring 6 or above on the GDS component of the 'Stress Test'. However, individual's income and age have a greater standardised association with depression compared to area-level deprivation. These findings are consistent with previous research (Walters et al., 2004), in which most of the association of deprivation and psychological distress was explained by individual rather than local area factors. Walters et al. (2004) argue that this does not necessarily mean individual factors are more important than area characteristics. However, this is to be expected because area-based and individual measures of deprivation are highly correlated. Due to this co-linearity, untangling the relationship between individual and area-level factors are likely to be complex (Walters et al., 2004).

To untangle the complex relationship between geography and place, this study argues similarly to a number of recent commentaries (Williams, 2003; Smith and Easterlow, 2005; Cummins et al., 2007; Morrissey et al., 2013) that the role of geography in health outcomes has become locked into a false dichotomy between area-level context and individual-level composition. This has undermined our ability to understand the interacting influences of social influences that are 'etched in space' (Soja, 1980; Harvey, 1989) on health (Smith and Easterlow, 2005; Morrissey et al., 2013). Previous research using the Stress Test by a team of psychologists (Kinderman et al., 2011) found that individual circumstance, a family history of mental health difficulties, social deprivation, and traumatic or abusive life experiences all strongly predicted higher levels of depression. However, these relationships were strongly mediated by psychological processes; specifically, lack of adaptive coping, rumination and self-blame (Kinderman et al., 2013). From their findings, Kinderman et al. (2013) propose that the psychological processes of the individual determine the causal impact of biological, social and circumstantial risk factors on mental health. Indeed, there is a strand of literature, particularly within the housing literature, that believes that an individual's ability to live successfully in difficult environs is partly contingent on their imaginative capacity to transcend what may be seen as oppressive and hostile environments (Jacobs, 2002). While previous health geography-based research with adolescents on mental health by Fagg et al. (2006) in London found that perceptions of area quality were linked to individual mental health. However, independently measured indicators of area deprivation were not so consistently associated with individual health outcomes. Thus, this research posits that the importance of geography in determining health outcomes may be reinforced via psychological processes rather than individual and neighbourhood characteristics.

Given the increasing prevalence of depression (Chang et al., 2012), this research concludes that future work on depression across space should leave the debate on the relative roles of individual versus contextual characteristics behind.

Instead, how individuals process their own social and environmental circumstances, rather than their actual circumstances, may be more important in determining whether, for example, an affluent individual in an affluent neighbourhood is depressed or not. Thus, geographical health research needs to begin to incorporate how individuals endogenously process their position within society as a means of deepening the understanding of the role of geography in health outcomes. Finally, the pathways to depression are of interest to health researchers across a variety of disciplines and policy makers in beginning to understand the determinants of depression. This study demonstrates the value gained in understanding the individual determinants of depression through interdisciplinary research (geography and psychology) and academic and non-academic collaboration (the BBC).

## References


Buber, I. and Engelhardt, H. (2011) The association between age and depressive symptoms among older men and women in Europe: findings from SHARE, *Comparative Population Studies*, 36(1), 103–126.

Cape, P.J. (2010) *Questionnaire Length, Fatigue Effects and Response Quality Revisited*, Survey Sampling International.

Chang, S., Hong, J. and Maeng, C. (2012) Economic burden of depression in South Korea, *Social Psychiatry Psychiatric Epidemiology*, 47, 683–689.

Christensen, H., Jorm, A., Mackinnon, A., Korten, A., Jacomb, P., Henderson, A. and Rodgers, B. (1999) Age differences in depression and anxiety symptoms: a structural equation modelling analysis of data from a general population sample, *Psychological Medicine*, 29, 325–339.

Congdon, P. (2006) A model for geographical variation in health and total life expectancy, *Demographic Research*, 14, 157–178.

Cummins, S., Curtis, S., Diez-Roux, A. and Macintyre, S. (2007) Understanding and representing 'place' in health research: a relational approach, *Social Science & Medicine*, 65(9), 1825–1838.

Department of Communities and Local Government (2011) *The English Indices of Deprivation 2010*, London: Department of Communities and Local Government.

Ellen, I., Mijanovich, T. and Dillman, K. (2001) Neighborhood effects on health: exploring the links and assessing the evidence, *Journal of Urban Affairs*, 23 (3–4), 391–408.

Fagg, J., Curtis, S., Congdon, P. and Stansfeld, S. (2006) Psychological distress among adolescents, and its relationship to individual, family and area characteristics: evidence from East London, UK, *Social Science & Medicine*, 63, 636–648.

Fone, D., Dunstan, F., Lloyd, K., Williams, G., Watkins, J. and Palmer, S. (2007) Does social cohesion modify the association between area income deprivation and mental health? A multilevel analysis, *International Journal of Epidemiology*, 36, 338–345.

Gatrell, A.C. and Elliott, S.E. (2009) *Geographies of Health* (2nd edition), Chichester, UK: Wiley.

Goldberg, D., Bridges, K., Duncan-Jones, P. and Grayson, D. (1988) Detecting anxiety and depression in general medical settings, *British Medical Journal*, 297: 897–899.

Harvey, D. (1989) *The Urban Experience*, Oxford: Basil Blackwell.

Health Survey of England (2011) *Health Social Care and Lifestyle*, London: Office for National Statistics, England.

Husain, N., Creed, F. and Tomenson, B. (1997) Adverse social circumstances and depression in people of Pakistani origin in the UK, *British Journal of Psychiatry*, 171, 434–438.

Jacobs, K.A. (2002) Subjectivity and the transformation of urban spatial experience, *Housing, Theory and Society*, 19(2), 102–111.

Jenkins, R., Lewis, G., Bebbington, P., Brugha, T., Farrell, M., Gill, B. and Meltzer, H. (1997) The National Psychiatric Morbidity surveys of Great Britain: initial findings from the household survey, *Psychological Medicine*, 27, 775–789.

Kinderman, P., Schwannauer, M., Pontin, E. and Tai, S. (2011) The development and validation of a general measure of well-being: the BBC well-being scale, *Quality Life Resource*, 20, 1035–1042.

Kinderman, P., Schwannauer, M., Pontin, E. and Tai, S. (2013) Psychological processes mediate the impact of familial risk, social circumstances and life events on mental health, *PLoS ONE*, 8(10), e76564.

Kinderman, P. and Tai, S. (2013) Stress Test, London: BBC Lab.

Mallinson, S. and Popay, J. (2007) Describing depression: ethnicity and the use of somatic imagery in accounts of mental distress, *Sociology of Health & Illness*, 29(6), 857–871.

McLennan, D., Barnes, H., Noble, M., Davies, J. and Garratt, E. (2011) *The English Indices of Deprivation 2010*. Available at: www.gov.uk/government/publications/english-indices-of-deprivation-2010.

Menard, S. (2011) Standards for standardized logistic regression coefficients, *Social Forces*, 89, 1409–1428.

Merlo, J., Yang, M., Chaix, B., Lynch, J. and Rastam, L. (2005) A brief conceptual tutorial on multilevel analysis in social epidemiology: investigating contextual phenomena in different groups of people, *Journal of Epidemiology and Community Health*, 59, 729–736.

Morrissey, K., Clarke, G., Hynes, S. and O'Donoghue, C. (2010) Examining the factors associated with depression at the small area level in Ireland using spatial microsimulation techniques, *Irish Geography*, 43(1), 1–22.

Morrissey, K., O'Donoghue, C., Clarke, G. and Li, J. (2013) Using simulated data to examine the determinants of acute hospital demand at the small area level, *Geographical Analysis*, 45(1), 49–76.

Nosek, B., Banji, M. and Greenwald, A. (2002) E-research: ethics, security, design and control in psychological research on the internet, *Journal of Social Issues*, 58(1), 161–176.

O'Connor, W. and Nazroo, J. (eds) (2002) *Ethnic Differences in the Context and Experience of Psychiatric Illness: A Qualitative Study*. London: The Stationery Office.

ODPM (2004) *The English Indices of Deprivation* (revised), London: Office of the Deputy Prime Minister.

Office for National Statistics (2012) *National Statistics Postcode Lookup (2011 Census) User Guide* (2012 edition), London: Office for National Statistics.

Peterson, L., Tsai, A., Petterson, S. and Litaker, D. (2009) Rural–urban comparison of contextual associations with self-reported mental health status, *Health & Place*, 15(1), 125–132.

Pontin, E. (2012) Research on pathways to mental health: testing the mediating psychological processes model of mental disorder. D.Clin. Thesis, University of Liverpool, UK.

Riva, M. and Smith, D. (2012) Generating small-area prevalence of psychological distress and alcohol consumption: validation of a spatial microsimulation method, *Social Psychiatry and Psychiatric Epidemiology*, 47, 745–755.

Shaw, C.M., Creed, F., Tomenson, B., Riste, L. and Cruickshank, J. (1999) Prevalence of anxiety and depressive illness and help seeking behaviour in African Caribbeans and white Europeans: two phase general population survey, *British Medical Journal*, 318, 302–305.

Singh, A.M., and Misra, N. (2009) Loneliness, depression and sociability in old age, *Industrial Psychiatry Journal*, 18(1), 51–55.

Skapinakis, P., Lewis, G. and Araya, R. (2005) Mental health inequalities in Wales, UK: multi-level investigation of the effect of area deprivation, *British Journal of Psychiatry*, 186, 417–422.

Smith, N. (2004) Goldberg anxiety and depression inventory, www.alswh.org.au/images/content/pdf/InfoData/Data_Dictionary_Supplement/DDSSection2GADS.pdf.

Smith, S.J. and Easterlow, D. (2005) The strange geography of health inequalities, *Transactions of the Institute of British Geographers*, 30, 173–190.

Soja, E.W. (1980) The socio-spatial dialects, *Annals of the Association of American Geographers*, 70, 207–277.

Sproston, K. and Nazroo, J. (eds) (2002) *Ethnic Minority Psychiatric Illness Rates in the Community (EMPIRIC)*, London: The Stationery Office.

Stafford, M. and Marmot, M. (2003) Neighbourhood deprivation and health: does it affect us all equally? *International Journal of Epidemiology*, 32(3), 357–366.

Surtees, P.G., Dean, C., Ingham, J.G., Kreitman, N.B., Miller, P.M. and Sashidharan, S. (1983) Psychiatric disorder in women from an Edinburgh community: associations with demographic factors, *British Journal of Psychiatry*, 142, 238–246.

Van de Velde, S., Bracke, P. and Levecque, K. (2010) Gender differences in depression in 23 European countries: cross-national variation in the gender gap in depression, *Social Science & Medicine*, 71(2), 305–313.

Walters, K., Breeze, E., Wilkinson, P., Price, G., Bulpitt, C. and Fletcher, A. (2004) Local area deprivation and urban–rural differences in anxiety and depression among people older than 75 years in Britain, *American Journal of Public Health*, 94(10), 1768–1774.

Weich, S., Twigg, L. and Lewis, G. (2005) Geographical variation in rates of common mental disorders in Britain: prospective cohort study, *British Journal of Psychiatry*, 187, 29–34.

Westling, E.L., Lerner, D.N. and Sharp, L. (2009) Using secondary data to analyse socio-economic impacts of water management actions, *Journal of Environmental Management*, 91, 411–422.

Williams, G. (2003) The determinants of health: structure, context and agency, *Sociology of Health & Illness*, 25(1), 131–154.

# Part V

# Environmental modelling

# 18 A case study of flooding in the Limpopo River Basin, Xai-Xai, Mozambique

*Robert Fligg and Joana Barros*

## 18.1 Introduction

This chapter presents an agent-based model for the Limpopo River Basin Area in Mozambique. More specifically, the study focuses on the evacuation procedure of a fast-flooding area in the vicinity of Xai-Xai, Province of Gaza.

The Xai-Xai area, located at the bottom of a vertical relief of 800 metres draining the land from four countries, Zimbabwe, Botswana, South Africa and Mozambique, is prone to very fast flooding. One of the worst floods in the area happened in 2000 due to the Cyclone Eline. Prior to the cyclone hitting the area, heavy rainfalls and a tropical storm brought 700mm of rain over two days, resulting in 4–8 metres of flooding. The flow of the Limpopo River at Xai-Xai at the time of the storm was estimated at 10,000 $m^3$/sec (10 times the normal rate), breaking through dykes and resulting in flash flooding (Christie and Hanlon 2001). It is estimated that 2 million people were affected by the floods (Christie and Hanlon 2001), but at the time the British Broadcasting Corporation reported 7,000 people near Xai-Xai were trapped in trees – some up to several days – and approximately 700 perished (BBC 2000). This was the worst flood since 1997, when the water reached 6.07m or 1.77m above flood level. Although a similar flood had occurred in 1915, which reached 5.7m or 1.4m above flood level, the area had experience bad floods quite frequently (1955, 1967, 1972, 1975 and 1981).

The reoccurrence of the natural phenomena that caused the magnitude of flooding in 2000 is estimated at once in 50 years and once in every 20 years for a flood of the 1977 magnitude (Christie and Hanlon 2001). A flooding event comparable to the one in 2000 occurred in January 2013 in southern Mozambique, killing 36 and displacing thousands (Appleton 2013; UNITAR 2013). The major flood episode in 2000 was actually produced by four successive flooding events. The first was an above average rainfall in October to December 1999. This was followed by the remnant of Cyclone Connie, which dropped a record rainfall between 4 and 7 February, resulting in flooding equivalent to the 1977 flood. On 13 February, a second crest of water came down the Limpopo River, once again matching the 1977 floods. Cyclone Eline then hit on 22 February, causing a crest on the Limpopo River, which resulted in a flood 3m above the levels reached in 1977 on 1 March.

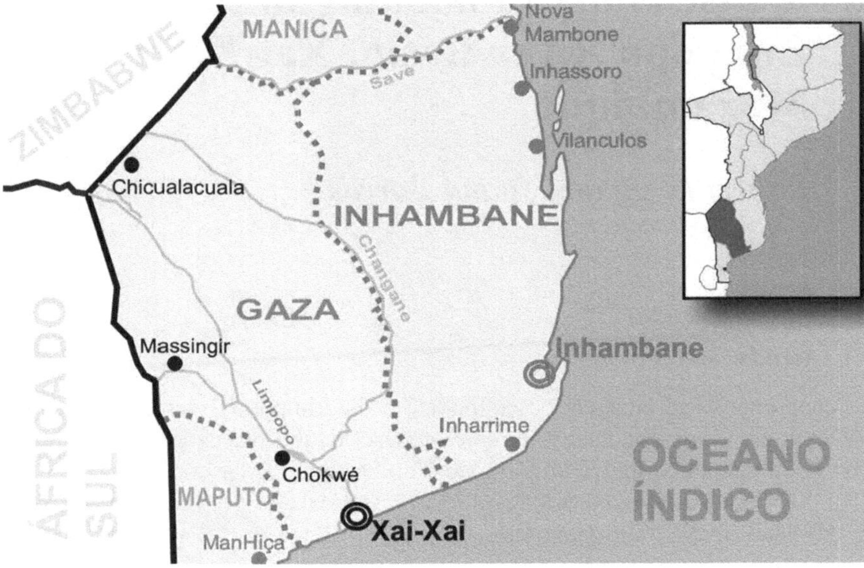


*Figure 18.1* Province of Gaza and location of Xai-Xai

Source: Koehne (2007)

The 2000 floods, however, were a result of a long-term chain of events. During the 1990s this area suffered from drought and the hard ground in catchment areas resulted in rapid run-off; the headwaters of the Limpopo River in northern South Africa, southeast Zimbabwe and southern Mozambique had three times as much yearly rain (approximately 1,800mm); and the dams in northern South Africa and southern Mozambique were near capacity after heavy rainfalls in October to December and could not absorb the record rainfall from subsequent cyclones, which resulted in the crests of flood water rushing down the Limpopo River (Christie and Hanlon 2001).

What transformed this event into a catastrophe was the fact the progressive flood events caught the population off guard, aided by the belief that flooding could not be worse than in 1977. So when the crest of water flowing down the Limpopo River reached Xai-Xai on 1 March, water levels rose rapidly to 5.7m above 'flood level', causing people to flee to high ground in the Xai-Xai area as well as to rooftops and trees (Christie and Hanlon 2001).

It is this reality of very fast flooding that the present research investigates using an agent-based model. The objectives of the project are: (a) to simulate the phenomena of a fast-flooding area and evacuation behaviour; and (b) to develop a neural network designed to simulate an agent's cognitive ability to sense, learn and adapt when travelling over a landscape during a flooding episode.

Agent-based models (ABM) have been largely applied to simulate disaster processes (Dawson et al. 2011; Lumbroso et al. 2013). ABMs are particularly suited to disaster modelling because they allow for an understanding of disaster processes from a bottom-up approach. ABMs provide applied geographers with an understanding of how disasters affect individuals and how individual behaviours shape a pattern of collective behaviour. Such understanding is essential for disaster management, in particular for setting evacuation procedures. ABMs of disasters can help the decision making of evacuation procedures (Brouwers and Verhagen 2003; Takahashi 2007) by testing evacuation times and routes in natural disasters (Lin and Manocha 2010; Dawson et al. 2011; Bakillah et al. 2012) or changing of routes and procedures in buildings or urban areas (Rahman et al. 2008; Su et al. 2011; Padgham 2015).

The next section introduces the model and discusses the choice of methodology employed. This is followed by sections on the model implementation, results, analysis and conclusion.

## 18.2 An agent-based model for Limpopo Basin flooding

The model presented here simulates a scenario of fast flooding along the lines of the one which occurred in 2000. The model is divided into two parts: the simulation of flooding and the simulation of the evacuation procedure taken by the local population.

The simulation of flooding is done in a very simplistic way, as will be detailed in section 18.4. The environmental information used to simulate the flooding are: rain rate, run-off and a digital elevation model (DEM). The DEM provides the elevation data that determines where the flooding takes place, while the rate of rain fall and run-off impacts the rate of flooding.

In areas of fast flooding like Xai-Xai, people often resort in seeking refuge on rooftops, in trees and on any higher ground (Christie and Hanlon 2001). Once flooding is imminent and an area starts to flood, people start travelling to the 'safe area'. This is the behaviour the present model attempts to simulate. For the purpose of this study, a 'safe area' or shelter is either an area located on the higher ground of Xai-Xai or any other area that is not prone to 'fast flooding'.

The model simulates the decision-making process of people when travelling towards a safe area, assuming that people evacuating do not have prior knowledge about the area and, thus, are learning about the terrain as they move. As such, the model does not include prior learning of the area as well as interaction with other people, which is another factor that can impact on people's evacuation behaviour (i.e. by going towards the neighbour's house because he knows a safe place/route).

In the model, each agent represents a single person. When moving towards a safe area, agents have the ability to sense the terrain and surrounding area they travel over. This is based on a pragmatic approach on how a person would travel over the terrain, for example, the best mode of travel is by road, followed by trails, farm land, and vegetation. The two main restrictions are that agents cannot travel across water or through buildings.

The decision-making process of agents is built upon concepts from cognitive science (Gonzalez et al. 2003). In the model, the cognitive ability of an agent is its ability to sense and recognise the characteristics of the landscape around them. As part of the decision-making process inbuilt in the model, this information is used by the agent to learn about the terrain and adapt their choices in order to make the best decision to move towards the 'safe area'. In other words, agents look for a safe route away from the flooding, making decisions based on their assessment of land cover at each step. This behaviour is illustrated by the flowchart in Figure 18.2.

Such a decision process is central to the model and incorporates an optimal operating range of weighted information that increases the agent's ability to adapt to change. This means that if an agent has used a specific land use that has delayed his journey or made it harder, it will learn that this is not a good option

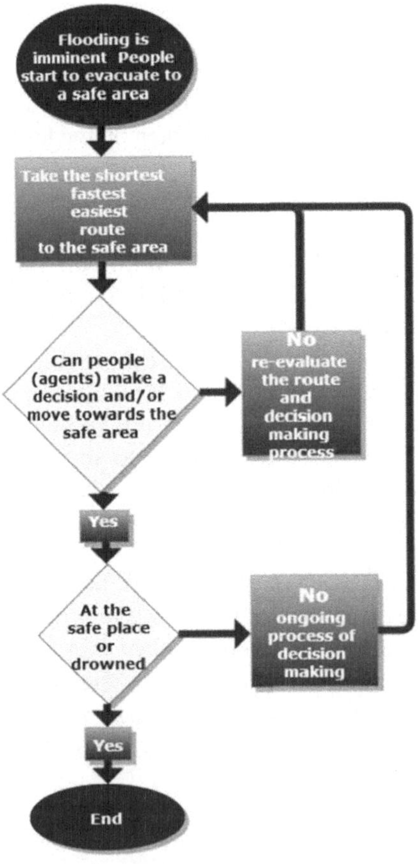


*Figure 18.2* Flowchart of agent's behaviour

and improve his choices in the remainder of his path. In the model this is simulated using a neural network, which incorporates the capability of learning and adapting onto the decision-making process.

An artificial neural network is designed to work similarly to a biological neural network having artificial nodes (neurons) 'linked to each other in a weighted way' forming a network (Cheung and Cannons 2002; Kriesel 2005). Figure 18.3 illustrates the structure as data being processed through layers; input neurons that are adjusted in the hidden layer neurons, until it emerges as output neurons (Colton and Gervás 2003; Kriesel 2005; Colton 2012). Sample data are often subdivided into training, validation and test sets. The system must first be trained with a training set – a subset of the data that will be used for learning. The validation set is used to adjust the weights and to assess the performance of the model (Ripley 1996). In this model, an agent gathers 'input' data about the terrain surrounding it. This information is processed through 'hidden layers' and emerges as 'output' to be used in the decision-making process as it travels over the terrain towards the 'safe area'.

The neural network developed for this study is a hybrid model, as the model required a spatio-temporal, dynamic, decision-making process. One of the requirements for this model was that training in one area should not apply to another, so the network had to be able to adapt to changing land use/land cover (LU/LC) during a flooding episode. As a result, the hybrid model included characteristics of five neural network types: feedforward (Ripley 1996); recurring (Kriesel 2005); stochastic (Turchetti 2004); neural-fuzzy (Kasabov 1996); and learning to rank (Li 2011). The implementation of the neural network is detailed in section 18.4.1.

## 18.3 Environmental data

The landscape used in the model was built using satellite imagery. A single tile of 50cm ground resolution GeoEye-1 satellite imagery dated of 2009 was acquired from GeoEye through Sani-International Technology Advisors Inc. covering a

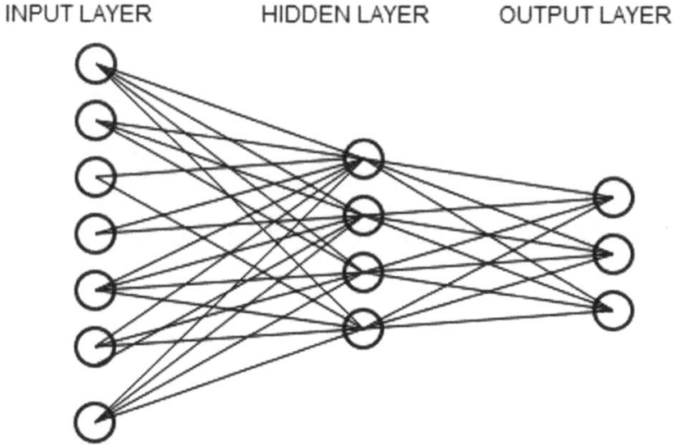


*Figure 18.3* Conceptual model of a neural network

Source: Adapted from Kriesel (2005)

140,000m square area in the vicinity of Xai-Xai, Mozambique. The raw imagery was acquired in four bands: red, green, blue and near infrared (NIR), and joined through an imagery process in ERDAS Imagine software. This imagery was used to derive the land use/land cover information for the project having a ground resolution that provided realistic information.

The imagery was categorized for features such as buildings, farm land, roads, vegetation, the Limpopo River, trails and hazards – the features that were used in the process of decision making by agents in the model. Remote sensing technology was used to assess the spectral signatures (red, green and blue) of each pixel within the imagery in order to classify the pixels for feature types and create a land use/land cover data file for the project. The following subsections detail the imagery processing and classification.

For the flooding behaviour in the model, a Digital Elevation Model (DEM) was used. The DEM obtained was an Aster GDEM 2011 Version2, 1 arc second or approximately 30m resolution. For a more realistic flooding model over the project area, the Aster GDEM was interpolated and a 6m resolution version produced.

### *18.3.1 Imagery processing and classification*

The 50cm GeoEye-1 satellite imagery was classified using the signature editor set to classified and supervised. Each feature was outlined on the imagery that best represented the feature to be captured and the spectral signature for the patch was recorded. The rest of the imagery was then assessed using the same spectral signature.

The spectral signatures of many features (red, green, blue) were assessed and categorized into 19 layers. It was expected that some features would have more than one spectral signature and that each one could be slightly different. For example, buildings could have three slightly different spectral signatures due to construction material and reflection. Within the 19 layers of feature types are: Building 1, Building 2, and Building 3. The GeoEye-1 satellite imagery pixels for these three layer types were combined into one feature category and assigned the attribute number 20. Similarly, Road-Urban, Road-Urban2 and Road-Highway were combined to create a feature category assigned number 40. After amalgamating similar features, 19 classifications were reclassified into nine feature categories, assigned attributes 10, 20, 30, 40, 50, 60, 70, 80 and 90.

### *18.3.2 Resampling the LU/LC file*

The LU/LC file for the 0.5m resolution GeoEye-1 satellite image was resampled from 0.5m to 15m using ERDAS (Plate 13). The resampled image (img) file was then converted to an ASC (ascii) file format, to be used in Repast Simphony Java platform in which the model was built.

## 18.4 Model implementation

The model was developed in Repast Simphony Java (Collier and North 2013) and implemented in five phases, as shown in Figure 18.4. These phases are: (1) Input, (2) Training/Backpropogation and Fuzzy Logic, (3) Learning to Rank and Output, (4) Recurrent, and (5) Results.

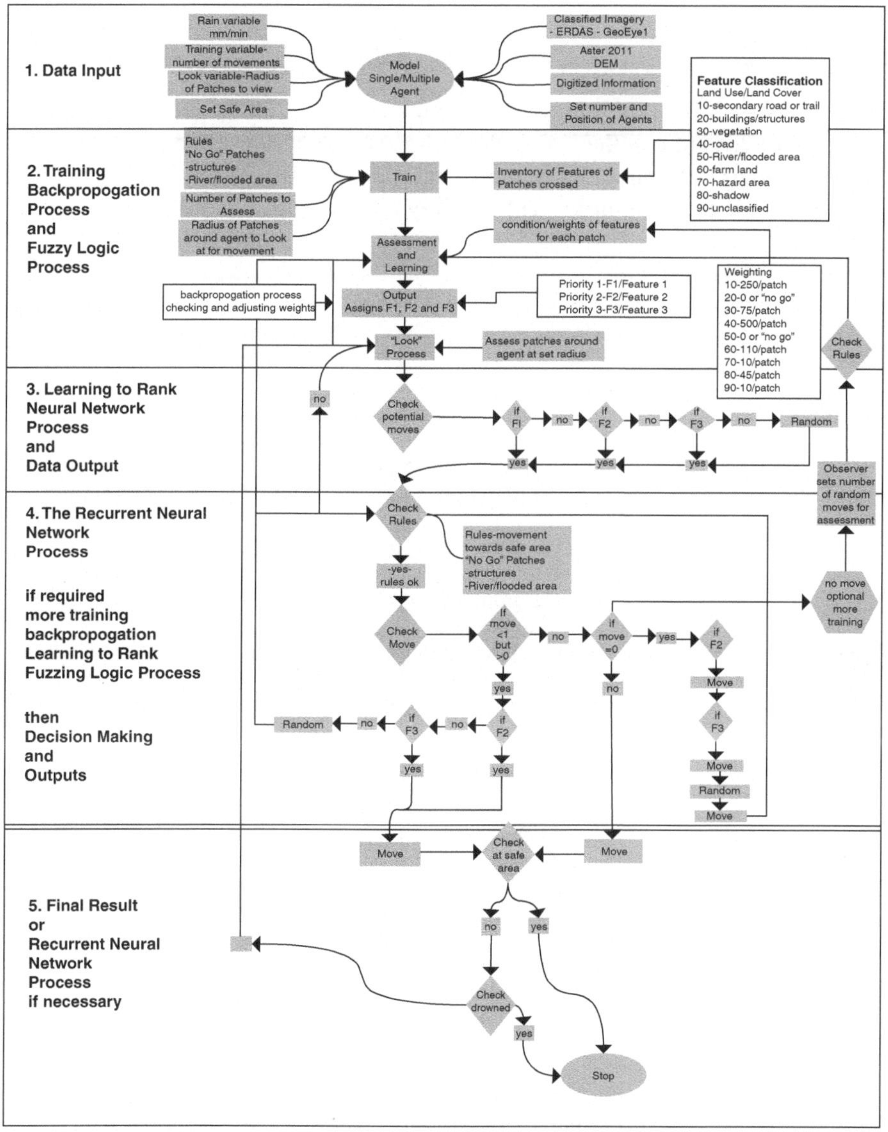


*Figure 18.4* Flowchart of model behaviour highlighting the different stages of implementation

In what follows, each phase will be detailed using the typical behaviour of a single agent as an example. As shown in Figure 18.4, the model behaviour on a typical run consists of all five phases. The model run will automatically stop when there are no agents present, that is, either when an agent is encompassed by flooding or when an agent has made it to the safe area. The neural network is called upon during phases 2 to 5 when land cover features are incorporated into the decision-making process. The specific behaviour of the neural network will be detailed in section 18.4.1.

### *Phase 1: Input*

When a scenario is initialized, the two main data files (DEM and the classified imagery) are loaded together using the setup settings summarized in Table 18.1. These variables determine the flooding rate, location of agents, location of the safe area, amount of training, agent mobility and sensing. Plate 14 illustrates the interface on setup.

*Table 18.1* Model variables

| | |
|---|---|
| Rainfall | The default is 1mm/minute. The rate can be set to any value per minute by the observer |
| Run-off<br>3 rates are available | The default setting is 2, but can be changed by the observer.<br>1 – simulates the ground is not saturated and flooding from run-off is set at none<br>2 – simulates the ground has some saturation and contributes to flooding at a low rate<br>3 – simulates the ground is highly saturated and contributes to flooding at a higher rate |
| Agent location | The x,y location of one or more agents is set before initialization by the observer |
| Agent rate of travel | Movement depends on the environment surrounding an agent. Rate of travel is pre-set at:<br>1 cell for feature 30 (dense vegetation), 70 (hazard), 80 (building shadows), 90 (unclassified)<br>2 cells for feature 60 (farm land)<br>3 cells for feature 10 (trail)<br>4 cells for feature 40 (road) |
| Safe area | A designated area known to be safe from flooding is set by the observer prior to initialization |
| Sensing (radius) | Values are pre-assigned according to the land cover the agent is on. The radial sensing values are:<br>up to 9 cells for features 30 (dense vegetation), 70 (hazard), 80 (building shadows), 90 (unclassified)<br>up to 25 cells for feature 60 (farm land)<br>up to 49 cells for feature 10 (trail)<br>up to 81 cells for feature 40 (road) |
| Training | Number of moves for training is set by the observer prior to initialization (the default is 5) |

*Phase 2: Training/Backpropogation and Fuzzy Logic Process*

Once a scenario is set up and the model starts running, an agent undergoes a training process gathering data about the terrain for the agent's first five moves (being the default). The brain (here represented as a neural network) of an agent undergoes a training process that sets up the initial weights for the land cover features. The knowledge base that an agent uses in this process was established by a fuzzy logic approach for feature values and rules about how an agent can move. At the end of this phase, an agent is ready to make intelligent decisions as it moves towards the safe area.

*Phase 3: Learning to Rank and Output*

After an agent is trained and is ready to move to a safe area, the neural network is called upon to assess an inventory of land cover information gathered by the agent. The neural network implements a process to optimize the weights and rank the inventory of land cover features, providing an agent with the three best moves for decision making (see details in section 18.4.1). If the three best moves (ranked from highest to lowest and assigned variables F1, F2 and F3) do not provide an agent with an acceptable move in accordance with model rules, a random feature will be selected and checked in accordance with model rules. A random feature is a cell that an agent can select in an arbitrary direction and distance of 1 or 2 cells depending on the difficulty encountered by the agent in moving.

*Phase 4: Recurrent Neural Network and Output*

While an agent is travelling over the terrain, the neural network is adapting and learning, constantly assessing and reassessing the land cover. If an agent is unsuccessful in finding a move, the neural network is called upon to undergo a more rigorous assessment that could include additional cells and/or incorporate more random moves. If a random selection does not help an agent move, the process has the option to revert back to phase 2 to undergo a reassessment that is similar to training in order to adjust the feature weights.

*Phase 5: Results*

If the process proceeds to phase 5, then a move is possible. The amount an agent moves is dependent on what features it can sense. The number of cells an agent senses is determined by the radius variable and this can change automatically depending on the land cover feature it is on. For example, a sensing variable or radius of 1 (searches 9 cells) and is used for feature 30 (dense vegetation), a radius of 2 (searches 25 cells) and is used for feature 60 (farm land), a radius of 3 (searches 49 cells) and is used for feature 10 (trails), and a radius of 4 (searches 81 cells) used for feature 40 (roads). The maximum an agent can move is determined by the radius variable; for instance, an agent can move faster on feature 40 (road) by advancing 4 cells if available. After each move, the location of an agent is assessed to determine if it is at the safe area or encompassed by flooding, at which time the agent is removed from the scenario. If not, the process loops back to collect another inventory of feature data, and re-implements phases 3, 4 and 5.

### *18.4.1 Developing the artificial neural network*

The design of the neural network is based on a process where an agent collects data to create an inventory of information. This information is then used in a series of operations that assesses feature weights, calculates the delta error that is used in a back propagation process, optimizes the weights, and then checks the model rules to produce a list of feature outputs that is ranked from highest to lowest. The network assigns the three highest land cover features with the output variables F1, F2 and F3 for agent decision making.

Figure 18.5 depicts the artificial neural network (ANN) with 21 inputs, hidden layers, three outputs, and describes it in seven processes.

In process 1, the number of cells an agent assesses depends on how far it can sense, which depends on the radius variable setting that fluctuates with the environment. As previously described in 18.4 Model implementation:

- a radius of 1 will sense up to 9 cells
- a radius 2, up to 25 cells
- a radius 3, up to 49 cells, and
- a radius 4, up to 81 cells

In process 2, the inventory of LU/LC data is assessed and categorized into the numeric variables shown in Table 18.2.

The categorized inventory of data will be used in the next process for weight assignment and movement associated with each cell. Features 20 (structures) and 50 (water) are not utilized for movement, as they are considered 'no go' cells.

The knowledge base on (LU/LC) feature weights is a pre-assigned percentage of 1,000 based on a pragmatic approach and trial runs. The percentage value is later used to optimize the weights in process 6 that determines the final output in process 7.

- feature 10 was set at 250 (25%)
- feature 30 at 75 (7.5%)
- feature 40 at 500 (50%)
- feature 60 at 110 (11%)
- feature 70 at 10 (1%)
- feature 80 at 45 (4.5%)
- feature 90 at 10 (1%)

In process 3, the weights are summed for of all the features (in the inventory) and stored. After the total weights for each feature are assessed during process 4, they

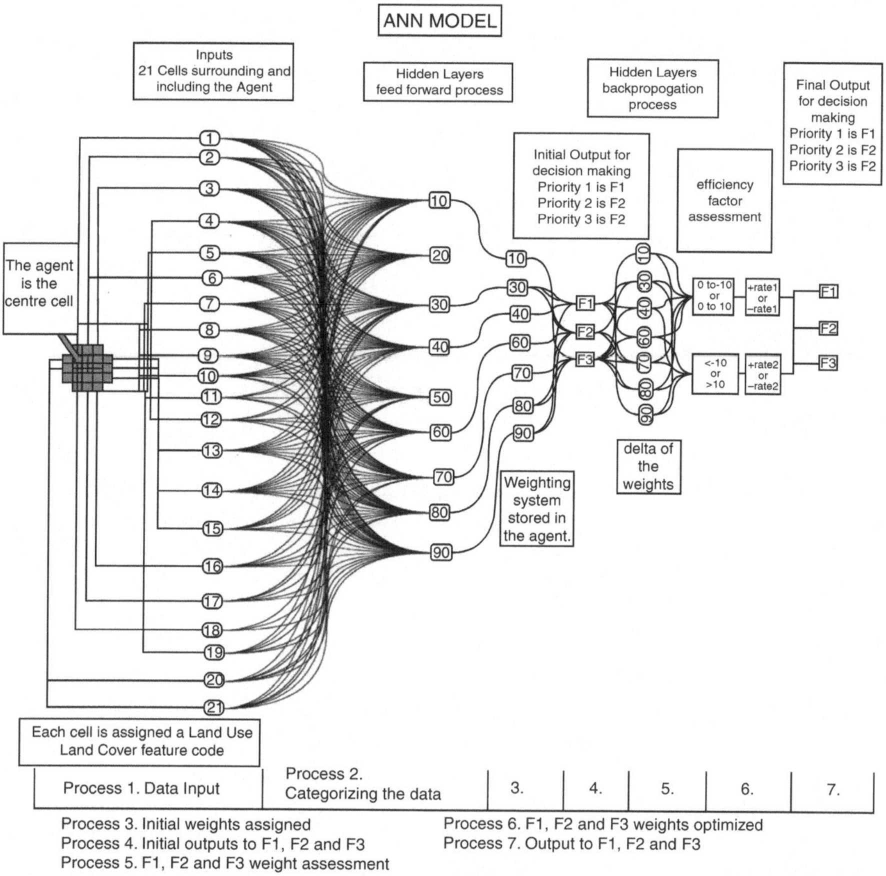


*Figure 18.5* Outline of the artificial neural network process

are ranked from highest to lowest, with the three highest features assigned variables F1, F2 and F3.

In process 5, the delta error is determined using the percentage of the total weights compared to the knowledge base for each feature. The delta error is used to determine the (optimization) efficiency value called ‘r’ and categorized in one of three ranges: (1) less than –10, (2) –10 to +10, or (3) greater than +10.

Depending on the category of the delta weights, each value has a learning factor of rate 1 or rate 2 to either reduce the weight or increase it to its optimal

*Table 18.2* Numeric variables assigned to features

| | | |
|---|---|---|
| 10 (trails) | 40 (roads) | 70 (hazards) |
| 20 (structures) | 50 (water) | 80 (building shadows) |
| 30 (dense vegetation) | 60 (farm land) | 90 (unclassified) |

value. In process 5, if the 'r' value was between –10 and 10, rate 1would be used, utilizing a small increase or decrease of the weights to approach the optimal value of 0. If the 'r' value has less than –10 or greater than 10, rate 2 would be applied to increase or decrease the weights at a greater rate into the range between -10 and +10. If more adaptation is required, the observer could increase the learning rate. The formula for adjusting the weights was +/– (rate1 or rate2) × (the features initial assigned weight).

When all the weights have been adjusted, in process 6 the output features are ranked from highest to lowest and once again the three highest are assigned the variables F1, F2 and F3.

## 18.5 Model interface

The landscape of the model is composed by a 470 × 470 matrix of 15m cells containing two layers of information: land cover and DEM, as illustrated in Plate 14.

The user can set the model parameters using the interface. These are: number of agents, rainfall in mm/minutes, and option for printing to an output file. The example shown in Plate 14 has one safe area, shown as a black star (bottom right).

## 18.6 Scenarios and analysis

Three scenarios are presented to demonstrate how the model worked. Scenario 1 tests the decision-making process over rural and urban land use/land cover. Scenario 2 tests the agent's ability to make difficult decisions, such as to find a bridge in order to get to a safe place. Finally, scenario 3, tests the ability of multiple agents to face different levels of difficulty in decision making.

In all scenarios, the rain rate is set at a value between 0.5 and 1mm per minute, a value that simulates the flood in 2000, and run-off rate that is moderate (a value of 2 – maximum value is 3) increasing the rate of flooding greater than the accumulation of rain. The location and number of agents was selected to provide a cross-section of information that would demonstrate how the model worked over three scenarios.

Data was collected for each scenario and used to evaluate the agent's ability to successfully travel the environment and find a safe place or inability to escape the flooding. The data are organized into four charts and one table that are described in Table 18.3. Charts A, B and C in Figures 18.6, 18.7 and 18.8, depict the values for (1) feature weights variables (wt10 to wt90), (2) optimization variables (r10 to r90), and (3) feature selection variables (F, F1, F2 and F3).

Chart D in Figures 18.6, 18.7 and 18.8 and Tables 18.4, 18.5 and 18.6 displays statistical analysis about the decision-making process used to assess how well the neural network worked and if there is evidence of adaption and path dependence. Linear regression calculations were performed on the data for the $R^2$ statistic (also called the correlation coefficient), providing a measure on how good the fit is or the correlation between the resultant values and the predicted values (Fox 2008). In this study there are no predicted values, therefore the results (feature selections 'F') are compared to the preferred values, which are the ranked feature

*Table 18.3* Description of figures for each scenario

| | |
|---|---|
| Figures 18.6, 18.7 and 18.8, Chart A | A line chart showing the weight variables (wt10, wt30, wt40, wt60, wt70, wt80 and wt90) value for each feature during a scenario. |
| Figures 18.6, 18.7 and 18.8, Chart B | A line chart showing the 'r' variable (r10, r30, r40, r60, r70, r80, r90) indicating how efficient the weighting process was during a scenario. The target value of 'r' is 0 and the optimal operating range is –10 to +10. |
| Figures 18.6, 18.7 and 18.8, Chart C | A line chart showing the ranked feature output, F1, F2 and F3 (described in section 4) and the feature the agent selected (F) during a scenario. |
| Figures 18.6, 18.7 and 18.8, Chart D | A bar chart showing the standard deviation (SD) and mean value of 'r' (the optimized value of each feature weight). |
| Tables 18.4, 18.5 and 18.6 | A table that depicts the $R^2$ (at the 95% confidence interval) for the correlation of features the agent selected 'F' with ranked feature values F1, F2 and F3, and the $R^2$ for the correlation of the optimization value of each feature 'r' with the weights for features 10 (trails), 30 (vegetation), 40 (roads), 60 (farm land), 70 (hazards), 80 (building shadows) and 90 (unclassified). The table also includes corresponding RMS values. |

selection variables F1, F2 and F3. The usual statistical value for $R^2$ is a value between 0 and 1, with a value closer to 1 indicating that a greater proportion of the variance is accounted for by the model (Mathworks 2012). For all the tick values, the $R^2$ assessment is based on the top three feature values (variables F1, F2 and F3) compared to the actual feature selected by the agent. $R^2$ values are also provided for the optimization value 'r' (for each feature) with respect to its corresponding feature weight, providing an indication on how well the agent could adapt and learn.

Root Mean Square error (RMS) (also known as the fit standard error, the standard error of regression and the standard deviation of the residuals) was calculated on the same data used for $R^2$. RMS is an estimate of the standard deviation of the random components in the data. A value closer to 0 indicates a fit that is more useful for prediction (Mathworks 2012). In this study, RMS was used to indicate how well scenario results compared with the results of the optimal operating range. RMS should be used in conjunction with the $R^2$ values to help assess how well the model performed.

In what follows, each of the scenarios will be described and analysed.

### *18.6.1 Scenario 1 – One Agent Rural/Urban Area*

In scenario 1, an agent was placed in a rural area just outside the lower urban area of Xai-Xai (Plate 15).

During scenario 1, the agent travelled across farm land, then various roads within lower Xai-Xai, and across a mix of rural/urban terrain to the 'safe area' in the upper area of Xai-Xai. The scenario time was 110 ticks (representing minutes)

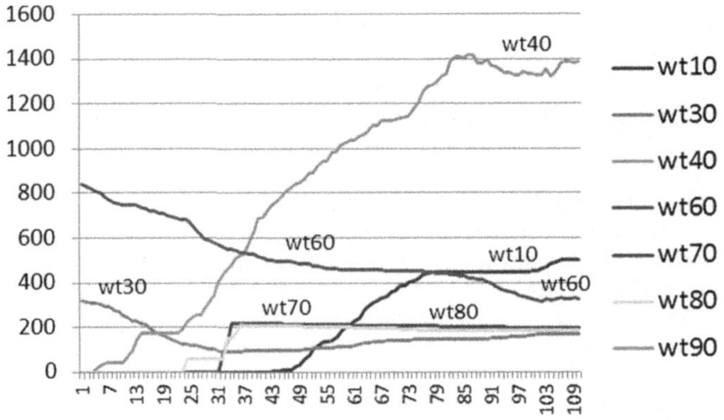


*Figure 18.6a* Illustration of the weights (wt) indicating how the agent interacted with the environment during the scenario and the values of the feature weights at each 'tick' interval that would be used in the ranking process

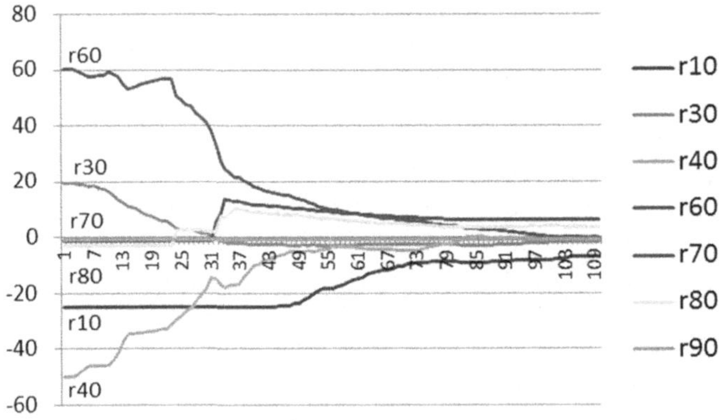


*Figure 18.6b* Optimization values (r) of the weights for each feature

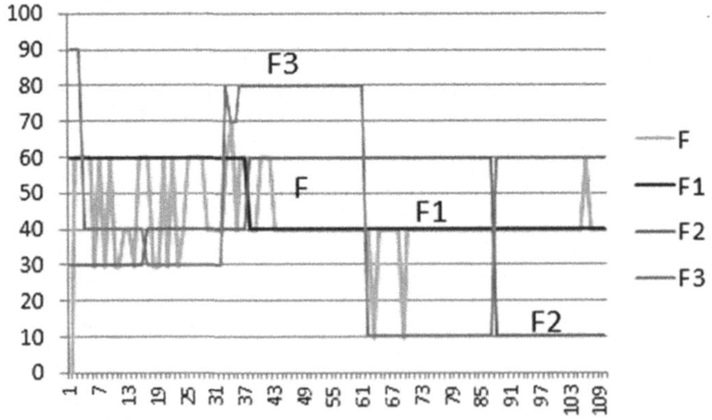


*Figure 18.6c* Feature selections F (actual) and ranked features F1, F2 and F3

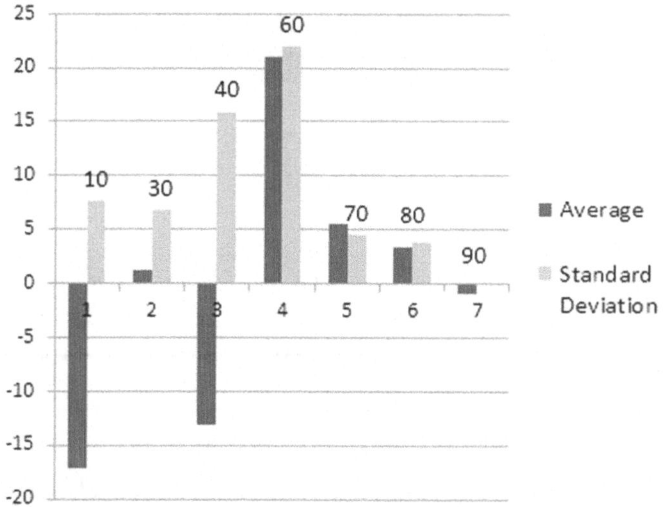


*Figure 18.6d* Standard deviation and mean of the 'r' values

with a rain fall of 1mm/minute, incurring a flood level of 3 metres. Although the flooding occurred quickly, the agent avoided the flooded areas and made it to the safe area.

Comparing Plate 15 to Figure 18.6a, the highest weight for the agent as it travelled through the rural area (from approximately 0 to 30 tick minutes) was feature 60 (farm land). Entering the urban area of Xai-Xai, it had to adapt as demonstrated by a spike in feature 40 (roads) and feature 10 (trails). In Figure 18.6a, the three highest feature weights for the agent at each 'tick' correspond to the feature variables F1, F2, and F3 that were used in the decision-making process (Figure 18.6c).

Figure 18.6b illustrates the optimization process and how it initially struggled to bring the weights for land cover features 60 (farm land) and 40 (roads) within the optimal values. Mid-way through the scenario all the weights were operating close to the preferred range of +/– 10.

In Figure 18.6c, the actual feature selection made by agents, indicated by the line 'F', depicts feature F1 was selected most of the time, with some F2, F3 and a few random selections. When the agent was in the rural area, the selection process was more erratic whereas in the urban area, feature 40 (road) was the main selection.

The standard deviation (SD) of optimization values 'r' for land cover features 60 (farm land), 40 (roads) and 10 (trails) showed they were slightly outside the preferred operating range, as illustrated in Figure 18.6d, and their corresponding mean values were close to or greater than 10. The mean values for all other features were within +/– 10. (The variable 'r' was described in the beginning of this section.)

*Table 18.4* Scenario 1 – Regression Analysis showing $R^2$ (at the 95% confidence interval) for the correlation of features the agent selected with ranked feature values F1, F2 and F3, and the $R^2$ for the correlation of the optimization value of each feature 'r' with the weights of features 10 (trails), 30 (vegetation), 40 (roads), 60 (farm land), 70 (hazards), 80 (building shadows) and 90 (unclassified). The table also includes corresponding RMS values

| | | | |
|---|---|---|---|
| Feature 1 F1 $R^2$ | 0.93 | RMS | 11.8 |
| Feature 2 F2 $R^2$ | 0.75 | RMS | 21.9 |
| Feature 3 F3 $R^2$ | 0.73 | RMS | 22.6 |
| $R^2$ 10 | 0.99 | RMS | 17.4 |
| $R^2$ 30 | 0.73 | RMS | 29.1 |
| $R^2$ 40 | 0.88 | RMS | 171.0 |
| $R^2$ 60 | 0.94 | RMS | 35.9 |
| $R^2$ 70 | 0.87 | RMS | 34.0 |
| $R^2$ 80 | 0.81 | RMS | 35.5 |
| $R^2$ 90 | 1.0 | RMS | 0 |

Table 18.4 indicates that the $R^2$ values for output features F1, F2 and F3 with the feature selection 'F' were approaching 1 in the order of F1, F2 and F3. The RMS values were approaching 0 in the order of F1, F3 and F2. This suggests the model was selecting features in the preferred order, with a stronger selection correlation for F1 features; F2 and F3 being approximately the same.

The $R^2$ for land cover features 10 to 90 (explained at the introduction of section 18.6) suggests the weights were adjusting and operating well (all over 0.7). The RMS values supported $R^2$ except for feature 40. Generally, the values show that the model was working well with the agent adapting to the environment.

### *18.6.2 Scenario 2 – One Agent Extreme Travel*

In this scenario one agent was placed in a rural area on the westerly side of the Limpopo River, presenting a challenge for the agent to find a way across the river in its effort to get to the 'safe area'. The scenario time was 311 tick/minutes with a flood level of 1.5m and rainfall of 0.5mm/minute. The flooding was moderate and the agent made it to the 'safe area' as shown by the agent's route taken in Plate 16.

Plate 16 illustrates that the agent travelled across farm land to the river. It then found the bridge, crossed over to Xai-Xai, taking various roads before reaching the 'safe area'.

Figure 18.7a illustrates the weight for feature 40 (roads) was highest for the entire scenario. The weight for feature 30 (vegetation) was higher in the rural area, but when the agent travelled from lower Xai-Xai through a transitional area (rural/urban) to the upper area, the weights for 10, 30, 60, 70 and 80 were converging.

In Figure 18.7b, the 'r' value for land cover features 10 and 30 indicates the process struggled to achieve the optimal operating range of +/– 10. Levels for all the weights improved as the scenario progressed.

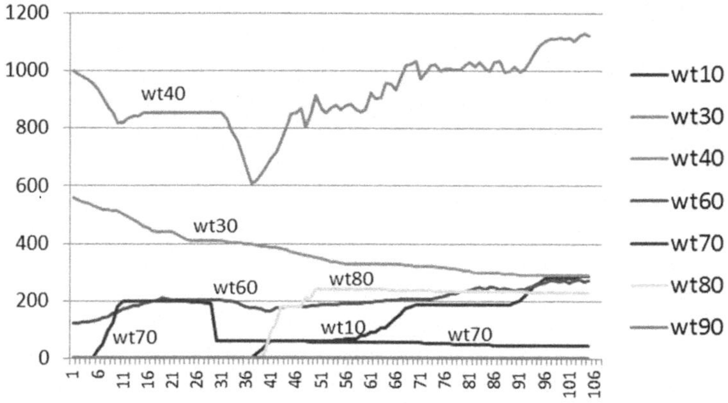


*Figure 18.7a* Weights (wt) of the features 10, 30, 40, 60, 70, 80 and 90

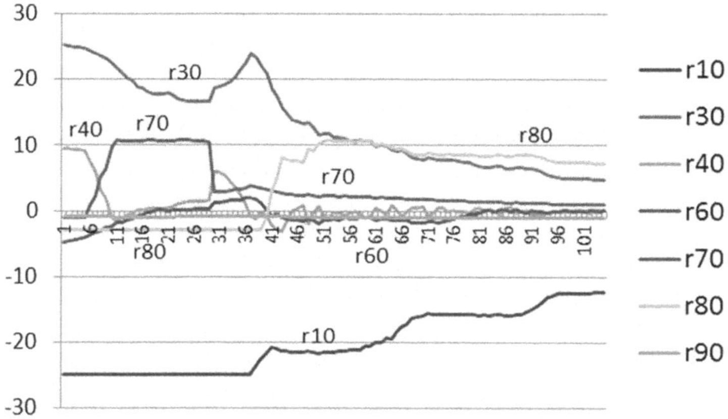


*Figure 18.7b* Optimization values (r) of the weights for each feature

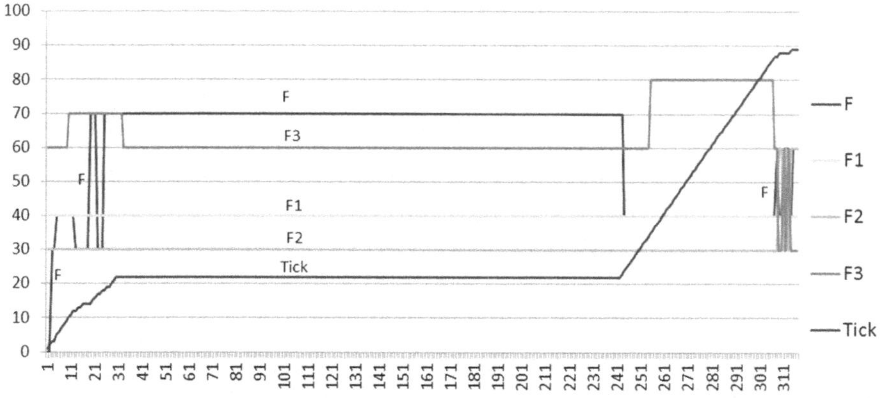


*Figure 18.7c* Feature selections F (actual) and ranked features F1, F2 and F3, and tick line for agent progress

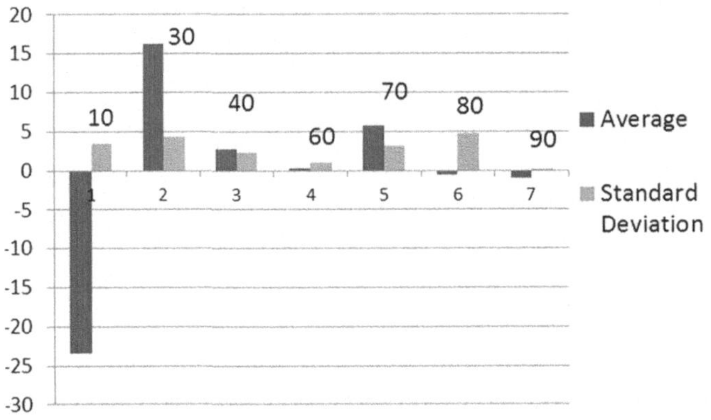


*Figure 18.7d* Standard deviation and mean of the 'r' values

Figure 18.7c illustrates selections were made on features F1, F2 and F3 as the agent travelled across the rural area until it discovered the river. Making a series of random selections while adapting to the terrain, the agent travelled along the river, found and crossed the bridge. It then travelled through lower Xai-Xai, with the main selection on F1, roads (40). Near the end of the scenario, while crossing over an area of rural and urban land cover, between lower and upper Xai-Xai, a series of selections on features F1, F2 and F3 were made. The selections by the agent suggest that it adapted to the terrain by selecting the best available land cover features while travelling to the 'safe area'. The majority of the land cover features with the exception of features 10 and 30 had a mean 'r' value within +/− 10. All the features had a SD within +/− 10, as shown in Figure 18.7d.

In Table 18.5, the $R^2$ values for the correlation of the features selected by the agent 'F' with F1, F2 and F3 were approaching 1 in the order of F1, F2 and F3, and the RMS values were approaching 0 in the same order. This suggests the decision-making process was working well as the agent made selections on features in the preferred order.

$R^2$ for the optimization value 'r' in relation to the weights for features 10, 30, 60, 70, 80 and 90 indicates the process was operating well for those features having values over 0.88. The $R^2$ value for feature 40 was low and RMS was much higher than the rest, suggesting improvement could be made to keep the 'r' value for 40 within the optimal operating range.

### *18.6.3 Scenario 3 – 25 Agents: 15 Urban/10 Rural*

In this scenario, 15 agents were placed in various locations in the lower urban area of Xai-Xai and 10 in the rural area. The whole area flash flooded to an elevation of 3m within 70 tick/minutes. The flooding trapped all the rural agents and 10 of the urban agents, only five made it to the safe area, as indicated in Plate 17.

*Table 18.5* Scenario 2 – Regression Analysis showing $R^2$ (at the 95% confidence interval) for the correlation of features the agent selected with ranked feature values F1, F2 and F3, and the $R^2$ for the correlation of the optimization value of each feature 'r' with the weights of features 10 (trails), 30 (vegetation), 40 (roads), 60 (farm land), 70 (hazards), 80 (building shadows) and 90 (unclassified). The table also includes corresponding RMS values

| | | | |
|---|---|---|---|
| Feature 1 F1 $R^2$ | 0.97 | RMS | 14.8 |
| Feature 2 F2 $R^2$ | 0.92 | RMS | 17.8 |
| Feature 3 F3 $R^2$ | 0.89 | RMS | 20.6 |
| $R^2$ 10 | 0.99 | RMS | 5.76 |
| $R^2$ 30 | 0.88 | RMS | 16.5 |
| $R^2$ 40 | 0.04 | RMS | 73.4 |
| $R^2$ 60 | 0.16 | RMS | 19.1 |
| $R^2$ 70 | 0.99 | RMS | 5.40 |
| $R^2$ 80 | 0.99 | RMS | 9.45 |
| $R^2$ 90 | 1.0 | RMS | 0 |

The results indicate that roads (40) and secondary roads/trails (10) were being utilized as the primary source of travel when it was available. However, the use of trails and rural features for travel was not fast enough to keep the agents ahead of the flooding.

Figure 18.8a shows the individual weights for the 25 agents. As with the two previous scenarios, agents displayed the capability of assessing the land cover and making decisions on an individual basis. Every agent in Figure 18.8a depicts a unique set of weights. Urban agents favoured feature 40 (roads) where feature 60 (farm land) was favoured for rural agents.

The optimization process appeared to work well for all the agents. As the scenario progressed, all 'r' values improved towards the optimal range of +/– 10 (see Figure 18.8b).

Figure 18.8c indicates feature F1 was selected most often, with fewer selections made on F2 and F3, and some random selections. Generally, the 25 agents performed similar to the individual scenarios.

Figure 18.8d indicates that the 'r' values for features 10, 30, 40 and 60, also being the features the agents selected most often, struggled to be within the preferred range of +/– 10. The values could be improved by adjusting the learning rate impacting how fast optimization of the weights occurs.

When an agent was removed from the scenario, a value of 10 or 20 was assigned to the file output indicating whether the agent made it to the 'safe area' or was trapped. For example, in Figure 18.8e, agents 1 to 20, were all trapped, indicated by the value 20 on the dark grey bar, and agents 21 to 25 made it to the safe area, indicated by the value 10. The corresponding light grey bar indicates at what point in the scenario the agent was removed. The last agent to be removed was number 17 at 70 ticks.

Figure 18.8f indicates the level of decision making required for each agent. Many agents travelled through the scenario using the first decision-making process, while many had to dig deeper to the max level of 47. Agent 20 got stuck

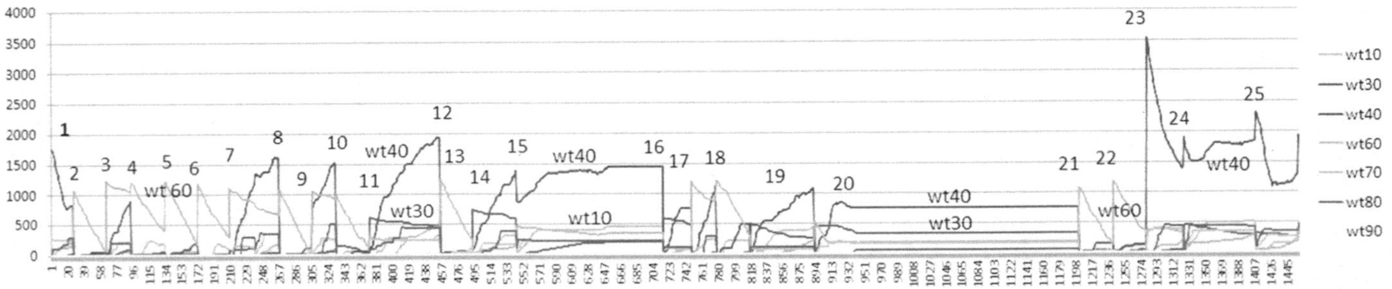


*Figure 18.8a* Weights (wt) of the features 10, 30, 40, 60, 70, 80 and 90 for 25 agents

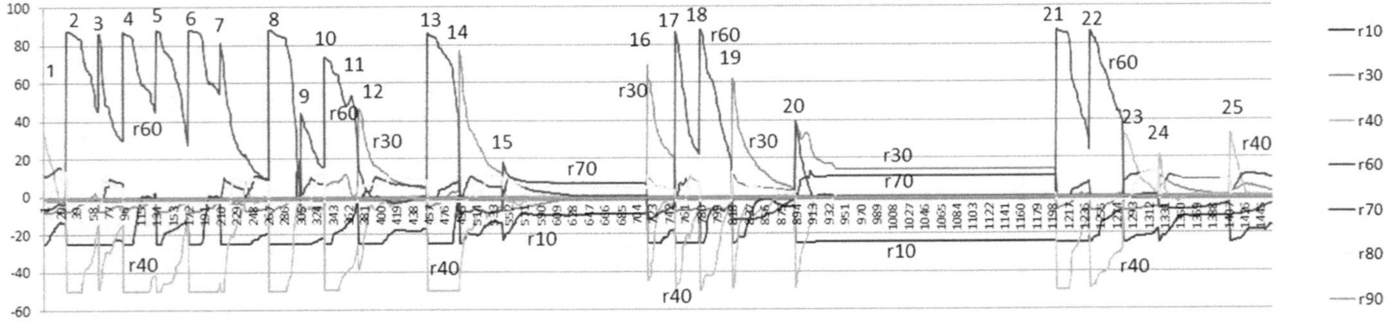


*Figure 18.8b* Optimization values (r) of the weights for each feature

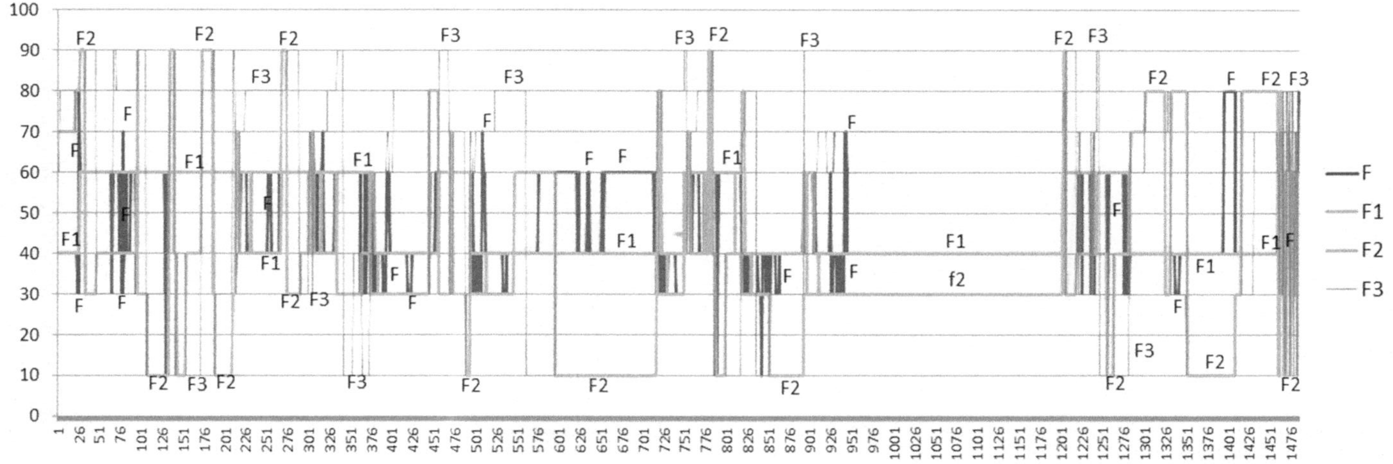


*Figure 18.8c* Feature selections F (actual) and ranked features F1, F2 and F3

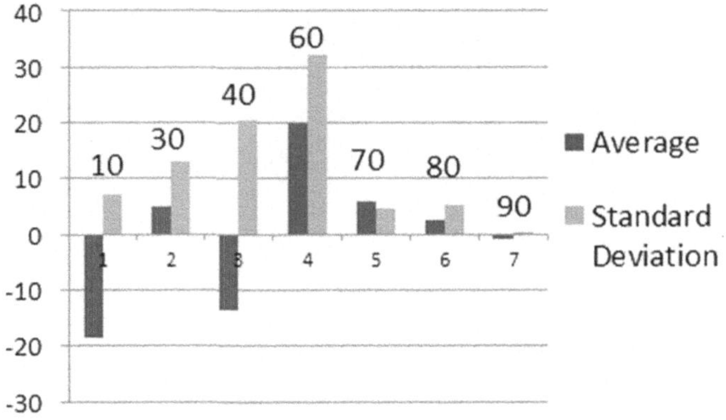


*Figure 18.8d* Standard deviation and mean of the 'r' values

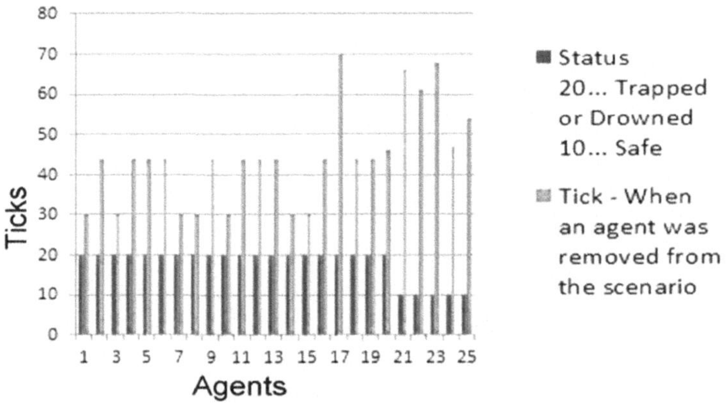


*Figure 18.8e* Illustration indicating when and why an agent was removed from the scenario

in a decision-making loop (indicated by the heavy grey bar) until it was able to move forward. This figure indicates that many agents struggled to find a viable decision at some point during the scenario.

The $R^2$ and RMS values in Table 18.6 indicate F1, F2 and F3 were approaching 1 and 0 respectively, in the order of F1, F3 and F2, suggesting agent selection on F2 should have been better. As a result, agents might have been able to move faster if output feature F2 was the preferred alternative when F1 was not available and possibly fewer agents might have been trapped by the flooding.

The $R^2$ for all the feature weights indicated the adjustment was working well, with values all over 0.61. The RMS for feature 40 (roads) was somewhat higher as it struggled to achieve the optimal operating range, suggesting this process affecting adapting and learning should be examined.

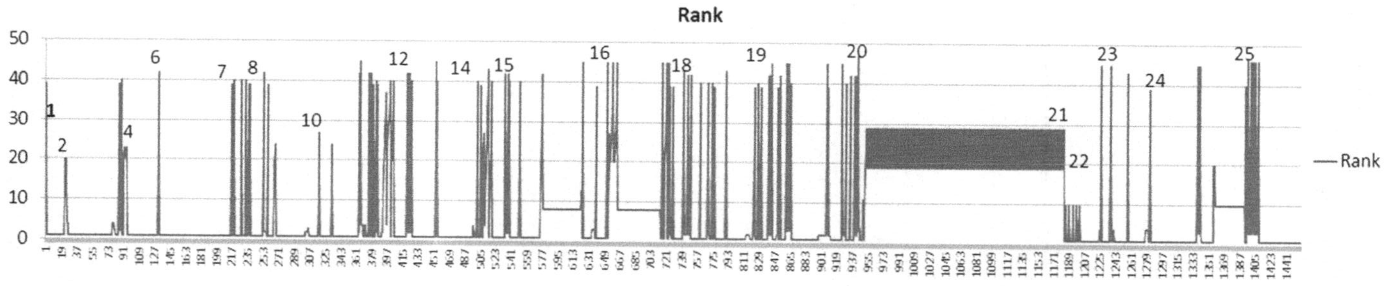


*Figure 18.8f* Illustration of the level of decision making each agent made during the scenario. The number on top represents the agents (1 to 25) with the level of decision making noted on the vertical axis

*Table 18.6* Scenario 3 – Regression Analysis showing $R^2$ (at the 95% confidence interval) for the correlation of features the agent selected with ranked feature values F1, F2 and F3, and the $R^2$ for the correlation of the optimization value of each feature 'r' with the weights of features 10 (trails), 30 (vegetation), 40 (roads), 60 (farm land), 70 (hazards), 80 (building shadows) and 90 (unclassified). The table also includes corresponding RMS values

| | | | |
|---|---|---|---|
| Feature 1 F1 $R^2$ | 0.92 | RMS | 13.3 |
| Feature 2 F2 $R^2$ | 0.65 | RMS | 28.8 |
| Feature 3 F3 $R^2$ | 0.76 | RMS | 23.7 |
| $R^2$ 10 | 0.63 | RMS | 98.4 |
| $R^2$ 30 | 0.70 | RMS | 104 |
| $R^2$ 40 | 0.69 | RMS | 342 |
| $R^2$ 60 | 0.61 | RMS | 201 |
| $R^2$ 70 | 0.64 | RMS | 67.4 |
| $R^2$ 80 | 0.84 | RMS | 55.8 |
| $R^2$ 90 | 1.0 | RMS | 0.0 |

## 18.7 Model's evaluation

The evaluation process of agent-based models is comprised of calibration, verification and validation (Manson 2003). Manson et al. (2012) describe *calibration* as the adjustment of model parameters to fit with desired results; *verification* as a way to ascertain whether the model runs according to design and intention, i.e. by running the model with data to determine if output data are in line with expectations; and, finally, *validation* as the process in which the model outputs are compared with real-world data. Although validation is an important part of an agent-based model's evaluation, it is also one of its greatest challenges (Crooks and Heppenstall 2012). In the case of the present study, validation would require real-world information which was not available for this project. Therefore, the evaluation of the model focuses on verifying the process of a calibrated ABM and how well the output data met with expectations.

The two parts of the model (flooding and evacuation behaviour) were evaluated separately. The flooding model was calibrated to simulate the flooding that took place in Xai-Xai in 2000 derived from the book *Mozambique and the Great Flood of 2000* by Christie and Hanlon (2001). In order to replicate this behaviour, the amount of rainfall used for the model scenarios (presented in section 18.6) was set between 0.5 to 1 mm per minute (6 cm/hr) in the medium to severe range.

By and large, the manner in which the flooding occurred in the model was consistent with the flooding experienced in the area, as outlined in the three scenarios presented in section 18.6. The flooding model could have been more realistic if a higher resolution and more accurate DEM had been used. The flooding model also lacked other considerations, such as hydraulics and mitigation measures, which would allow for a more realistic outcome. Still, the model did provide a realistic flooding environment for the agents to interact with, which satisfied the objectives of the project.

The evacuation behaviour was evaluated using a combination of three different methods: visual assessment of the agent's route, statistical analysis, and assessment of the output from the neural network. Those were used to evaluate the decision-making process for each scenario presented in section 18.6.

For conducting the visual assessment, the routes produced by agents in the model were reproduced using environmental data in ArcGIS. This was done by converting the coordinates of the routes and overlaying them on GeoEye1 0.5m imagery using ArcGIS. The routes taken by agents in the Repast Simphony Java model were then visually compared with the same routes produced in ArcGIS.

The assessment of the routes selected by the agents were performed for all three scenarios (see Plates 15, 16 and 17) and presented significant similarities to the ArcGIS routes. Thus, the results of visual assessment suggest that agents make decisions about the terrain in a similar manner to a person. The fastest features to travel over, such as roads, are taken when available, and cells with structures and water are avoided. All the scenarios verified that the process was meeting expectations as agents made decisions similar to humans as they moved towards the safe area. An example is the agent's behaviour in scenario 2, which demonstrates characteristics of human decision-making when the agent found and crossed the bridge as it travelled to the safe area, although some decisions were poor ones, in part due to a lack of information about the area and cells that could not be sensed. In addition, the resampling process degraded features that contributed to agents getting trapped.

The statistical analysis evaluation was conducted using linear regression by calculating the $R^2$ and RMS statistic. $R^2$ provided information to evaluate the correlation between feature outputs and features the agent selected, while RMS or root mean square error provided an estimate on how well scenario results of the feature weights compared with the optimized results.

The statistical analysis yielded interesting results that provide us with an insight into human decision making using a hybrid model of neural networks. When the neural network was working well, $R^2$ and RMS could be both high and low. An $R^2$ value that was closer to 1 suggested the decision-making process was operating well, whereas a low (closer to 0) statistic suggested adapting and learning was taking place. The reverse was seen for RMS and therefore both values were evaluated together. Scenarios that required lots of adapting, often accomplished by random moves, showed a lower correlation of optimal weights 'r' with the actual weights. Scenarios that did not require as much adapting and less random moves showed a higher value. Correlation of the features indicated F1 was the preferred selection, with F2 and F3 being a close second and third option. In scenarios where little adaptation was required and few random selections were made, the $R^2$ value was 1 and RMS 0.

In each scenario the operation of the neural network was assessed (presented in section 18.6) using Charts A to D, in Figures 18.6, 18.7 and 18.8. Chart C shows the output of the neural network, being the three features (variables F1, F2 and F3) that were provided to each agent for selection at each tick interval, and the actual selection (variable F) that was made. Chart C shows that the decision-making process often resembled the human characteristic of 'panic'. This occurs when an agent is trapped by flood water or by buildings and the feature selection line (F) 'zig-zags' between various selections. For example, in Scenario 1 (see Figure 18.6c), between ticks 1 and 37, the feature selection line jumps between different features as the agent moved through the rural area as it encountered various land cover. In all scenarios, agents experienced areas that had fewer problems in decision making, thus having a 'calm' appearance as a result of less feature selection activity (see Figure 18.7c, between ticks 241 and 301). In this example, the period of calm occurred after a series of random selections and weight adjustment as the agent searched for and found the road to cross the bridge over the Limpopo River. Scenario 3, provides additional information on the level of decision making (Figure 18.8e) encountered by each agent. If no problems were encountered by an agent the first level would be sufficient. However, if an agent encountered severe difficulty in making an acceptable decision, up to 47 decision levels could be tried. The chart indicated each agent experienced a unique level of difficulty in making decisions.

Travel time by an agent to the safe area, using an average walking rate of 4km/hr (according to Cavagna et al. 1976), was considered reasonable. The distance from lower Xai-Xai to the safe area (in the upper area) is approximately 3 km. If an agent did not get trapped, the travel time was approximately 60 tick minutes to the safe area and approximately 110 tick minutes when commencing in the rural area, about 4 km from the safe area. Travel time in each scenario took into consideration what type of terrain the agent was travelling over as the rate of travel over roads was approximately 2½ times faster than over rural areas.

## 18.8 Conclusions

The model was successful in simulating a fast-flooding area, thus providing a framework to study evacuation behaviour as well as developing a neural network to simulate an agent's cognitive ability to sense, learn and adapt when travelling over a landscape during a flooding episode.

The evaluation of the model (presented in section 18.7) suggested that agents did demonstrate cognitive abilities and human characteristics, but it was not possible to perform a full validation of the model due to lack of data. It is well known that validating human behaviour is very difficult to achieve and, in fact, it is considered as one of the main challenges of ABMs, as highlighted by Kennedy (2012). Thus, further calibrating the learning rate to improve the decision making and progress of an agent during a flooding episode may not necessarily improve the results as not all people learn at the same rate. In other words, without obtaining data on real human behaviour further calibration of the model is likely to be a fruitless effort to improve decision-making results.

The evaluation process demonstrated there is potential for using this design of hybrid model of neural networks on further research about the evacuation behaviour of people in the fast-flooding area of Xai-Xai, which can possibly be extended to other geographic areas where models integrate human decision making and land use/land cover. The architecture in Repast Simphony Java is flexible and adaptable, such that further steps could be taken to improve the model with more land cover/land use features, improved DEM, a larger knowledge base of information, other decision-making processes, and increased agent movement. Further development of the model could also involve the participation of local groups providing essential data for model calibration and validation.

While there are challenges in agent-based modeling when integrating human decision making with the complex phenomenon of fast flood and changing land use/land cover, this study highlights the potential of using a hybrid model of neural networks for decision making in ABMs.

The model presented here is the first steps towards a tool which can effectively inform decision making on evacuation procedures in the Limpopo River Basin Area in Mozambique. By allowing for a better understanding of how individuals behave when evacuating the area during fast-flooding events, the model can inform practitioners in the development of an effective disaster management plan. The model can be particularly helpful in informing the stages of preparedness and response. Once the model is fully developed and validated, different evacuation routes can be tested and the safer routes identified. This can inform the development of a safe evacuation procedure as part of the preparedness plan for future events to be implemented by local authorities together with the local population. Similarly, based on the evacuation behaviour of individuals, critical areas for rescue can also be identified that can be part of the response plan for the disaster. Priority areas for rescue can be identified that allow emergency services to have an effective action plan. This knowledge can hopefully be transferable to other areas of fast-flooding and inform evacuation procedures and policies elsewhere.

## References

Appleton, R. (2013) '"Mozambique Flooding Kills 36, Displaces Thousands": Earth Changes and Poles Shift', http://poleshift.ning.com/profiles/blogs/mozambique-flooding-kills-36-displaces-thousands (accessed 5 February 2014).

Bakillah, M., Dominguez, J.A., Zipf, A., Liang, S.H.L. and Mostafavi (2012) 'Multi-agent Evacuation Simulation Data Model with Social Considerations for Disaster Management Context', in S. Zlatanova, R. Peters, A. Dilo and H. Scholten (Eds.), *Intelligent Systems for Crisis Management* (pp. 3–16). Dordrecht: Springer.

BBC (2000) 'Mozambique: How Disaster Unfolded', *BBC News*, Thursday, 24 February 2000.

Brouwers, L. and Verhagen, H. (2003) 'Applying the Consumat Model to Flood Management Policies', *Agent-Based Simulation*, April 2003, Agent-Based Simulation, Montpellier, France.

Cavagna, G.A., Thys, H. and Zamboni, A. (1976) 'The Sources of External Work in the Level Walking and Running', Istituto di Fisiologia Umana, Universital di Milano and Centro di Studio per la Fisiologia del Lavoro Muscolare del CNR, Milan, Italy.

Cheung, V. and Cannons, K. (2002) 'An Introduction to Neural Networks', Signal and Data Compression Laboratory Electrical and Computer Engineering, University of Manitoba Winnipeg, Manitoba, Canada.

Christie, F. and Hanlon, J. (2001) *Mozambique and the Great Flood of 2000*. Cumbria, UK: Long House Publishing Services.

Collier, N. and North, M. (2013) *Repast Java Getting Started*, http://repast.sourceforge.net/docs/ (accessed 22 April 2014).
Colton, S. (2012) Artificial Intelligence (PowerPoint presentation), Imperial College, London, UK, www.doc.ic.ac.uk/~sgc/teaching/pre2012/v231/lecture12.ppt (accessed 9 September 2012).
Colton, S. and Gervás, P. (2003) 'Creativity in Arts and Science', *The Interdisciplinary Journal of Artificial Intelligence and the Simulation of Behaviour*, 1(4), December 2003, www.aisb.org.uk/publications/aisbj/issues/AISBJ%201(4).pdf (accessed 9 September 2012).
Crooks, A.T. and Heppenstall, A.J. (2012) 'Introduction to Agent-based Modelling', in A.J. Heppenstall, A.T. Crooks, L.M. See and M. Batty (Eds.), *Agent-Based Models of Geographical Systems* (pp. 85–105) Dordrecht: Springer.
Dawson, R.J., Peppe, R. and Wang, M. (2011) 'An Agent-based Model for Risk-based Flood Incident Management', *Natural Hazards*, 59(1), 167–189.
ERDAS Imagine, Expert Classifier Overview (Software), www.gis.usu.edu/unix/imagine/ExpertClassifier.pdf (accessed 9 September 2012). www.erdas.com/products/ERDASIMAGINE/ERDASIMAGINE/Details.aspx (accessed 22 August 2012).
Fox, J. (2008) *Applied Regression Analysis and Generalized Linear Models*. Thousand Oaks, CA: Sage.
Gonzalez, C., Lerch, J. and Lebiere, C. (2003) 'Instance-Based Learning in Dynamic Decision Making', *Cognitive Science*, 27, 591–635.
Kasabov, N.K. (1996) *Foundations of Neural Networks, Fuzzy Systems, and Knowledge Engineering*. Cambridge, MA: MIT Press.
Kennedy, W.G. (2012) 'Modelling Human Behaviour in Agent-Based Models', in A.J. Heppenstall, A.T. Crooks, L.M. See and M. Batty (Eds.), *Agent-Based Models of Geographical Systems* (pp. 166–179). Dordrecht: Springer.
Koehne, A. (2007) 'Gaza Mozambique', http://commons.wikimedia.org/wiki/File:Mozambique_Gaza_destaque.png (accessed 2 May 2002). Reproduced under GNU Free Documentation License.
Kriesel, D. (2005) 'A Brief Introduction to Neural Networks', www.dkriesel.com (accessed 2 May 2014).
Li, H. (2011) *Learning to Rank for Information Retrieval and Natural Language Processing*. San Rafael, CA: Morgan and Claypool Publishers.
Lin, M. and Manocha, D. (2010) *Simulation Technologies for Evacuation Planning and Disaster Response*, Research Brief, Institute of Homeland Security Solutions, administered by RTI International.
Lumbroso, D., Davison, M. and Tagg, A. (2013) 'Agent-based methods in emergency management', www.hrwallingford.com/projects/agent-based-methods-in-emergency-management (accessed 25 June 2015).
Manson, S.M. (2003) 'Validation and Verification of Multi-Agent Models for Ecosystem Management', in M. Janssen (Ed.), *Complexity and Ecosystem Management: The Theory and Practice of Multi-Agent Approaches* (pp. 63–74). Northampton: Edward Elgar.
Manson, S.M., Sun, S. and Dudley Bonsal, D. (2012) 'Agent-Based Modeling and Complexity', in A.J. Heppenstall, A.T. Crooks, L.M. See and M. Batty (Eds.), *Agent-Based Models of Geographical Systems* (pp. 124–139). Dordrecht: Springer.
Mathworks (2012) The MathWorks Inc., www.matworks.com/help/toolbox/curvefit/bq_6zzm.html (accessed 28 August 2012).

Padgham, L. (2015) *Urban Decision-making and Complex Systems*, Global Cities Research Institute, www.global-cities.info (accessed 26 June 2015).

Rahman, A., Mahmood, A. and Schneider, E. (2008) 'Using Agent-based Simulation of Human Behavior to Reduce Evacuation Time', in T. Bui, T. Ho and Q. Ha (Eds), *Intelligent Agents and Multi-Agent Systems* (pp. 357–369). Berlin and Heidelberg: Springer.

Ripley, B.D. (1996) *Pattern Recognition and Neural Networks*. Cambridge: Cambridge University Press.

Su, Z., Jiang, J., Liang, C. and Zhang, G. (2011) 'Path Selection in Disaster Response Management Based on Q-learning', *International Journal of Automation and Computing*, February.

Takahashi, T. (2007) 'Agent-based Disaster Simulation Evaluation and its Probability Model Interpretation', in B. Van de Walle, P. Burghardt and C. Nieuwenhuis (Eds.), *Proceedings ISCRAM 2007*. Delft, The Netherlands.

Turchetti, C. (2004) *Stochastic Models of Neural Networks*. Amsterdam: IOS Press.

UNITAR (2013) *Flood Waters over Chokwe, Guika, Bilene, and Xai-Xai Districts, Gaza Province, Mozambique*. Châtelaine, France: United Nations Institute for Training and Research, www.unitar.org.

# 19 A computational framework for mitigating land degradation

## A synergy of knowledge from physical geography and geoinformatics

*Tal Svoray*

### 19.1 Introduction

Deterioration in soil fertility caused through soil misuse by humans is usually termed as land degradation (e.g. Lal and Stewart, 1990). Among the various factors that affect land degradation worldwide, water-erosion and runoff flow are the most destructive processes (De Santisteban et al., 2006; Amundson et al., 2015). These hydrological processes lead to the formation of gullies, which in turn cause soil loss and exposure of infertile layers (Poesen et al., 2003), water ponding and increased salinity in lower areas (Chu and Marino, 2005), sediment logging downslope and buildup of fine material off-site (Mandal et al., 2005). Studies show that soil loss processes in the US (Trimble and Crosson, 2000) and Europe (Boardman, 1998) cause serious destruction that current research cannot fully quantify and predict. Water ponding and salinity can decrease crop yield substantially (McFarlane and Williamson, 2002; Bahceci et al., 2006), leading to economic losses and exclusion of large areas from the circle of cultivation (Manjunatha et al., 2004). Sediment logging can cause extensive damage by covering of seeds and even small seedlings, which eliminates the use of the damaged part of the field for the current season (Gyssels et al., 2002; Flugel et al., 2003). An additional problem is the damage caused by the sediments to infrastructure, including the cover of roads, sewage systems and bridges, and the decrease in the capacity of water reservoirs (Van Rompaey et al., 2003). The various aspects of land degradation are of worldwide concern, while, in many agricultural catchments, intensive human activity has increased pressure on land resources to the point of no return (Holy, 1980; Boardman, 2006). The influence of land degradation on the economies of developed countries is far-reaching (Thibier and Wagner, 2002), but in underdeveloped countries, the threat to human lives is even more critical, as large populations suffer malnutrition, hunger and even death as a result of diminished crop production (Smith, 1998).

The impairment caused by land degradation processes to agricultural fields requires a solution in the magnitude provided by state-of-the-art technology to guide decision makers, practitioners and farmers on how to, where and when to tackle the possible threat of soil movement by water. Identification of areas prone to gullying can help in prioritizing soil conservation practices that can be extremely expensive when management of an entire catchment is the problem at hand. The main problem is that in relation to the wide areas of agricultural fields, areas prone to gully initiation are very small, and their detection can be like searching for a needle in a haystack or in other words, they are what is known in information systems engineering as the 'rare cases problem' (Svoray et al., 2012).

The field of physical geography covers the study of earth surface processes that determine soil and water movement in agricultural catchments and can help in predicting the gully head movement. Remote sensing can serve as an excellent tool to derive data on field conditions in the catchment and regional scale. Thus, the combination of these two disciplines can assist in combating soil erosion processes and land degradation through the development of spatially and temporally explicit mapping systems of areas prone to soil degradation. In addition, the application of geographical information systems (GIS), coupled with expert-based and decision supporting systems, could act as an efficient tool to manage agricultural catchments in a sustainable manner.

As land degradation due to soil erosion by water is a widespread phenomenon, previous studies of identifying risks and assessing erosion damage were applied in numerous countries and for diverse crop areas. Table 19.1 shows that most of the studies have used various empirically-based techniques, including multivariate statistical models, Monte Carlo simulations and decision trees. Some of these studies apparently achieved good results, yet most of them have not made the further step needed for using the risk maps to recommend actions. Furthermore, most of the systems developed are heuristic in nature and although their results can be accurate, they do not use knowledge of the mechanisms that drive the erosion process itself, and thus cannot be easily generalized over wide regions.

On the other hand, decision support systems (DSS) that were developed to provide planners and practitioners with recommendations on the appropriate agrotechnology or conservation practices were usually limited to improve crop yields or reduce watering and fertilization costs and were less oriented towards an attempt to reduce soil erosion and soil loss. Table 19.2 shows a selection of recent research studies that developed spatial decision support systems (SDSS) for agricultural purposes. The table clearly shows that the focus of these studies is on agricultural productivity and development of infrastructure such as dams, while less effort has been dedicated to develop SDSSs to prevent soil erosion by water and land degradation.

The *aim* of this chapter is therefore to suggest a computational framework that allows us: (a) to study the main erosion process – gully development – using spatially and temporally explicit modelling; (b) to identify areas under threat using the knowledge acquired in the previous stage; and (c) to use the risk layer as a basis for a spatial decision support system for soil conservation. The method is demonstrated on wheat fields in a semiarid zone in Israel, an area suffering from intensive water erosion processes.

*Table 19.1* Comparative table for risk assessment of soil loss processes in agricultural catchments

| *Source* | *Method* | *Site* | *Tool* | *Crop type* |
|---|---|---|---|---|
| Le Gouée et al., 2010 | Large-scale assessment and mapping soil erosion hazards based on a decision tree coupled with a GIS. Parameters used are soil erodibility, agricultural practices, topography and rainfall characteristics. | France | GIS | Many (large scale model) |
| Maeda et al., 2010 | Using a landscape dynamic simulation model, an erosion model and synthetic precipitation datasets generated through a Monte Carlo simulation, created a future agricultural expansion and climate change scenarios that evaluate their potential impacts on soil erosion. | Kenya | GIS | Maize, beans, peas, potatoes, cabbage, tomato, cassava and banana |
| Lucà et al., 2011 | GIS-aided procedures for the evaluation of gullying susceptibility on a statistical basis, using four statistical models, differed in types of model, data sampling and variables. | Italy | GIS | Mainly field crops and olive groves |
| Shruthi et al., 2011 | Object-oriented image analysis (OOA) to extract gully erosion features from satellite imagery, using a combination of topographic, spectral, shape (geometric) and contextual information. | Morocco | RS | Rain-fed wheat, barley and maize |
| Navas et al., 2013 | Creating a spatial distribution map of estimated soil erosion and deposition rates, using estimates of soil redistribution rates derived from $^{137}Cs$ measurements, and data such as relief, topography, hydrological network and DEM. | Spain | GIS | Winter barley |
| Conoscenti et al., 2014 | GIS and multivariate statistical analysis (with variables such as bedrock lithology, land use, slope aspect, distance from roads, elevation, WI, etc.) including logistic regression, to assess gully erosion susceptibility. | Italy | GIS | Permanent crops |

*Table 19.2* Comparative table for spatial decision support systems to reduce soil loss in agricultural catchments

| *Source* | *Method* | *Site* | *Tool* | *Crop type* |
|---|---|---|---|---|
| Lorenz et al., 2013 | An approach for transferring crop rotations and related soil management strategies using statistical and spatially explicit data. Applied with a spatial decision support tool GISCAME, to assess and visualize the impact of alternative agricultural land-use strategies on soil erosion risk and ecosystem services provision. | Germany | GIS | Cereals, corn, legumes, oilseeds, root crops |
| Patel et al., 2015 | By using hydrological, geological and geomorphological data, authors identified areas that experience extreme runoff and soil erosion. Using a DSS based on RS data and GIS, the watersheds were ranked and a priority was given to build dams to prevent runoff and soil erosion. | India | GIS / RS | Large-scale model |
| Macary et al., 2014 | An MCDA (multiple criteria decision analysis) model based on ELECTRE TRI-C and GIS for erosion risk assessment in agricultural areas. Using hydrological data, land use, DEM and slopes, road networks, etc., and classification of this data to varying levels of risk by cross-referencing, the model produces a map of levels of risk for the farming parcels, to help the farmers in decision-making. | France | GIS | Mainly maize and wheat |
| Jaiswal et al., 2015 | A MCDA tool based on AHP, for prioritizing vulnerability levels of drainage basins, using hazard parameters affecting soil erosion, such as topography, soil, geomorphology, conservation and management practices and more. | India | GIS/RS | Large-scale model |
| Shahbazi and Jafarzadeh, 2010 | Using MicroLEIS, DSS system for agro-ecological land evaluations, based on soil data, climate, crops, etc. By implementing the recommendations of the system, a sustainable soil management can be achieved, improving soil quality through the use of soil-specific practices, adapted to local soil, terrain and climatic conditions. | Iran | GIS | Wheat, alfalfa, sugar beet, potato and maize |
| Frank et al., 2014 | Making use of the ecosystem services concept in regional planning. Using GISCAME system based on variables such as soil, DEM, Precipitation, land use, road systems, etc., Possible scenarios to reduce water erosion are examined, and recommendations for regional planning are given. | Germany | GIS | Mostly short crop rotations, such as 'Rape–Wheat–Rye' and 'Rape–Wheat–Corn silage–Rye' |

## 19.2 Methodology

### *19.2.1 Overview*

The computational framework described in this chapter is aimed at supplying the practitioner with a means to decide on the most suitable soil conservation practices based on field conditions and history of cultivation (Svoray et al., 2015), in an optimized, efficient and economic manner. The suggested framework is based on three successive steps to be taken by the expert: (1) knowledge on erosion processes in agricultural catchments is extracted from physically-based and space-time dynamic models; (2) this knowledge is translated to decision rules that can be applied to develop an information layer determining the risk level of gully development in the catchment in a spatially and temporally explicit manner; and, finally, (3) the risk layer from step 2 acts as a basis for an SDSS to prioritize necessary actions to reduce erosion and soil loss damage (Figure 19.1).

### *19.2.2 Extraction of knowledge from dynamic soil erosion models*

Modeling of soil erosion processes can be used as an important means of acquiring a better understanding of how to apply conservation practices of

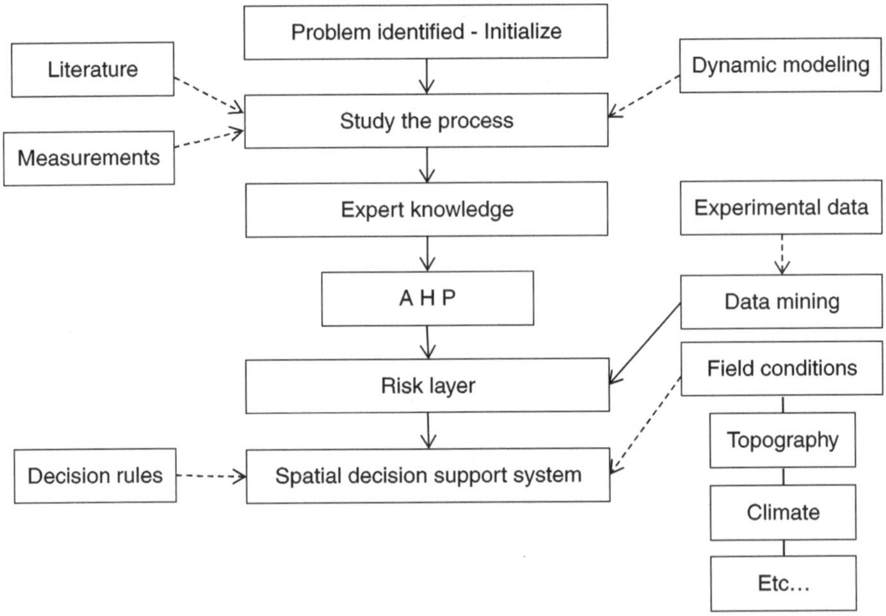


*Figure 19.1* A flowchart of the proposed framework for mitigating land degradation. Actions start when a problem is identified, the governing processes are studied with physically based models or similar. In the next step the knowledge acquired is translated into computer language using AHP (Analytic Hierarchy Process). Next, a risk layer is compiled and used as an input to a decision support system

natural, agricultural and built environments. Catchment-scale erosion modeling is particularly sought, since it allows efficient soil conservation planning by providing spatial data over large areas that may be used to decrease erosion-related problems. The potential of such models for environmental management and planning organizations is clear, but most of the state-of-the-art soil erosion models are difficult to apply over large areas due to intensive labor and detailed data requirements. Several models can be demonstrated here as a useful tool to extract the necessary information on space-time dynamics in gully development. To exemplify, the use of FuDSEM (Fuzzy Dynamic Soil Erosion Modeling; Cohen et al., 2008) is presented here. This model predicts soil erosion on a daily basis using a rainfall database and a set of topographic and environmental data. The model simulates the infiltration excess runoff mechanism (Hortonian) in the sub-catchment scale, highlighting the temporal dynamics of this process. The model divides the erosion process into a sequence of four sub-routines including: (i) simulating soil water profile and antecedent conditions of soil moisture; (ii) simulating runoff generation and water flow; (iii) calculating transport capacity function; and, finally, (iv) predicting soil erosion. This model uses fuzzy algebra while each sub-routine is calculated by an individual joint membership function, which combines the relevant parameters. These include: explicit calculation of soil moisture potential; calculation of runoff potential by considering the soil moisture profile; and accumulation of runoff in space based on digital elevation model (DEM) data. The transport capacity of the runoff is calculated based on the accumulated and *in situ* runoff potential; soil erosion potential is calculated based on the transport capacity potential; and the model proceeds to the next day in the rainfall database until reaching the last day in the wet season. The functions and weights used in FuDSEM are the outcome of generalized interpretation of common knowledge of erosion processes. Unlike standard, physically-based models, the weights do not intend to represent an accurate quantitative relationship between the parameters, but to provide a general interpretation of the processes as foreseen by the modeler. This is adequate, since the model predicts the potential of the parameters which intend to represent its relative spatial and temporal distribution and not to provide a quantitative prediction of erosion yield. Therefore, the relationships between the parameters (i.e. functions and weights) are not directly linked to a specific study, but were chosen by fusion of the relevant literature and expert knowledge. Figure 19.2 shows the fuzzy rule-based model outcome with predicted variations due to the local topographic and environmental conditions.

Most importantly with regards to the proposed framework, the FuDSEM model allows us to identify areas under stronger erosion threats and to relate the observed erosion rate to underlying mechanisms and topographic and environmental conditions. This efficiently serves the experts in the next step of the task, namely the creation of a risk assessment map for gully development in the catchment.

### *19.2.3 Risk assessment*

Once the knowledge has been extracted from the model, the next step is to identify areas under risk. The proposed framework creates the risk assessment map through

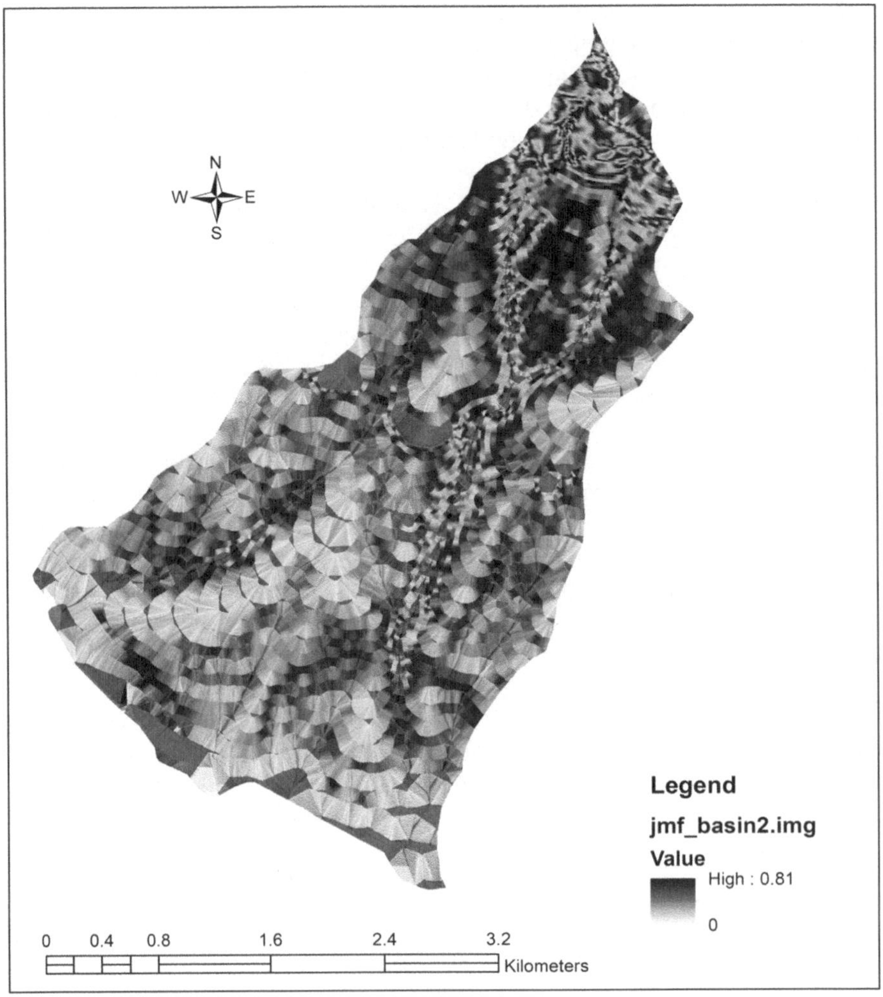


*Figure 19.2* Values of soil loss potential for the Yehezkel catchment in Northern Israel. The values are predicted by the fuzzy rule-based model. The output value range is between 0, the lowest potential for soil loss, and 1, the highest potential

three consecutive steps: (i) gathering field and remotely-sensed data to simulate the climatic, environmental and human-induced conditions in the catchment; (ii) use of pairwise comparison matrices to assign the criteria from step (i) with weights; and (iii) use of weighted linear combination (WLC) to compile an erosion risk map. In the example provided here, the experts recommended a database of ten criteria to represent the most important factors influencing erosion processes in the studied catchment. All ten criteria were mapped as GIS layers, including Topographic slope, aspect and upslope contributing area, which were calculated from a 2×2-$m^2$

spatial resolution DEM (Svoray and Markovitch, 2009). Vegetation, rock and soil cover of the study area were mapped from a sub-meter resolution orthophoto using supervised Maximum Likelihood Classification (Svoray and Ben-Said, 2010). Rainfall intensity was mapped using meteorological radar data and tillage direction was digitized for all fields of the catchment. A land-use map with spatial resolution of 2×2-m$^2$ was compiled, based on data from the National GIS of Israel, made by the Survey of Israel. The data were verified using updated data from visual interpretation and field visits. The land-use classes in the entire catchment included mainly orchards as well as field crops, and at a much lower coverage percentage – woodlands of two sub-classes, dense and sparse; and grassland. Unpaved roads were manually digitized from the 2006 orthophoto, based on visual interpretation. Based on the experts' recommendations, the roads layer was divided into two criteria: (i) roads-as-runoff-contributors, enhancing the effect of roads on the contributing area downslope; and (ii) roads-as-barriers to water ponding and sediment deposition, enhancing the effect of roads as barriers to water and sediment flow from upslope. The two layers were coded with classes based on the distance from the road.

The pairwise comparison method was used to assign the weight of influence of the ten criteria on soil erosion in the catchment with the help of four experts (two drainage engineers and two soil pedologists). The experts compared every possible pair of criteria in the 10×10 criteria matrix using the Saaty influence scale, ranging between 1 to 9 (Saaty, 1977). This was carried out as follows: the expert compared the criterion in the first row against the criterion in the first column. The score he assigned to this comparison depended on the level of relative influence on the relevant process of the criterion in the row against that in the column. The value 1 stands for equal importance between the two indices, the value 9 stands for the criterion in the row being extremely more important than that in the column, and the value 1/9 stands for the criterion in the column being extremely more important than that in the row. The expert then repeats the process for each cell in this matrix. Each expert filled out the pairwise comparison matrix separately, according to his understanding. Once the weights for each criterion were obtained, they were implemented over the entire study area, using WLC. To normalize the criteria, they were divided into categories by the experts. The application of WLC resulted in a raster layer storing the score for the occurrence of, or risk of, soil loss, in each cell. The validity of the risk layers was tested against field data that included 88 plots of severe gullying which were physically observed. Figure 19.3 shows the risk map that was extracted as an outcome of this process.

### *19.2.4 Creating a spatial decision support system (SDSS)*

The risk map, although a useful tool to identify areas under risk, does not provide a recommendation of what to do in the areas under threat in order to reduce land degradation. For this purpose one needs a software tool that will provide the necessary tools to decide on the best suitable soil conservation practice for each field or part of it based on the local conditions. Thus the next step of the proposed framework is to create a decision support system which could provide the solution for this

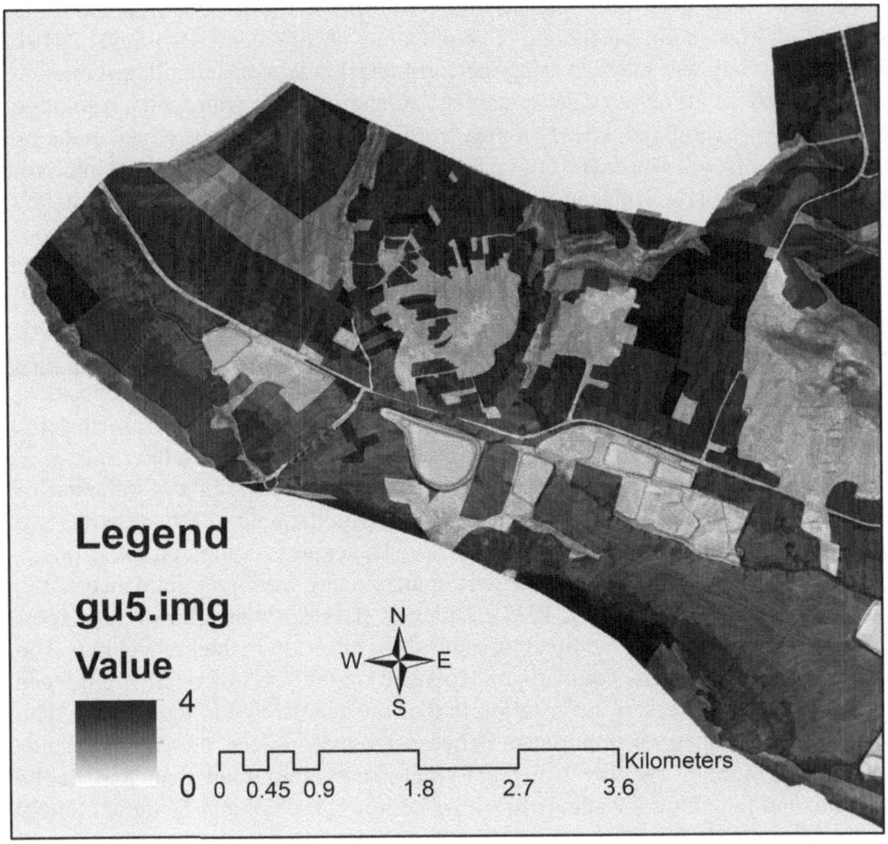


*Figure 19.3* Risk-level values for soil loss in part of the Harod catchment in Northern Israel. Risk layer is predicted using weighted linear combination while the weights are assigned based on the pairwise comparison method

need. This system is an interactive tool which combines useful information from a collection of data with expert knowledge, to detect and solve problems and to make a more established decision. An SDSS is a decision support system that is especially suitable for dealing with different kinds of problems in a spatially and temporally explicit manner, as it includes functions of GIS and remote sensing. SDSS is not only economic from a planning standpoint, but also provides a reliable method to efficiently use financial resources, since it can be used both to target erosive hotspots and select appropriate soil conservation practices.

The development of an SDSS to recommend soil conservation actions is of central importance for the entire framework presented here. The risk map is an important input for the SDSS. The system could take several strategies, but as an example it could work based on the following six steps (Figure 19.4): (a) The system identifies areas with no risk, selects them using an Structured Query

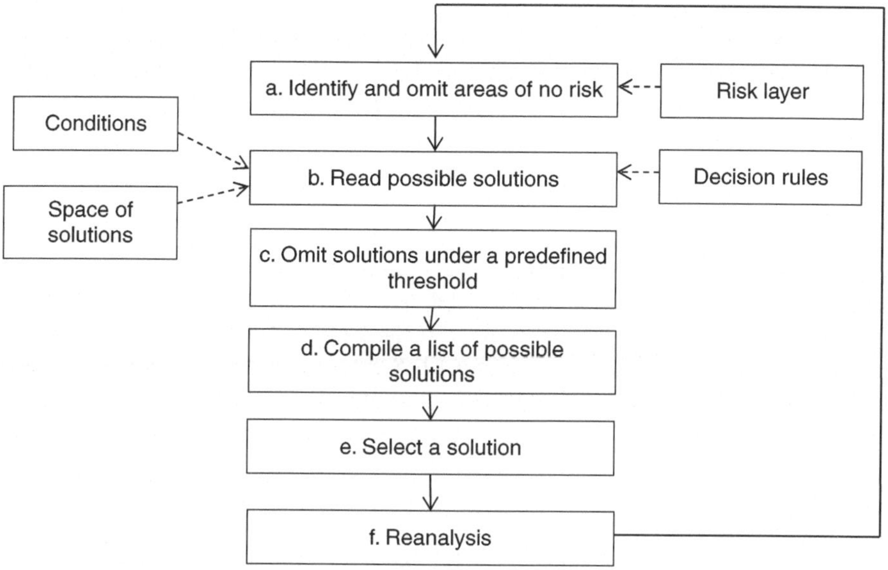


*Figure 19.4* A flowchart describing the six stages required to apply the SDSS to a catchment under risk of land degradation through water erosion processes

Language (SQL) operator and omits these areas from the analysis. As the areas of no risk are relatively large, they save substantial computation requirements and time from the system. (b) The system reads the possible solutions from the solutions space and rates them based on a set of decision rules and the conditions read from the given cell treated. (c) Based on the score attained for each solution for each cell, the system omits solutions under a predefined threshold. These solutions are not provided to the user as their efficiency is indisputably low. (d) The system lists the possible solutions with scores higher than the threshold and provides the user with a set of possible solutions rated according to the scores attained previously. (e) The user either uses the solution given the highest score (naive user) or makes up his mind based on his own experience or limitations that are not expressed in the system (i.e. this specific farmer will never apply a no-tillage approach due to personal reasons). As a sixth stage (f) for further development of the system, the user can feed the system with a layer that describes the solution assigned to each field or sub-field and the implication of each solution to the risk map. In that sense, the system becomes dynamic and can support a general trend or feedback of continuous improvement of the catchment state in terms of land degradation.

## 19.3 Summary

The computational framework presented here provides the necessary basis for a more sustainable use of the land in agricultural fields extensively cultivated in areas suffering from a strong threat of erosion. The methodology suggested is based on a mechanistic understanding of erosion and deposition processes and an educated use of the synergy between spatial databases and expert-based and decision support systems to identify areas under risk. Perhaps the most important and relevant to this book is that the methodology actually reflects the knowledge that geographers acquire through their academic studies.

## Acknowledgements

I thank my students at Ben-Gurion University who executed into computer codes some of the ideas presented in this chapter. Continuous support by the Soil Sciences Branch and the Chief Scientist of the Israeli Ministry of Agriculture is highly acknowledged. I thank Professor Eli Stern and Professor Graham Clarke, who invited me to write this chapter as an output to a talk I gave at the IGU-AGC Meeting in Bangkok.

## References

Amundson, R., Berhe, A.A., Hopmans, J.W., Olson, C., Sztein, A.E. and Sparks, D.L. (2015) Soil and human security in the 21st century, *Science 348*, 1261071.

Bahceci, I., Dink, N., Tari, A.F., Agar, A.I. and Sonmez, B. (2006) Water and salt balance studies, using SaltMod, to improve subsurface drainage design in the Konya-Cumra Plain, Turkey, *Agricultural Water Management 85*, 261–271.

Boardman, J. (1998) An average soil erosion rate for Europe, myth or reality? *Journal of Soil and Water Conservation 53*, 46–50.

Boardman, J. (2006) Soil erosion science, reflections on the limitations of current approaches, *Catena 68*, 73–86.

Chu, X.F. and Marino, M.A. (2005) Determination of ponding condition and infiltration into layered soils under unsteady rainfall, *Journal of Hydrology 313*, 195–207.

Cohen, S., Svoray, T., Laronne, J.B. and Alexandrov, Y. (2008). Fuzzy-based dynamic soil erosion model (FuDSEM): modelling approach and preliminary evaluation. *Journal of Hydrology 356*(1), 185–198.

Conoscenti, C., Angileri, S., Cappadonia, C., Rotigliano, E., Agnesi, V. and Marker, M. (2014) Gully erosion susceptibility assessment by means of GIS-based logistic regression: a case of Sicily (Italy), *Geomorphology 204*, 399–411.

De Santisteban, L.M., Casali, J. and Lopez, J.J. (2006) Assessing soil erosion rates in cultivated areas of Navarre (Spain), *Earth Surface Processes and Landforms 31*, 487–506.

Flugel, W.A., Marker, M., Moretti, S., Rodolfi, G. and Sidrochuk, A. (2003) Integrating geographical information systems, remote sensing, ground truthing and modelling approaches for regional erosion classification of semi-arid catchments in South Africa, *Hydrological Processes 17*, 929–942.

Frank, S., Fürst, C., Witt, A., Koschke, L. and Makeschin, F. (2014) Making use of the ecosystem services concept in regional planning – trade-offs from reducing water erosion, *Landscape Ecology 29*, 1377–1391.

Gyssels, G., Poesen, J., Nachtergaele, J. and Govers, G. (2002) The impact of sowing density of small grains on rill and ephemeral gully erosion in concentrated flow zones, *Soil and Tillage Research 64*, 189–201.

Holy, M. (1980) *Erosion and Environment*. Pergamon Press, Oxford.

Jaiswal, R.K., Ghosh, N.C., Galkate, R.V. and Thomas, T. (2015) Multi Criteria Decision Analysis (MCDA) for watershed prioritization, *Aquatic Procedia 4*, 1553–1560.

Lal, R. and Stewart, B.A. (1990) *Advances in Soil Science, Soil Degradation*. Springer, New York.

Le Gouée, P., Delahaye, D., Bermond, M., Marie, M., Douvinet, J. and Viel, V. (2010) SCALES, a large-scale assessment model of soil erosion hazard in Basse-Normandie (northern-western France), *Earth Surface Processes and Landforms 35*, 887–901.

Lorenz, M., Fürst, C. and Thiel, E. (2013) A methodological approach for deriving regional crop rotations as basis for the assessment of the impact of agricultural strategies using soil erosion as example, *Journal of Environmental Management 127*, S37–S47.

Lucà, F., Conforti, M. and Robustelli, G. (2011) Comparison of GIS-based gullying susceptibility mapping using bivariate and multivariate statistics, Northern Calabria, South Italy, *Geomorphology 134*, 297–308.

Macary, F., Dias, J.A., Figueira, J.R. and Roy, B. (2014) A multiple criteria decision analysis model based on ELECTRE TRI-C for erosion risk assessment in agricultural areas, *Environmental Modeling & Assessment 19*, 221–242.

Maeda, E.E., Pellikka, P.K., Siljander, M. and Clark, B.J. (2010) Potential impacts of agricultural expansion and climate change on soil erosion in the Eastern Arc Mountains of Kenya, *Geomorphology 123*, 279–289.

Mandal, U.K., Rao, K.V., Mishra, P.K., Vittal, K.P.R., Sharma, K.L., Narsimlu, B. and Venkanna, K. (2005) Soil infiltration, runoff and sediment yield from a shallow soil with varied stone cover and intensity of rain, *European Journal of Soil Science 56*, 435–443.

Manjunatha, M.V., Oosterbaan, R.J., Gupta, S.K., Rajkumar, H. and Jansen, H. (2004) Performance of subsurface drains for reclaiming waterlogged saline lands under rolling topography in Tungabhadra irrigation project in India, *Agricultural Water Management 69*, 69–82.

McFarlane, D.J. and Williamson, D.R. (2002) An overview of water logging and salinity in southwestern Australia as related to the 'Ucarro' experimental catchment, *Agricultural Water Management 53*, 5–29.

Navas, A., López-Vicente, M., Gaspar, L. and Machín, J. (2013) Assessing soil redistribution in a complex karst catchment using fallout 137 Cs and GIS, *Geomorphology 196*, 231–241.

Patel, D.P., Srivastava, P.K., Gupta, M. and Nandhakumar, N. (2015) Decision Support System integrated with Geographic Information System to target restoration actions in watersheds of arid environment: a case study of Hathmati watershed, Sabarkantha district, Gujarat, *Journal of Earth System Science 124*, 71–86.

Poesen, J., Nachtergaele, J., Verstraeten, G. and Valentin, C. (2003) Gully erosion and environmental change, importance and research needs, *Catena 50*, 91–133.

Saaty, T.L. (1977) A scaling method for priorities in hierarchical structures, *Journal of Mathematical Psychology 15*, 234–281.

Shahbazi, F. and Jafarzadeh, A.A. (2010) Integrated assessment of rural lands for sustainable development using MicroLEIS DSS in West Azerbaijan, Iran, *Geoderma 157*, 175–184.

Shruthi, R.B., Kerle, N. and Jetten, V. (2011) Object-based gully feature extraction using high spatial resolution imagery, *Geomorphology 134*, 260–268.

Smith, L.C. (1998) Can FAO's measure of chronic undernourishment be strengthened? *Food Policy 23*, 425–445.

Svoray, T. and Ben-Said, S. (2010) Soil loss, water ponding and sediment deposition variations as a consequence of rainfall intensity and land use, a multi-criteria analysis, *Earth Surface Processes and Landforms 35*, 202–216.

Svoray, T., Levi, R., Zaidenberg, R. and Yaacoby, B. (2015) The effect of cultivation method on erosion in agricultural catchments, integrating AHP in GIS environments, *Earth Surface Processes and Landforms 40*, 711–725.

Svoray, T. and Markovitch, H. (2009) Catchment scale analysis of the effect of topography, tillage direction and unpaved roads on ephemeral gully incision, *Earth Surface Processes and Landforms 34*, 1970–1984.

Svoray, T., Michailov, E., Cohen, A., Rokah, L. and Sturm, A. (2012) Predicting gully initiation, comparing data mining techniques, analytical hierarchy processes and the topographic threshold, *Earth Surface Processes and Landforms 37*, 607–619.

Thibier, M. and Wagner, H.G. (2002) World statistics for artificial insemination in cattle, *Livestock Production Science 74*, 203–212.

Trimble, S.W. and Crosson, P. (2000) Land use – US soil erosion rates – myth and reality, *Science 289*, 248–250.

Van Rompaey, A., Bazzoffi, P., Dostal, T., Verstraeten, G., Jordan, G., Lenhart, T., Govers, G. and Montanarella, L. (2003) Modeling off-farm consequences of soil erosion in various landscapes in Europe with a spatially distributed approach, Proceedings of the OECD Expert Meeting on Soil Erosion and Soil Biodiversity Indicators Conference, 25–28 March, Rome, Italy.

# Index